International Series in Operations Research & Management Science

Volume 330

The book series **International Series in Operations Research and Management Science** encompasses the various areas of operations research and management science. Both theoretical and applied books are included. It describes current advances anywhere in the world that are at the cutting edge of the field. The series is aimed especially at researchers, advanced graduate students, and sophisticated practitioners.

The series features three types of books:

• Advanced expository books that extend and unify our understanding of particular areas.

• Research monographs that make substantial contributions to knowledge.

• Handbooks that define the new state of the art in particular areas. Each handbook will be edited by a leading authority in the area who will organize a team of experts on various aspects of the topic to write individual chapters. A handbook may emphasize expository surveys or completely new advances (either research or applications) or a combination of both.

The series emphasizes the following four areas:

Mathematical Programming: Including linear programming, integer programming, nonlinear programming, interior point methods, game theory, network optimization models, combinatorics, equilibrium programming, complementarity theory, multiobjective optimization, dynamic programming, stochastic programming, complexity theory, etc.

Applied Probability: Including queuing theory, simulation, renewal theory, Brownian motion and diffusion processes, decision analysis, Markov decision processes, reliability theory, forecasting, other stochastic processes motivated by applications, etc.

Production and Operations Management: Including inventory theory, production scheduling, capacity planning, facility location, supply chain management, distribution systems, materials requirements planning, just-in-time systems, flexible manufacturing systems, design of production lines, logistical planning, strategic issues, etc.

Applications of Operations Research and Management Science: Including telecommunications, health care, capital budgeting and finance, economics, marketing, public policy, military operations research, humanitarian relief and disaster mitigation, service operations, transportation systems, etc.

This book series is indexed in Scopus.

Dalė Dzemydienė • Aurelija Burinskienė •
Kristina Čižiūnienė • Arūnas Miliauskas

Development of Smart Context-Aware Services for Cargo Transportation

An Operational Management Approach

Springer

Dalė Dzemydienė
Department of Business Technologies and Entrepreneurship,
Faculty of Business Management
Vilnius Gediminas Technical University
Vilnius, Lithuania

Aurelija Burinskienė
Department of Business Technologies and Entrepreneurship,
Faculty of Business Management
Vilnius Gediminas Technical University
Vilnius, Lithuania

Kristina Čižiūnienė
Logistics and Transport Management Department,
Faculty of Transport Engineering
Vilnius Gediminas Technical University
Vilnius, Lithuania

Arūnas Miliauskas
Department of Business Technologies and Entrepreneurship,
Faculty of Business Management
Vilnius Gediminas Technical University
Vilnius, Lithuania

ISSN 0884-8289 ISSN 2214-7934 (electronic)
International Series in Operations Research & Management Science
ISBN 978-3-031-07198-0 ISBN 978-3-031-07199-7 (eBook)
https://doi.org/10.1007/978-3-031-07199-7

This Springer imprint is published by the registered company Springer Nature Switzerland AG
The registered company address is: Gewerbestrasse 11, 6330 Cham, Switzerland

Preface

The universe of discourse of this research is concerned with the development of intelligent systems for contemporary management processes of transportation of goods. To understand how context recognition systems can be integrated into intelligent transport management systems and operate in freight transport processes, it is important to understand and evaluate the factors influencing the interaction of transport processes. The aim of this research is to develop a system for the provision of intelligent services for the operational management of freight transportation. The objectives to achieve this aim concerns the development of a multi-layered methodology that would allow describing complex transportation processes, contextual information and their management in a computer-based system. Such an approach requires the analysis of a fairly wide range of knowledge-based methods and technologies. The methodology describes the stages of analysis and application of methods and ICT. The review of tools and methods is based on comparative analysis techniques, other methods and system architecture components are developed originally by the authors. The authors examine quite important areas for the development of such systems, which include:

- An analysis of the context concept and representational structure that is important in freight transport management processes;
- An analysis of forwarding sector activities;
- The identification of variables that will form the information base required for the delivery of goods, taking into account the entire cycle of freight transportation and logistics;
- The assessment of requirements for describing the algorithms, which ensure context recognition service engineering

The topics are concerned in the monograph in order of using appropriate ICT infrastructure for the analysis of the observed data, for right designing of system's components on the basis of the analysis of the literature sources:

- The evaluation and selection of the sets of wireless communication network tools.
- The classification of services and priorities for provision.

The integration of all necessary components according to the specifics of the context-aware services is analysed.

After accumulating and comparing the necessary methods in the course of research:

- The architecture of the intelligent service delivery system is designed, which integrates context-aware services that help increase the efficiency of the transport system.
- Evaluate the cargo delivery management process: risks, solutions, and optimization.
- The chosen methods of data collection, aggregation, and dissemination and their application are evaluated by continuing the prototype of the developed simulation system.
- Transport infrastructure is described by semantic models and its capabilities are evaluated by means of simulation modelling.
- The need for information systems in freight transport is identified and assessed.
- The road, rail, and water transport sectors and the information systems applied in them in transport and forwarding activities are analysed.
- Intermodal and multimodal transport are valued.

Multi-criteria decision support methods are integrated into the intelligent service infrastructure to ensure sustainable multimodal freight transportation management.

This monograph is written by authors: Dalė Dzemydienė, Aurelija Burinskienė, Kristina Čižiūnienė, and Arūnas Miliauskas. Short description of activities of authors is provided.

Dalė Dzemydienė (Švanytė) is the leading professor and the chief research fellow working at the Department of Business Technologies and Entrepreneurship of the Faculty of Business Management of the Vilnius Gediminas Technical University and at the Institute of Data Science and Digital Technologies of the Faculty of Mathematics and Informatics of the Vilnius University (Lithuania). She holds a diploma with the honor of applied mathematics in the specialization of software engineering in 1980, achieved PhD of mathematics—informatics sciences in 1995, a diploma of habilitation doctor procedure in the field of social sciences of management and administration in 2004. She holds the academic rank of a professor gained in 2009. She published more than 300 research articles, 3 manual books, and 3 monographs with co-authors. She is the head of the Legal Informatics Section of Lithuanian Computer Society (LIKS), a member of the LIKS Board, a member of the European Coordinating Committee for Artificial Intelligence (ECCAI), and a Board member of the Lithuanian Operation Research Association (LITORS) and a member of European Operational Research Society (EURO). She is a member of the Editorial boards of International Scientific Journals: Technological and Economical Development of Economy (2008–2017); Strategic Property Management; Electronic Security and Digital Forensics. She is leading in scientific research work at the programme of Social Technology and teaching at Curriculum of Bachelor, Master, and Doctoral study programmes. She was a supervisor of doctoral studies and the research programmes of Social Technology, Modelling, and Development of Social and Technological Systems. She was the supervisor of doctoral studies of Gediminas Gricius (Informatics sciences), Mindaugas Kurmis (Technological Informatics Engineering sciences), Giedrius Romeika (Social Management and Administration sciences), Darius Drungilas (Informatics Engineering sciences), Ramūnas Dzindzalieta (Informatics Engineering sciences), Ramutė Naujikienė (Social Management and Administration sciences), and others. She is an organizer of international scientific conferences in the field of databases and information systems. She was the member of Programme Committees of International Baltic Conferences of DB and IS, ALTA'2022, 13th Intern. Conf. "Business and Management 2022", ADBIS, CIGSUD'2006, coordinator of International Doctoral Consortiums, etc. She is a member of Editorial Boards of International Scientific Journals: "Technological Economical Development of Economy" (2008–2017); "Strategic Property Management"; "Electronic Security and Digital Forensics"; "Applied Computer Systems". Her research interests include artificial intelligence methods, decision support systems, evaluation of sustainable development processes, and smart services.

Aurelija Burinskienė is a chief researcher at the Institute of Dynamic Management of the Vilnius Gediminas Technical University and an associate professor at the Department of Business Technology and Entrepreneurship. She has 12 years of experience as a logistics project manager. She is the manager of the scientific project "Smart context-aware services for operational freight transport management" (2020–2023) and participant in the DIGINNO-digitization project of land freight transport. She was a Board member of the international cooperation project COST CA15105 from 2018.

She is the head of a team of researchers of the Faculty of Business Management engaged in distribution and logistics research since 2016. She has published over 70 publications in scientific journals and books. Recently she published a scientific study "Determining the trade-offs of inventory management approaches in the face of COVID-19". She has 10 years of experience in lecturing (teaching such courses as logistics, warehouse, and inventory management, e-logistics, and logistics process optimization systems). She is a member of the editorial board of the Logistics Scientific Conference and the editorial board of a periodic international journal published abroad. She made a presentation at the MIM 2019 conference in Berlin, leading the part "Intelligent Methods and Systems Supporting Supply Chain Decision Making—II". The researcher has also been a member of committee of the Business Logistics study programme.

Kristina Čižiūnienė is an associate professor working at the Department of Logistics and Transport Management, Faculty of Transport Engineering, Vilnius Gediminas Technical University. She defended her doctoral dissertation and hold a doctoral degree diploma in Social Sciences and Economics. She holds an excellent track record as an expert in lecturing and training specialists in logistics, service logistics, transport, human resource management, organization of industrial practices, and marketing in logistics. She was a team member and a senior expert in project "Promoting innovative business establishment" (No. VP2-1.4-ŪM-05-V-01-006). She gained solid knowledge and experience in preparing and developing teaching materials for the study programmes and modules taught at the Faculty of Transport Engineering. In 2019, she was the executor of the outsourced project at Lithuanian Railways. As a senior expert of the project, she ensured the planning, coordination, delegation of functions, and performance monitoring; assisted students throughout events and consultations, and cooperated with business partners. Kristina Čižiūnienė was a manager of the project "ATVERK" organized by Lithuanian Engineering Industry Association LINPRA. The training "Non-formal development of entrepreneurship based on Demol model" was held in the course of the project in Lithuania and Finland. Kristina Čižiūnienė is an author and co-author of over 80 scientific publications, several monographs and handbooks. She is a member of the Editorial board of Journal Special issue in Logistics that is a part of the section "Maritime and Transport Logistics".

Arūnas Miliauskas is a specialist engineer working at the Department of Business Technology and Entrepreneurship of the Faculty of Business Management of the Vilnius Gediminas Technical University, and an assistant research fellow at the Institute of Data Science and Digital Technologies of the Faculty of Mathematics and Informatics of the Vilnius University (Lithuania). He earned a master's degree in Informatics at Vilnius University. He published about 15 scientific articles. His research interests include artificial intelligence methods and smart service development platforms.

Vilnius, Lithuania

Dalė Dzemydienė
Aurelija Burinskienė
Kristina Čižiūnienė
Arūnas Miliauskas

Author Contributions

The Preface is written by Dalė Dzemydienė, the Chap. 1 is devoted for introduction into the research area and is delivered by Kristina Čižiūnienė and Aurelija Burinskienė; the issues of development of context recognition methods, communication infrastructure, and algorithms of smart service delivery system are presented by Dalė Dzemydienė (Chaps. 2–4); the analysis of cargo delivery management process and context-aware services support means is described by Aurelija Burinskienė (Chaps. 5–6, 8–12); the issues of context service engineering for automated web-based freight brokerage is offered by Arūnas Miliauskas (Chap. 7); the description of information technology used in various transport sectors, the analysis of intermodal transport and multimodal transport are provided by Kristina Čižiūnienė (Chaps. 13–18); the representation of decision support for multimodal transportation of cargo in smart service delivery system is presented by Dalė Dzemydienė (Chap. 19). The glossary of terminology and list of abbreviations is provided by all authors at the end of the monograph.

Acknowledgements

This research is supported by the project that received funding from the European Regional Development Fund (project No 01.2.2-LMT-K-718-03-0030) under a grant agreement with the Research Council of Lithuania (LMTLT).

The authors are thankful for opportunities of research work and experiments on base of contemporary ICT means at the Department of Business Technology and Entrepreneurships of the Faculty of Business Management of Vilnius Gediminas Technical University—Vilnius Tech, as well as the authors are thankful for valuable consultations of colleagues from the Institute of Data Science and Digital Technologies of the Faculty of Mathematics and Informatics of Vilnius University, from the Department of Informatics and Statistics of the Faculty of Marine Technologies and Natural Sciences of Klaipėda University and from the Department of Logistics and Transport Management, Faculty of Transport Engineering of Vilnius Gediminas Technical University—Vilnius Tech.

2014-2020 Operational
Programme for the
European Union Funds
Investments in Lithuania

Contents

Chapter 1
Introduction to Research Area

Kristina Čižiūniene and Aurelija Burinskiene

Abstract This chapter presents research background and research areas that are going to be investigated in this monograph. The authors revise the dynamism of the environment which is important for operational freight transportation. Also, herein the separate attention is given to the ongoing digitalization initiatives and their new opportunities for transport management. Later on, in the chapter some ongoing researches in their area and the overview of European Commission focus directions are presented. Further, the development of contextual systems and the discussion on context meaning are introduced. Finally, the chapter ends with the presentation of research object.

Keywords Background of research · Context-sensitive systems · Environment in freight delivery processes

1.1 Importance of the Research

The gross domestic product of the country is largely dependent on possibilities of freight transport. About three-quarters of all land freight is carried on European Union (EU) roads. One of the factors that affect the development of the international market is the ability to transport the goods to other markets quickly, safely, and at the lowest possible cost. The rapid growth of trade flows between countries and the implementation of best practices means relatively rapid growth of delivery volumes.

There are many freight delivery solutions on the market. Digitization, including digital information flows, adoption of new technologies and context-aware services, offers many opportunities to improve efficiency, reduce financial costs, and improve the quality of road freight transport service. Digitization also offers a potential for a radical change in the business ecosystem, including any changes in freight transport, such as improved traffic flows, reduced fuel consumption, and higher efficiency of

K. Čižiūniene (✉) · A. Burinskiene
Vilnius Gediminas Technical University, Vilnius, Lithuania
e-mail: kristina.ciziuniene@vilniustech.lt; aurelija.burinskiene@vilniustech.lt

D. Dzemydienė et al., *Development of Smart Context-Aware Services for Cargo Transportation*, International Series in Operations Research & Management Science 330, https://doi.org/10.1007/978-3-031-07199-7_1

the transport system. Digitization has restructured the current benefit network. The operation of digital disconnection and interconnection allows increasing the efficiency of multimodal transport and carriers. In addition, smart context-aware support systems help to establish new forms of mobility and reduce freight transportation costs. These systems are a combination of transport and technology. Uniquitous connectivity and attitudes are generated forward configuration. However, realizing this potential is an issue of appropriations, not just design and implementation. Everyday users, from engine manufacturers to public and private implementation partnerships, service providers, and individual users, put some effort into the implementation and the development of context-aware support systems.

In recent decades, the increasing use of smart devices and sensors has made it possible to accurately detect changing environmental conditions, which has led to the development of contextual systems where contexts, such as the current location or the user, affect the functioning of apps. The opportunities and advantages of such systems have been extensively researched in various areas in search of various scenarios where contextual perception can render beneficial. The fact that each research area has its requirements makes defining the context particularly difficult. There is still too little focus on contextual systems. This is for a variety of reasons. The most important is the wide variety of data sources and the need to ensure privacy.

Costs of freight transport fall within costs of goods sold or operating costs of companies. Transport is often considered to be one category of logistics costs along with warehousing. During the period from 1980 to 2015, transportation costs decreased by 25%, which stimulated the growth of freight delivery volumes.

The complexity of management of freight transport, especially when it comes to ensuring conditions for multimodal transportation processes, encumbers the search for more optimal management solutions to meet the green development requirement. The complexity of transport management processes, the multi-dimensional nature of data warehousing (DWs), as well as many elements in the process of decision-making, hinder the development of operational systems for online decision-making. Management processes face challenges related to the communication and information infrastructure to ensure freight transport operations and to manage freight transport processes efficiently.

The process includes the provision of intelligent service infrastructure for the operational management of freight transport, the analysis of infrastructure and multimodal freight transport in the context of heterogeneity of communication channels and multi-criteria support of decisions.

The purpose of the study—to form a method for assurance the infrastructure of smart intelligent services for the implementation of on-line multi-modal freight transport management processes. Objectives include the development and integration of multi-criteria solutions for hosting components into multi-layered management of computer intelligent systems information.

Below are described the steps of the possible extension of the structure of knowledge-based systems by the methods of decision support methods, the possibilities of working with wireless networks, and the recognition of contextual situations. The results are based on an analysis linked to alternatives evaluating multimodal (Eastern and Western Europe) transport corridors and the requirements of sustainable development in search of more efficient management perspectives.

1.2 Topicality of the Research Problem

The identification of problems with the development of advanced cargo transportation systems is related to many issues according to the complexity of such difficult process characteristics and ICT. Extensive recognition of the spectrum and the specificity of their provision under the cordless networks are the issues that raise concern. A study was also carried out to find appropriate contextual data provision, addressing the definition of context and context-awareness, and the possibility of applying formalized techniques used for the identification of objects in transport. This stage of the advanced system solves the problem of prioritizing potential services, ensuring the maximum data transmission channels quality and preventing the influx of wireless channels. This methodological framework integrates algorithms used to determine the contextual description of a situation and provides insights useful for decision-making and conflicts resolution. The safe service system components have been developed for safer transport management. The obtained research results are based on the development of priority schemes of service provision, analysis of intelligent services heterogeneity, and the addition of intelligent transport potential in the process of information transportation in the context of intelligent recognition.

Researchers are now addressing the digital transformations integration into the expansion of intelligent systems that can help manage the freight process (Du et al., 2021; Kaffash et al., 2021). To this day, transportation management processes are very complex. The flow of freight transport by various modes of transport has grown significantly. Developing advanced tools based on future communication and information technologies (ICT) that can adapt to a variety of vehicle automation tools and communication infrastructures (Dzemydienė et al., 2020; Osmundsen et al., 2018) has faced significant challenges. These challenges are concerned with the development of smart services for the deployment of data from wireless sensor networks (WSNs) (Sharma et al., 2021; Zagrouba & Kardi, 2021). The means for the development of intelligence transport system (ITS) require the extension of adaptability of heterogeneous WSNs infrastructure, expand the capabilities of IoT technology with next-generation smart services.

Our research is designed for the development of the advanced systems that meet the performance of the operating regime, adaptation, safety, and efficiency requirements without flood wireless channels. This should include the provision of an intelligent service system with tracking and context functions that can respond in real time to the needs of processes for freight transport. The application of an intelligent system (integrating contextual services) under complex processes for transportation has been problematic because the development of a platform is based on a wireless sensor network (WSN) infrastructure. The information obtained must be recognized in the exercise of the transport stages and the determination of sustainable freight transport management. This method is based on the methods of developing the component architecture of the Intelligent Adaptive Service Delivery System (IASDS). The results of previous studies were used to develop the system (Dzemydienė et al., 2020; Dzemydienė; & Radzevicius, 2020, Kurmis et al., 2015) in the hope that it will be useful in developing infrastructure for self-contained transport with expanding intermodal communication equipment and channels that will support safer processes.

The new Digital Agenda for Europe 2030 (COM (2020) 67 final) sets out several important objectives that have an impact on the development of ITS, cargo transport processes, and connections to the targets of the 2030 Agenda for Sustainable Development (Regulation (EU) No 2018/1724). The new ICT development strategy to 2030 (COM (2020) 67 final) introduces several significant innovations, namely recent ambitions focusing on artificial intelligence (IoT), cloud computing and blockchain technologies for supercomputers, IoT, and quantum computing technologies.

The Digitalization strategy sets out key commitments and opportunities:

- The efforts will be forwarded for the development of a Single Digital Gateway (SDG) platform (Regulation (EU) No 2018/1724). This SDG Regulation obliges all countries of the EU to provide the entry to the data supporting service (21 of them); e-mail services will have to become fully digital in seven life events (i.e. cross-border and online).
- Unified efforts will be forwarded for the protection of personal data and the free data movement (COM (2020) 67 final). The Open Data Directive and the Regulation of Data Protection (Regulation (EU) No 2016/679) have imposed strict obligations on the Member States of the EU.
- Objectives of the Information Systems and Networks Directive (Directive (EU) 2019/1024) have led to new measures to develop cybersecurity. The increased focus on cybersecurity has obliged the achievements in preparing means against cybercrimes and cyberattacks.

All these means affect the development of the potential of smart service delivery infrastructure. The safety of drivers and all traffic participants on the road during transport cycles is a key requirement for the development of ITS. The systems help to provide the necessary services for users on the base of just-in-time regime. The efforts for adapting them accordingly to the needs of users are vital. Some operational services, which are necessary for the interventions in the transportation process have to be developed for ensuring the safety of the transport process.

The interoperability of all necessary components has to be achieved for service provision systems. We are trying to achieve the efficiency of user–vehicle interaction and to provide structural support for the assessment of existing and past situations of transportation. With operative control procedures, which are obtained from the smart system we are trying to influence future situations. ICT infrastructure will include various new types of wireless sensor networks (WSNs) and wireless channels. The ability to integrate heterogeneous services is coordinated in the plane of moving objects by the functionality of a geographic information system (GIS) and can also help develop smart systems. The integration of data of sensors into the moving vehicle monitoring process helps to predict processes and help in the assessment of environmental data, thus promoting the creation of intelligent system structures. However, there are problems in prioritizing the service delivery process, limiting inappropriate information and data traffic over wireless to safe driving.

Advances in the Internet of Things (IoT) technology, mobile computing, and big data analytics have led to the need for smart services that allow us to adapt to changing circumstances and understand the context. Today, developing systems for the provision of smart services is a challenging task, because it is missing a suitable model for awareness support (Le Dinh et al., 2021).

The development of systems for context-aware intelligent services is one of the main areas of research, especially in the field of IoT industrial engineering, intelligent solutions, and context recognition (Lim & Maglio, 2018). However, there is a lack of such type studies in the transport and logistics sector. The context-aware systems for intelligent service is still a new field, encompassing many new topics for research.

In general, two research areas can be distinguished: the development of ITS and context-aware web services (Truong & Dustdar, 2009; Chang et al., 2017; Kurmis et al., 2016; Dzemydienė & Burinskienė, 2021).

We can analyse the following typical expression of a context: *"A «stakeholder» (know-who) performs «operations» (know-how) on «objects» (know-what) at «time» (know-when) in «a location» (know-where) because of «contract» (know-with) to be consistent with «rule of business» (know-why)…"*, according to Le Dinh et al. (2021).

Le Dinh and Pham Thi (2012) argue that service consists of three elements of service: provision service, consumption service, and operation service. Customers always want accurate and timely business support when they encounter product/service issues, especially when it comes to complex products/services.

Intelligent service delivery systems are helpful in management processes of freight transport. There are many challenges in developing these systems by analysing the many factors that affect such complex and dynamic systems (Dzemydienė & Burinskienė, 2021). The methodology is based on the methods for intelligent system architecture development with integration of methods, which integrate context-aware, real data identification. As well as the methods for description of data warehouse conceptual structures, and AI-based transportation situation recognition algorithms are evaluated for integration.

1.3 Integration of Context Recognition in Freight Transport Management

Exploring the possibilities of context integration recognition programmes in the transport sector has focused on the widespread use of various sensors, the processing and generation of their rich data into new services and useful information. Such systems are improved road safety and take benefits for drivers.

According to Chang et al. (2017), safety programmes allow vehicles to continuously monitor the surroundings (especially the conditions of other vehicles or the road) and to prevent accidents by taking timely actions. In the field of transport, current research is mainly focused on the observation phase (Golestan et al., 2016), driver (status), vehicle (condition) (Al-Sultan et al., 2013; Sun & Yu, 2014; Wartzek et al., 2011), and environmental (situation) monitoring (Boyraz et al., 2012; Ohn-Bar et al., 2015; Sathyanarayana et al., 2011; Sun et al., 2010; Yang et al., 2013). Different sensors are used for monitoring of movement (Al-Sultan et al., 2013; Sathe & Deshmukh, 2016; Sun et al., 2009; Zhang et al., 2011), including location sensors (like radar, GPS, infrared), imagine sensors (like cameras), bio-sensors (like ECG, EGG, alcohol sensors), network of controller area (CAN)—bus-based integrated diagnostics (OBD), and other data resources (like roadside infrastructure).

Drivers can face with unusual situations while driving. Some of such unusual situations can be qualified as very serious and the recognition of the levels of danger is required. The stages of recognition of unsafety of situations have influenced the triggering of the alarms. Many researchers have analysed transport service systems, which are included in transport management for passenger information, business vehicle management needs, public transport management, and other services. By receiving large amounts of data from a variety of resources, drivers can create and interactively access the set of relevant services and useful information (such as route navigation, planning, dynamic parking, vehicle diagnostics, and tracking). There is no comprehensive context-based service system developing approaches, which can combine situational awareness and business services for the provision of services for drivers with the right business services and help to meet their needs accurately and efficiently.

Intelligent service delivery systems are designed for better transport safety management, as these systems can help to manage the freight process and ensure its security. They apply contextual data for various freight delivery processes.

For understanding our research object we shortly describe the context's definitions. Context has become increasingly important because it allows customizing application features and user interfaces according to travel conditions. The usage of context is analysed in mobile users' applications for development of models of type the business to customs (B2C) service segment. These service applications in the transport sector require discussion. When the system identifies the most common circumstances such conditions became important for the implementation of contextual data. The presentation of definitions concerning the presentation of data objects, and classification examples from business applications business-to-business (B2B market), the identification of active and passive context-aware data is important (Swarnamugi & Chinnayan, 2019).

In recent decades, the increasing use of smart devices and sensors has made it possible to accurately detect changing environmental conditions (Gungor & Hancke, 2009). This has led to the development of contextual systems where application behaviour is affected by contexts, such as the geographical position or the users' context (Schilit et al., 1994a; Want et al., 1992).

Regarding different context scenarios that may be beneficial, the opportunities of such applications are researched in various areas (Bazire & Brézillon, 2005; Dey, 2001; Dey et al., 2001). Since each situation has its requirements, defining a context is particularly difficult.

Primary data by nature require different collection and processing approaches. Collecting data from various sensors and other equipment is challenging. For solving this problem, the system classifying context into types has been developed, which takes into account the peculiarities of the data itself. Collecting information and analysing data is a standard task in cargo delivery, due to focus on the work environment for drivers, trucks, freight, and papers as well as data processing and recognizing. Helping potential programmers to implement contextual systems, solutions that use many sensors to determine current circumstances are needed.

Analysing context definitions, we found many terms which are proposed and adapted for specific areas (Bazire & Brézillon, 2005; Dey, 2001). Based on Dey's (2001) work, Perera et al. (2014a) analysed the most common definitions in contextual computing and found that they can be all separated into such layers:

- Definitions formulated by using example cases (Dey et al., 2001; Ryan et al., 1999). Although the terms can be useful for the users which lack experience, harmonizing all cases could be a problematical task because nobody could indicate the spectrum of all possible cases (Aggeliki et al. 2016).
- Definitions are formulated by dimensions. The terms describe the contextual data according to its characteristics, indicating the object and position of that object (Liu et al., 2011; Schilit et al., 1994b).

Definitions by aspects are most useful for freight transport purposes, however, they would require discussions as to whether or not particular circumstances could be considered contextual (Mo et al. 2010).

As previously mentioned, the authors (Dey, 2001; Perera et al., 2014b) define context data as "information which could be collected to report the situation linked to the object (freight, resources, service points, etc.)". According to the definition of Bazire and Brézillon (2005), context means "a series of circumstances that frame an incident or an object" and one proposed by Dey et al. (2001) defines context as "data which describes the situation in terms of interaction between users, applications and the environment". Any reported situation touches not only the economic operators but also undertakings that are relevant for the contextual perception of the transaction describing the location of the entity itself rather than the entity. In the case of a multi-cultural environment, the context system should be easily adapted to the customer and their language. When it comes to tangible and intangible contexts, which are even more important for freight delivery purposes, this context class should be emphasized in the definition, defining it as "any data that could be used to describe the freight, its state, or its environmental situation, if the information is considered relevant to the interaction between the user (driver, planner, consignor, consignee, etc.) and the application".

Several researchers identified different types of context (Chen & Kotz, 2000; Chong et al., 2007; Fan et al., 2014; Yanwei et al., 2011). One popular classification was proposed by Dey (2001), which distinguishes between primary and secondary types of contexts. The author said that the primary contexts of time, identity, place, and activism can fully grasp any situation. He also emphasized that this context data is secondary and indicates that is better to start from the primary one. A controversial example is requirements for trucks, such as freight delivery temperature, which is treated as the secondary one. He suggests distinguishing between more and less important circumstances, that it is not always easy as this largely is linked with the usage situation.

Other authors have proposed the checking of the context data that can actively identify the dynamism that is related to the object (Chen & Kotz, 2000). The active context is used for the identification of the current status of the object and its condition. Passive context information is information that is not actively used in decision-making. For example, it indicates that contexts change but only describes the changes over time. An overview of various contextual applications shows the structure that illustrates how likely a context-sensitive system has used the types of active and passive context, and the probability that indicates the occurrence of one of the importance of the situation assessing the importance: high, medium, or low (Fuchs et al., 2008; Ward et al., 1997).

Although key circumstances are general in various applications, certain areas mainly manifest in certain environments. Three main categories can be distinguished, namely, the user, the environment, and the system:

- In the stage of analysing of the category "Users", for each user a separate context model is required. The several models could be overlapping, some of them are dedicated to a larger group of users, but others—have individual needs.
- In the stage of analysing of the category "Environment", the environment is divided into multi-dimensional parts of groups and the context is analysed by the assessment of surrounding of the freight.
- In the stage of analysing category "System", the system is developed and validated by assessing of context data and multi-criteria parameters are prepared for decision-making.

These systems have specific categories of information collection and model recognition in common. Data are collected from different data sources and summarized following the historical data or data which is extracted directly from the last update received from the data source (Vahdat-Nejad et al., 2016).

Having loaded the truck, the driver gets a shipping task (task). The system should support the driver, providing him with information about the route (how to get to the service point (place)). During delivery, the system constantly gets information about

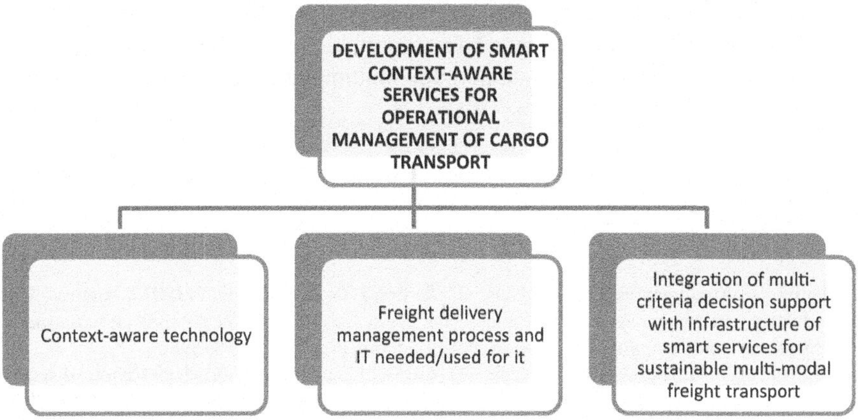

Fig. 1.1 Short description of research object

freight temperature, etc. Regardless of the driver, the system could search for stop places. Search results (search) are displayed according to search categories which are depending on the position (place) of a particular driver and his current needs. Later on, upon arrival at the location, the driver is informed about the details of the shipping task (task).

In the case of big data, contexts and other information are sorted, presented to the driver, reported to the service point and the transport planner. Similar information could be presented on resources and service points, while the research could be extended by including more modes of transport in it.

To understand how context recognition systems should be integrated and operate in the freight transport processes, the understanding of the factors of description of the research object is important (Fig. 1.1):

The analysis of important research areas is included in monograph:

- Infrastructure of smart service provision for operative management of transportation of goods;
- Forwarding sector;
- Processes of freight delivery and context-aware service engineering;
- Network design methods and specifics of context-aware services;
- Context-aware services that support efficiency improvement in the transport system;
- Freight delivery management process: risks, decisions, and optimization;
- Methods of data collection, aggregation, and dissemination, and their application;
- Transport infrastructure and its capacity;
- Needs for information systems in freight transport;
- Information technology (IT) used in the road, railway, and water transport sectors;

- Information systems in transport and forwarding activities;
- IT used intermodal and multimodal transport;
- Representation of decision support for multimodal transportation of cargo in smart service delivery system.

References

Aggeliki, V., Stratogiannis, G., Caridakis, G., & Siolas, G.,& Mylonas, P. (2016, August). User adaptive and context aware smart home using pervasive and semantic technologies. *Journal of Electrical and Computer Engineering*. https://doi.org/10.1155/2016/478980

Al-Sultan, S., Al-Bayatti, A.-H., & Zedan, H. (2013). Context-aware driver behaviour detection system in intelligent transportation systems. *IEEE Transactions on Vehicular Technology, 62*(9), 4264–4275.

Bazire, M., & Brézillon, P. (2005). Understand the context before using it. In *Modelling and contextual use* (pp. 113–192).

Boyraz, P., Yang, X., & Hansen, J.-H. (2012). Computer vision systems for "context-aware" active vehicle safety and driver assistance. In J. Hansen, P. Boyraz, & K. Takeda (Eds.), *Digital signal processing for in-vehicle systems and safety* (pp. 217–227). Springer.

Chang, J., Yao, W., & Li, X. (2017). The design of a context-aware service system in intelligent transportation system. *International Journal of Distributed Sensor Networks, 13*(10), 1–8. https://doi.org/10.1177/1550147717738165

Chen, G., & Kotz, D. (2000). *Study of contextual mobile computer research. Technical report TR2000–381*. Department of Computer Science, Dartmouth College.

Chong, S. K., McCauley, I., Loke, S. W., & Krishnaswamy, S. (2007). Context sensors and data muffling. In *Contextual understanding of self-managing systems (devices, applications, and networks) that are taking place* (pp. 103–117).

Dey, A. K. (2001). Understand and use context. *Personal and Everywhere Arithmetic, 5*(1), 4–7.

Dey, A. K., Abowd, G. D., & Salber, D. (2001). Conceptual framework and toolkit to support a rapid prototype of contextual applications. *Human-Computer Interaction, 16*(2), 97–166.

Du, Y., Wang, J., Rizos, C., & El-Mowafy, A. (2021). Vulnerabilities and integrity of precise point positioning for intelligent transport systems: Overview and analysis. *Satellite Navigation, 2*(3). https://doi.org/10.1186/s43020-020-00034-8

Dzemydienė, D., & Burinskienė, A. (2021, July). Integration of context awareness in smart service provision system based on wireless sensor networks for sustainable cargo transportation. *Sensors (Basel), 21*(15), 5140. https://doi.org/10.3390/s21155140

Dzemydienė, D., Burinskienė, A., & Miliauskas, A. (2020). An assessment of provision of heterogeneous services for sustainable cargo transportation process management by roads. *Sustainability, 12*(20), 8405. https://doi.org/10.3390/su12208405

Dzemydienė, D., & Radzevičius, V. (2020). An approach for networking of wireless sensors and embedded systems applied for monitoring of environment data. In G. Dzemyda, J. Bernatavičienė, & J. Kacprzyk (Eds.), *Data science: New issues, challenges and applications* (Studies in computational intelligence) (pp. 61–82). Springer. https://doi.org/10.1007/978-3-030-39250-5_4

Fan, Y., Wolfson, J., Adomavicius, G., Das, K. V., Khandelwal, Y., & Kang, J. (2014). *SmarTrAC: A smartphone solution for context-aware travel and activity capture. Research paper*. University of Minnesota.

Fuchs, S., Rass, S., & Kyamakya, K. (2008). A constraint-based and context-aware overtake assistant with fuzzy probabilistic risk classification. In *IADIS international conference wireless application and computing* (pp. 93–100).

Golestan, K., Soua, R., & Karray, F. (2016). Situation awareness within the context of connected cars: A comprehensive review and recent trends. *Inform Fusion, 29*, 68–83.

Gungor, V. C., & Hancke, G. (2009). Industrial networks of wireless sensors: Challenges, design principles and technical methods. *IEEE Industrial Electronics Transactions, 56*(10), 4258–4265.

Kaffash, S., Nguyen, A. T., & Zhu, J. (2021). Big data algorithms and applications in intelligent transportation system: A review and bibliometric analysis. *International Journal of Production Economics, 231*, 107868. https://doi.org/10.1016/j.ijpe.2020.107868

Kurmis, M., Andziulis, A., Dzemydienė, D., Jakovlev, S., Voznak, M., & Gricius, G. (2015). Cooperative context data acquisition and dissemination for situation identification in vehicular communication networks. *Wireless Personal Communications*. [New York] Springer LLC, *85* (1), 49–62. https://doi.org/10.1007/s11277-015-2727-1

Le Dinh, T., & Pham Thi, T. T. (2012). Information-driven framework for collaborative business service modelling. *International Journal of Service Science, Management, Engineering, and Technology (IJSSMET), 3*(1), 1–18. https://doi.org/10.4018/jssmet.2012010101

Le Dinh, T., Thi, T. T. P., Pham-Nguyen, C., & Nam, L. N. H. (2021). A knowledge-based model for context-aware smart service systems. *Journal of Information and Telecommunication*. https://doi.org/10.1080/24751839.2021.1962105

Lim, C., & Maglio, P. P. (2018). Data-driven understanding of smart service systems through text mining. *Service Science, 10*(2), 154–180. https://doi.org/10.1287/serv.2018.0208

Liu, W., Li, X., & Huang, D. (2011). Study on contextual perception. In *Computer science and services system (CSSS), international conference 2011* (pp. 144–147). IEEE.

Mo, T., Li, W., Chu, W., & Wu, Z. (2010). CABS3: Context-awareness based smart service system. In *2010 6th International Conference on Wireless Communications Networking and Mobile Computing (WiCOM)* (pp. 1–4). IEEE.

Ohn-Bar, E., Tawari, A., & Martin, S. (2015). On surveillance for safety critical events: In-vehicle video networks for predictive driver assistance systems. *Computer Vision and Image Understanding, 134*, 130–140.

Osmundsen, K., Iden, J., & Bygstad, B. (2018). Digital transformation: Drivers, success factors, and implications. In *Proceedings of the Mediterranean Conference on Information Systems* (pp. 1–15), Corfu, Greece, 28–30 September 2018.

Perera, C., Zaslavsky, A., Christen, P., & Georgakopoulos, D. (2014a). Context aware computing for the Internet of Things: A survey. *IEEE Communications Surveys and Tutorials, 16*(1), 414–454.

Perera, C., Zaslavsky, A., Christen, P., & Georgakopoulos, D. (2014b). Context knows computing for the Internet of Things: Survey. *IEEE Communication Surveys and Textbooks, 16*(1), 414–454.

Ryan, N., Pascoe, J., & Morse, D. (1999). Improved fieldwork of reality: The context is perceived by the assistant of archaeology. *Bar International Series, 750*, 269–274.

Sathe, A. D., & Deshmukh, V. D. (2016). Advance vehicle-road interaction and vehicle monitoring system using smart phone applications. In *2016 online international conference on green engineering and technologies, Coimbatore, India, 19 November 2016* (pp. 1–6). IEEE.

Sathyanarayana, A., Boyraz, P., & Hansen, J.-H. (2011). Information fusion for robust "context and driver aware" active vehicle safety systems. *Inform Fusion, 12*(4), 293–303.

Schilit, B., Adams, N., & Want, R. (1994a). Contextual computer programs. In *Mobile computer systems and applications, 1994 MCSA 1994 first seminar* (pp. 85–90). IEEE.

Schilit, B. N., Teimer, M., & Hrsg, G. (1994b). Dissemination of active map information for mobile hosting devices. *IEEE Network, 8*(5), 22–32.

Sharma, H., Haque, A., & Blaabjerg, F. (2021). Machine learning in wireless sensor networks for smart cities: A survey. *Electronics, 10*, 1012. https://doi.org/10.3390/electronics10091012

Sun, Y., & Yu, X. (2014). An innovative nonintrusive driver assistance system for vital signal monitoring. *IEEE Journal of Biomedical and Health Informatics, 18*(6), 1932–1939.

Sun, J., Wu, Z., & Pan, G. (2009). Context-aware smart car: From model to prototype. *Journal of Zhejiang University-Science A, 10*(7), 1049–1059.

Sun, J., Zhang, Y., & He, K. (2010). Providing context-awareness in the smart car environment. In *Proceedings of the IEEE 10th international conference on computer and information technology, Bradford, 29 June–1 July 2010* (pp. 13–19). IEEE.

Swarnamugi, M., & Chinnayan, R. (2019). Context-aware intelligent reliable service model for an intelligent transport system based on ontology. *Lecture Notes on Electrical Engineering, 597*, 22–30.

Truong, H. L., & Dustdar, S. (2009). A survey on context-aware web service systems. *International Journal of Web Information Systems, 5*(1), 5–31. https://doi.org/10.1108/17440080910947295

Vahdat-Nejad, H., Ramazani, A., Mohammadi, T., & Mansour, V. (2016). Review of context-aware applications of automotive networks. In *Automotive communications* (pp. 1–15).

Want, R., Hopper, A., Falcao, V., & Gibbons, J. (1992). Active bullet tracking system. *ACM Information Systems Transactions (TOIS), 10*(1), 91–102.

Ward, A., Jones, A., & Hopper, A. (1997). New technology for setting up active office locations. *IEEE Personal Communications, 4*(5), 42–47.

Wartzek, T., Eilebrecht, B., & Lem, J. (2011). ECG on the road: Robust and unobtrusive estimation of heart rate. *IEEE Transactions on Biomedical Engineering, 58*(11), 3112–3120.

Yang, Y., Chen, B., & Suetal, L. (2013). Research and development of hybrid electric vehicles CAN-bus data monitor and diagnostic system through OBD-II and android-based smartphones. *Advances in Mechanical Engineering, 5*(3), 741240.

Yanwei, S., Guangzhou, Z., & Haitao, P. (2011). Research on the context model of intelligent interoperability systems in the Internet of Things. In *It in medicine and education (ITME), international symposium 2011* (Vol. 2, pp. 379–382).

Zagrouba, R., & Kardi, A. (2021). Comparative study of energy efficient routing techniques in wireless sensor networks. *Information, 2021*(12), 42. https://doi.org/10.3390/info12010042

Zhang, J., Tan, B., & Sha, F. (2011). Predicting pedestrian counts in crowded scenes with rich and high-dimensional features. *IEEE Transactions on Intelligent Transportation Systems, 12*(4), 1037–1046.

Chapter 2
Structure of Context Recognition in Computer-Based Systems for Transportation Management

Dalė Dzemydienė

Abstract The process of context recognition in dynamically changing transportation processes is quite complex. The complexity arises from many aspects of understanding the concept of "context". When developing computer-based systems, it is important to define what is included and how the concept of context is understood. Of course, a fairly wide range of ICT infrastructure components can be integrated into the digitization of contextual data. Quite a variety of devices and software can be included to help capture environmental data: different wireless channels, heterogeneous wireless sensor networks (WNSs), different flow management and e-document management systems, and various monitoring systems. Heterogeneity arises when we want to include different sensors and different means of communication. Vehicles can also be equipped with a wide variety of specialized apparatus and systems. Roads are equipped with special tools and road infrastructure software components. One of the goals of this part of the study is the desire to convey a wide range of context recognition processes. Artificial intelligence (AI) methods can be used for context recognition and such components form the basis of intelligent service systems. The provision of smart services should be based on a wide range of management needs in freight transport processes. How primary data sources are incorporated into all possible interconnected infrastructures is analysed in this section.

Keywords Context awareness · Information Communication Technology (ICT) · Information Management System (IMS) · Processing methods · Data aggregation

D. Dzemydienė (✉)
Vilnius Gediminas Technical University, Vilnius, Lithuania

Vilnius University, Vilnius, Lithuania
e-mail: dale.dzemydiene@mif.vu.lt

D. Dzemydienė et al., *Development of Smart Context-Aware Services for Cargo Transportation*, International Series in Operations Research & Management Science 330, https://doi.org/10.1007/978-3-031-07199-7_2

2.1 Related Works in the Research Area of Context Analysis for Transportation

The wide spectrum of equipment and communication channels can be included in context-aware process. Wide spectrum of types of technologies, like Global Positioning System (GPS), Geographical Information Systems (GIS), Internet of Things (IoT) with different possibilities of wireless communication channels, wireless sensor networks (WSNs) are integrated for context analysis. The video recording and recognition of objects from these records can be included as well. For the right development of a smart service provision system, all these things can be characterized and classified.

Several types of sensors are included in communication infrastructure and one of them belongs to the set of road side units (RSUs). RSUs are forming a large part of the needful infrastructure with wireless communication possibilities. The question—how to include in the system all possibilities of RSUs for traffic management, hazard warnings, and systems of e-calls, etc., became an important integrational task. In the types of WNSs are included equipment for measures of speed, temperature, humidity, chemical materials, power, and others.

Regarding the development of advanced systems, the authors have proposed several solutions for the development of adaptive service systems (Dzemydienė et al., 2021; Bellavista et al., 2012; Bu et al., 2006). Heterogeneous sensors can be integrated into the system for monitoring processes of freight transportation. We can define such sets of physical devices by including:

- The set of sensors for environmental monitoring;
- The set of sensors for monitoring the mode of operations of vehicles;
- The set of sensors for human health monitoring.

Concrete possibilities of integration of functionality of sensors depend on the development of adapters for integrating intelligent subsystems with the functionality of GPS and GIS.

The problem of collection of information from the environment during transportation and its dissemination in dynamic transport means have some proposals for solutions in Buchholz et al. (2003) and de Oliveira et al. (2013). The analysis of development and application of the ontology for the representation of monitoring data can help in a clear understanding of the monitoring data. The application domain ontology is directly linked to a higher level of representation of semantic models in terms of concepts in the domain and can be helpful in computer-based applications of meta-models and structures of repositories of data warehouses. The domain ontology with other mechanisms for obtaining knowledge (i.e. rules, decision trees, etc.) is needful for designing the recognition algorithms of context and situations.

Many research works raise questions of understanding the context (Buchholz et al., 2003; Bu et al., 2006; Bellavista et al., 2012). A few definitions of a context and a variety of interpretations are presented in Table 2.1. Some authors provide

Table 2.1 Analysis of various aspects of context understanding in related research works

Aspects of context analysis	Definitions of context	Scientific works
Domain includes surroundings, environment, and people	The context is understood as the location and identity of the surroundings, people and objects, and their changing	Schilit and Theimer (1994)
Localization and recognition of meteorological data	Location and identity of people around the user, time of day, time of year, temperature, etc.	Brown et al. (1997)
Conceptual definition	Set of entities interested in physical conceptual states	Pascoe et al. (1998)
Application factors	Application settings	Rodden et al. (1998)
Environmental surroundings	Reification of the environment or anything that describes the surrounding environment where the system performs	Andrea and Savigni (2001)
Environmental factors	Time, environment, products, and company information	De la Vara et al. (2010)
Circumstance factors	A set of circumstances or facts around a particular event or situation	Buchholz et al. (2003)
Circumstance triggers	Circumstances that trigger an event, a claim, or the idea, which allows to understand it	Bu et al. (2006)

similar definitions of a context, with some interpretations. The word "context" has a very broad definition.

Assessing the quality and usefulness of data collection for operational traffic management has become a problem. Wireless channel flooding problems can arise with the implementation of a wide range of communication services based on both V2V and V2I communication types. The high number of data transmissions can affect the contingency of communication network channels. If the data is not filtered and sampled, we can get large amounts of useless information and unnecessarily overload large data repositories. Opportunities offered by communication channels focus on solving these problems. In the area of V2V communication arise the questions:

- What kinds of communication protocols can be implemented in the V2V communication infrastructure?
- How to collect primary data and transmit and disseminate information?

When the range of devices selected is limited, communication protocol standards and their capabilities should be considered. One of the solutions for this problem-solving is presented in IEEE Trial-Use Standard for Wireless Access (2006) and other standards (IEEE Std WAVE, 2010; IEEE, 2021).

Our proposals for the solving of such problems are concerning the development of the mechanisms for restriction of service provision by including the algorithms, which enable the prioritization of services and data delivery processes. The priorities are related to the goals of drivers at real-time moments, transport ergonomics, and especially with safety needs.

The concept of quality became important for understanding the context. The quality concept was introduced by Buchholz et al. (2003) as "a set of parameters that reflects the quality requirements and properties of context data". The research focuses on the interdisciplinary concept and its interpretation for the development of AI systems, by providing the possibilities of a combination of different methods. Some methods of behavioural science, operating and embedded systems, communication networks, and other fields are included. Such requirements create a huge gap between top requirements for the context of the complexity of operating mobile systems, services and operations, and the context that is derived from the environment. The adaptability, ingenuity, and flexibility were introduced as requirements for the platform of forming IoT by Bu et al. (2006).

Heterogeneity of devices, high level of mobility, and supplementing the topology of the system with the ability to understand context require high quality of management of context information, which is extremely complicated, where the assessment of errors and partial information can play a significant role.

Application possibilities of the specially created context-aware modules, which are adapted in changing context situations are described by Zhang et al. (2011). The management of context defined by a fast-changing topology in the environment, such as vehicle communication networks, is a challenge due to several factors by Wibisono et al. (2009):

- "Context becomes important due to high mobility of nodes. Context information is intricately linked to physical locations but updating this information may be challenging due to repeated disconnection of nodes from the source of context.
- Relevance of temporary context is subject to dynamic changes. Dynamic situations give temporary importance to the context".

Context can become ambiguous and redundant in V2V communication services in particular when Ad Hoc networks are created. The regulation of the flow of context information is difficult in the service provision process at the conditions of the Ad Hoc network. Such problems were described in Padovitz (2006). The author analyses the context state as the collection of current sensor readings.

The reasoning behind the recognition of context information has properties of multi-attribution and multi-dimensionality. The multi-attribute utility theory (MAUT) proposed by De Oliveira et al. (2013) for integration of heuristics rules and expert experience can be distinguished. The implementation of methods of the weighting of parameters became popular in decision support systems, allowing the description of dynamic situations.

The usage of the theory of the MAUT allows one to organize the space of situations in the model distinction. Information is then provided from the model (e.g. the state of the context and a description of the situation space) that calculates the degree of confidence that the situation has occurred. It is possible to calculate the reliability and their appearance in a certain interval of confidentiality. Such an approach enables the comparison of the calculated reliability with a specific situation of a certain threshold, which allows you to compare the results calculated for different situations.

The recording of primary data from dynamically moving transport means and roadside equipment is dealing with heterogenic structures of communication (Ye et al., 2012). An assessment of the situation attempts to the assessment of the wider spectrum of infrastructure and the given data rather than assessing one individual sector.

2.2 Understanding the Context of Information According to Scenario-Based Approaches

The concept of context can be examined by analyzing certain scenarios. The types of scenarios are classified such as the internal system scenario, interaction scenario, and/or contextual scenario. Research results in works of Padovitz (2006) and de Oliveira et al. (2013) help in the analysis of some types of scenarios. Pohl and Haumer (1997) made an introduction in modelling contextual information about scenarios.

For example, business goals are laid down and relate to services provided by a system, representing external relations of system stakeholders, using the information obtained by a system service or stating organizational policies.

The structuring of the context in information systems according to the scenario-based approach (Table 2.2) allows distinguishing the following:

- An internal scenario of the system to examine the operation of the system itself (e.g. an interaction scenario between the objects of the system itself);
- An interaction scenario depicting knowledge of the interaction of a system with its context;

Table 2.2 Review of approaches for description of context-aware scenarios

Approaches based on context-aware scenarios	Internal system scenario is analysed	Interaction scenario is analysed	Contextual scenario is analysed
Armour et al. (2009)	x	x	x
Benner (2019)	x	x	x
Holbrook (1990)		x	x
Jacobson (1995)		x	
Kyng (1995)	x	x	x
do Prado Leite et al. (2000)		x	x
Booch et al. (1996)		x	
Karculias (1997)	x	x	x
	x	x	x
Galán-Jiménez et al. (2019)	x	x	x

- Interaction scenarios help in the representation of context with the system. The diagrams which help to represent the sequences of messages (traces) are used for the representation of such scenarios;
- Contextual scenarios are quite difficult for the representation of bidirectional interaction between system and context.

The type of internal system scenarios is focused on the processes that especially happened inside of the system and do not detailed the processes in embedded systems. The type of interaction scenarios represents mainly processes on the system itself.

2.3 Methods for Linking of Contextual Data with Components of Information Management System

The linking of contextual data in the entire freight transport recognition process requires some kind of linking it to components of the scenario, rather than the whole scenario itself. The definition of related contextual knowledge requires a more detailed description. Context is very important in the recognition of the data obtained and affects not only the whole scenario but specific components of the scenario as well, e.g. a single or sets of interactions between the system and the system user.

Two models of semantical description of context are analysed. The models are presented by using the Unified Modeling Language (UML) and visual form of class diagram notation. One of such models is designed by paying more attention for transportation scenario (Fig. 2.1). The semantical model is designed by paying more attention for interaction between participating components in the description of transportation interaction scenarios (Fig. 2.2). Both models emphasize the main

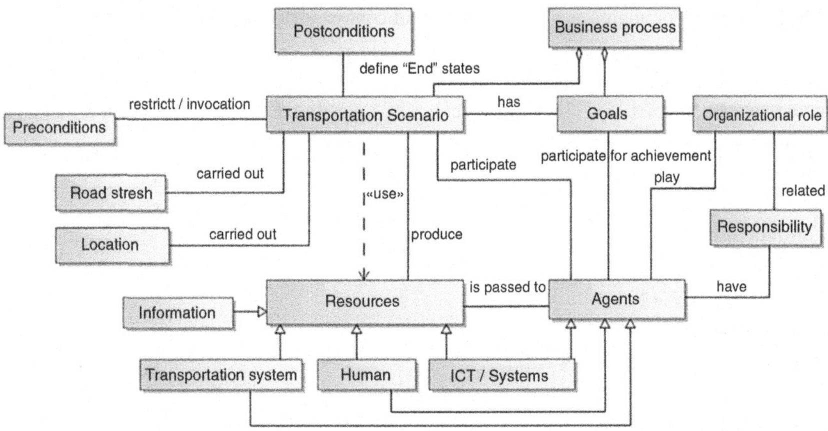

Fig. 2.1 Conceptual scheme of context of participating objects in the description of context in transportation scenario

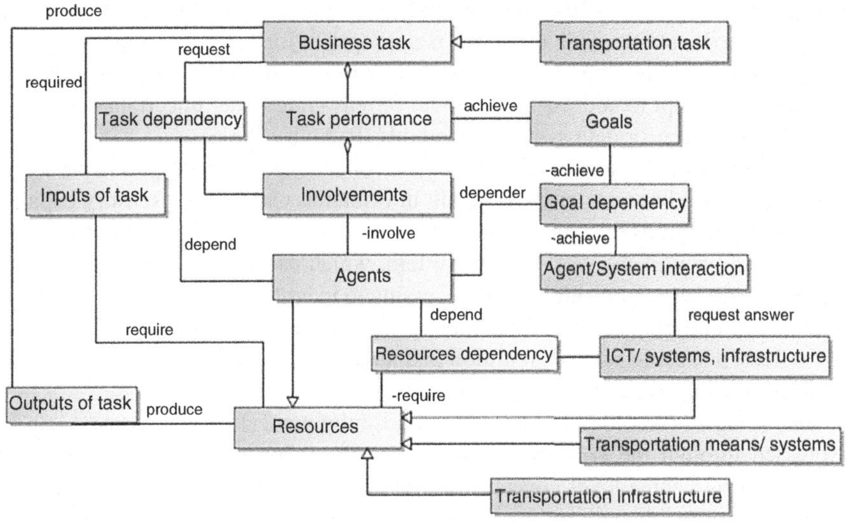

Fig. 2.2 Conceptual scheme of components comprising the interaction context

components for designing the ontology of context in the transportation. The relations between concepts are dealing with multi-level interactions of transport processes as well.

With the help of conceptual models, we can depict the interaction of components involved in business processes (Booch et al., 1996). When it is necessary to visualize scenarios of transportation processes in an intelligent system, we use analogous modeling tools.

The goals of transportation have preconditions of the implemented transportation scenario. Other important concepts can be included, for example, locating object's coordinates, initializing a session according to the geographic situation, recognizing sensor position data, and making different decisions in situation recognition.

Contextual knowledge can be related not only to the whole scenario but also to individual scenario components, e.g. a single interaction or sets of interactions between the system and the system user (Fig. 2.2).

All these components are described in detail through the representation of specific service provision structures.

2.4 Examples of Context Data Processing Methods

Recently research has focused on a variety of contextual aspects, including context middleware and toolkits for obtaining information, and ontologies that provide dictionaries for contextual description. The focus was on context abstractions and sharing to identify important context abstractions and characteristics. However, the issue of uncertainty, which is one of the fundamental ones in a sensor-based mobile environment, has been addressed very little. Also, attempts have been made to detect and reason data integrity, using methods based on ontologies to describe context-dependent applications. This approach is appropriate for finding inconsistencies in information and applying subject-specific rules for reasoning. Alternative approaches attempt to address the problem of reasoning under uncertain conditions by focusing on sensor data synthesis methods (Padovitz, 2006).

For analysing the context space model, the context-sensitive computing paradigm can be interpreted as an attempt to obtain information with limited sensory capabilities under conditions relevant to the scope. The nature of the context can be treated as a limited view of the world to the system, which can be used immediately (for trigger actions) or require additional processing to identify event situations (for more detailed reasoning). Thus, we fundamentally draw the line between a set of events that reflects a specific system condition and an understanding of a situation that can be derived using contextual knowledge.

According to these descriptions, the philosophy associated with context can be expressed through the context-situation pyramid, which defines the hierarchy of abstraction of a three-level concept. The first is the basic level of raw data. This information (possibly with some processing) is used to create the concept of context, e.g. context is the information used in a model to describe real-world situations.

Geometric metaphors can be used for the description of the context space, the context and situations as first-class model objects. Let us start with the description of the application space—a set of discourses, according to the available information in the application context. The type of information is defined as a context attribute and denoted as a_i. The value of the sensor scan at time moment t_i is the value of the context attribute at time t_i and is denoted as $a\_i \wedge t_i$. A scope is a multi-dimensional space made up of a domain of values for each context attribute in which a context can be captured. It helps to perceive sub-spaces (possibly described in fewer dimensions) that reflect real-world situations. These sub-spaces are called situational spaces. Situation spaces are described as areas of acceptable values in selected dimensions and represent sets of values that reflect real-world situations. The acceptable range of values is denoted as $A\_i \wedge j$ and is described as a set of elements V that satisfies the predicate:

$$A\_i{\wedge}j = \{V \mid P(V)\}.$$

The situation space is presented as an ordered list of those domains and is denoted as $S\_j = (A\_1 \wedge j, A\_2 \wedge j, \ldots, A\_n \wedge j)$ and is consisting of n acceptable domains of these attributes. The actual values of the information received by sensors are defined by the state of the context, e.g. a set of readings for current sensors.

Situation space is made up of three-dimensional context attributes that satisfy individual predicates. The state of the context is plotted at different time moments $(C\_i \wedge t1$ & $C\_i \wedge t2)$, representing a set of different specific values. At time t_1, the state corresponds to (or is included in) the description of the situation space, and at time t_2, the position of the context state is beyond the description of the situation space.

Different methods of reasoning can be used for the context and situations represented through context spaces with uncertainty, first using new reasoning based on Multi-attribute Utility Theory (MAUT), then moving to other reasoning methods, such as the Dempster–Shafer theory.

Different heuristics have been used for effective reasoning about the context. In artificial intelligence, heuristics depict rules based largely on expert experience or common sense and are used as guidelines for problem-solving. Using the MAUT is recommended as a way to integrate such heuristics into a measure of confidence that reflects the degree of confidence in the occurrence of a situation.

For the expression of information, significance is introduced to the significance function. To model these differences in the significance of context attributes for a given situation, a significance function is defined, assigning weights w_1, $w\_2, \ldots,$ $w\_n,$ $(w\_i \in [0,1], \sum w\_(i = 1) \wedge n \ [\![w\_i = 1]\!])$ for context attributes. The weights reflect the importance of each attribute (compared to other attributes) in describing the situation.

The contribution (i.e. utility value) function can be described as the significance function model for expressing the relative importance of the attributes of the situation space. The contribution function models have the individual contribution of elements from a specific domain that describe the situation. Instead of just knowing: is the value in the field or not? The value itself is estimated. The set of values, which belong to the range can become an indication of a certain situation. The value being in a certain range (i.e. determining the value of $c \in [0,1]$ for that element) can be described as a contribution function.

The contribution of the value of the context attribute may be affected by the accuracy of the sensor readings. Facts from an inaccurate sensor can be incorporated into a contribution function that can provide low levels of support for values that are similar but fall outside the scope. This is similar to membership functions in fuzzy sets, where functions calculate the degree of membership between a value and a known set. The input function allows modelling uncertainty about sensor readings and incorporating them as a part of the reasoning process.

Applying previous definitions of context data quality and based on the modelled context information, data with uncertainty or low accuracy can be used in the system to determine the occurrence of relevant situations. Thus, situation occurrence indicators (or evidence support) are expressed as values of context attributes in the context data state under consideration, according to their importance in acceptable areas of the situation. These values have their ranges, and in the analysis of a specific situation, they indicate the place of each attribute within a certain range. The fact that some of the values of the context attributes are beyond the scope of situations in its domain weakens the likelihood of the occurrence of that situation.

The state of the context is related and can be expressed through the description of the situation space and assess the degree of confidence in the likelihood of the occurrence of that situation. The calculated confidence is compared to the confidence threshold, thus facilitating the decision regarding the likelihood of the occurrence of the situation (comparing the calculated confidence with the individual threshold of a specific situation, which allows comparing the results calculated for different situations).

MAUT provides a convenient way to combine seemingly different contributions into a single measurement, expressing the result as a benefit. A benefit and contribution to the goal of identifying the likelihood of the occurrence of the situation are provided as evidence supporting the hypothesis of occurrence of the situation when the values of the context attributes are in the relevant domain. Indicators can show the state of the context and express the matches to the description in the situational space.

The MAUT is considered to be an evaluation scheme that provides an overall evaluation function *v(x)* for object *x* to indicate the overall benefit of the object. The valuation function is traditionally defined as the weighted sum of the dimensions of the objects being valued, which reflect different contributions relevant to the object (2.1). The result of the calculation in one numerical measurement varies between 0 and 1. The following is the expression of the function used to calculate confidence in the occurrence of a situation, taking into account the ideas discussed, the result of ranges between 0 and 1:

$$\text{Confidence} = \sum_{i=1}^{n} w_i c_i, \tag{2.1}$$

where w_i denotes the weight assigned to the contribution function area i in the situation space, and c_i indicates the contribution level of the value of the context attribute i in the situation space description.

Having come across any sensor inaccuracies, the reasoning procedure is extended to include heuristic methods in the confidence measurements, which allows integrating additional knowledge about sensor inaccuracies as a part of the reasoning process. This heuristic approach is based on the reasoning that the higher the probability that a context attribute belongs to a domain, the greater the contribution of the context attribute in decision-making, and vice versa. This heuristic approach provides capabilities and a way to calculate the level of a sensor display input during service execution rather than modelling it during development. For example, dim light in a room can be a strong indicator of a presentation being made. However, if a light sensor is inaccurate, then the contribution of the detected dimmed light to the likelihood of a presentation must be reduced. This heuristic method makes it possible to distinguish more precisely the information available for reasoning under uncertain conditions.

The measurement of confidence using this heuristic method is expressed as

$$\mu(S) = \sum_{i=1}^{n} w_i \times \Pr\left(\widehat{a}_i^t \in A_i\right), \tag{2.2}$$

where $\Pr\left(\widehat{a}_i^t \in A_i\right)$ represents the reliance on the fair value reading in the relevant range of acceptable values.

Two types of context attributes are distinguished according to the definition of a situational space, which has different effects on reasoning outcomes:

- *Symmetric contribution*: A context attribute that increases confidence in the likelihood of the occurrence of a situation if its values fall within a given range and reduces the reliance thereon if they are outside this range.
- *Asymmetric contribution*: Contextual attributes that increase confidence in the likelihood of the occurrence of a situation if its value falls within the appropriate range but the sensor values are outside the range, do not reduce the calculated confidence.

Applications of the heuristic approach for reasoning purposes require constraining asymmetric attributes in the corresponding acceptable range of values. Asymmetric attributes outside of their respective fields are ignored, recalculating the weights (in case of symmetric contribution) of the remaining context attributes to maintain their relative importance.

2.5 Needs for Aggregation of Contextual Data and Possible Solutions

Vehicles can receive, collect, and transmit contextual information. Vehicles, which can obtain contextual data from the surrounding network nodes, can transmit such information to other neighbouring network nodes within the coverage area or transmit data packages to the central data collection server. Such data transmission increases the number of transmitted packets if they contain similar or identical information. Data aggregation methods are used to solve these inefficient data transmission problems. The essence of these methods is to reduce the amount of data transmitted by combining the relevant information into an aggregate data cluster. This allows vehicles to combine, update, or delete certain information, as it may be duplicated, similar, or outdated. For example, two data samples for two cars can be replaced by one sample with a small error, if the cars are next to each other and moving at a similar speed (Kakkasageri & Manvi, 2014b).

The disadvantage of aggregation mechanisms is the decrease in the quality of contextual information. Aggregation methods, in some way, optimize data flow/storage needs by integrating different information from different cars. Due to the specific requirements of automotive communication networks discussed, the development of such methods is a particularly challenging task.

The main characteristics which are taken into account when developing of the aggregation methods for automotive communication networks according to Dietzel, Kargl, et al. (2011); Dietzel, Schoch, et al. (2010); Dietzel et al. (2011) are the following:

- The aggregation process must not be demanding for computing and communication resources.

- Loss of context quality due to aggregation must be kept to a minimum.
- Aggregated information must not contain duplicate or redundant information.
- The aggregation process must integrate contextual information, according to the location of the vehicle and the time of receiving the information.
- It must be possible to exchange aggregated information with other nodes.

2.6 Centralized Aggregation of Context Data

In centralized aggregation, one node aggregates data centrally. One of the most recent works in this area has been proposed by Zhu et al. (2015). The authors offer a method called a Tree that aggregates data based on the shortest tree path, then assigns a wait time to each node using dynamic programming methods. A comparison of taxi route information found that this method generated less redundant information compared to other solutions. One of the best-known centralized aggregation methods is VESPA (Defude et al., 2008; Delot et al., 2010). The method has been designed to process and transmit any type of event-based information, such as traffic accidents, emergency stops, free parking spaces, and more. The method is based on the concept of Encounter Probability, which exchanges data between cars and generates information that can then be used by drivers. When the data becomes obsolete, it is deleted.

Algorithms of this type lack solution for not increasing the latency of context data exchange required for security services and for not reducing the quality of context data during the aggregation process.

2.7 Distributed Context Data Aggregation

Applying the distributed aggregation methods, each car aggregates data locally. These methods are very reliable, but require large wireless communication resources and generate significant data traffic for service information as the number of network nodes increases. Freschi et al. (2014) have offered such a solution. This method has been designed to reduce data amount by monitoring pavement damage. Data are collected using smartphones and then aggregated according to characteristic readings based on spatial/temporal characteristics. Another method has been offered by Yu et al. (2012). This is an adaptive forwarding control method called Catch-Up. It dynamically changes the transfer rate, making it more likely to be received and aggregated. This method is based on the Distributed Learning Algorithm. Each car pays off using local knowledge and chooses the optimal delay time.

The systems and methods analysed lacked solutions to avoid duplication and overlap. In the transmission of event-based messages, this problem becomes particularly pronounced with increasing network density, so that a large number of nodes transmit the same information. Intelligent adaptive solutions to these problems are needed to filter out duplicate data, thus achieving efficient network scalability.

2.8 Cluster-Based Aggregation of Context Data

In cluster-based aggregation, nodes aggregate data into groups. The amount of official information is reduced because the data is aggregated in parallel. Such a method increases the efficiency of the system by reducing the dynamics of the VANET topology. VeSCA (Ucar et al., 2014) have proposed this type of algorithm. In the algorithm, mobile nodes are clustered based on relative mobility to minimize the amount of service information. Cluster members apply data aggregation before transmitting data to the main cluster node. In terms of data aggregation ratio, latency and data delivery ratio, the proposed method outperforms other previous methods of cluster-based aggregation.

A solution for semantic data aggregation and the method for division of the path into segments that are based on the cluster-ID is offered by Koyamparambil Mammu et al. (2015). The data is stored in data structures consisting of a supercluster, a cluster, and cluster member nodes. This method uses an algorithm to decide on the data to be used in aggregation, determining its level by the density of cars and the occupancy rate of the channel. Kakkasageri and Manvi (2011, 2013, 2014a, b) have widely explored these problems. The aggregation solutions proposed by the authors are based on multi-agent systems that take into account the speed, direction, mobility, and communication quality of a car, deriving the network stability indicator based thereon. Gricius et al. (2015) further explore multi-agent systems.

Most of the cluster-based aggregation methods reviewed have been adapted from the methods used in MANET networks, but they do not take into account important aspects of automotive communication networks, such as highly dynamic environments, special mobility models, and car movement scenarios.

2.9 Conclusions

The analysis of definitions of context-aware information is important for the development process of smart service provision systems. The description of possibilities to provide the conceptual structures of the context is important in developing computer-based ontology.

The provided comparative review of methods for context understanding in literature helps us in the construction of a wider spectrum of requirements for description of context in the system of context recognition and define multi-spectral understanding of cargo transportation processes. The multi-dimensional structures of ISs are revealed, which can be implemented in real processes of management as well in the smart service provision system development. The semantical models of the descriptional layers form the ontological views of the domain and their structure of context. Different aspects of proposals for definition of context are described and help in understanding of forming of the approach for description of context in the smart service delivery system. By revealing of main properties of the context, the provided analysis show that the context became an important component in understanding of different situations and can help us in the development processes of more adaptable service provision systems.

References

Andrea, A. F., & Savigni, A. (2001). A framework for requirements engineering for context-aware services. In *Proc. of 1st international workshop from software requirements to architectures (STRAW 01)* (pp. 200–201).

Armour, M., Rivaux, S. L., & Bell, H. (2009). Using context to build rigor: Application to two hermeneutic phenomenological studies. *Qualitative Social Work, 8*(1), 101–122. https://doi.org/10.1177/1473325008100424

Auld, J., Mohammadian, A. K., & Doherty, S. T. (2009). Modeling activity conflict resolution strategies using scheduling process data. *Transportation Research Part A: Policy and Practice, 43*(4), 386–400.

Bellavista, P., Corradi, A., Fanelli, M., & Foschini, L. (2012). A survey of context data distribution for mobile ubiquitous systems. *ACM Computing Surveys, 44*(4), 1–45. https://doi.org/10.1145/2333112.2333119

Benner, M. (2019). Smart specialization and institutional context: The role of institutional discovery, change and leapfrogging. *European Planning Studies, 27*(9), 1791–1810. https://doi.org/10.1080/09654313.2019.1643826

Booch, G., Rumbaugh, J., Jacobson, I. (1996). *The unified modeling language for object-oriented development.* Rational Software Corporation.

Brown, P. J., Bovey, J. D., & Chen, X. (1997). Context-aware applications: From the laboratory to the marketplace. *IEEE Personal Communications, 4*(5), 58–64. https://doi.org/10.1109/98.626984

Buchholz, T., Küpper, A., & Schiffers, M. (2003). Quality of context: What it is and why we need it. In *Proceedings of the work shop of the HP Open View University Association* (pp. 1–14). http://doi.org/10.1.1.147.565

Bu, Y., Gu, T., Tao, X., Li, J., Chen, S., & Lu, J. (2006). Managing quality of context in pervasive computing. In *Sixth International Conference on Quality Software*. QSIC. https://doi.org/10.1109/QSIC.2006.38

Defude, B., Delot, T., Ilarri, S., Zechinelli, J.-L., & Cenerario, N. (2008). Data aggregation in VANETs: The VESPA approach. In *Proceedings of the 5th Annual International Conference on Mobile and Ubiquitous Systems: Computing, Networking, and Services* (pp. 13:1–13:6). ICST. https://doi.org/10.4108/ICST.MOBIQUITOUS2008.3897

De la Vara, J., Ali, R., Dalpiaz, F., Sánchez, J., & Giorgini, P. (2010). Business processes contextualisation via context analysis. In J. Parsons, M. Saeki, P. Shoval, C. Woo, & Y. Wand (Eds.), *Conceptual Modeling – ER (SE – 37)* (Vol. 6412, pp. 471–476). Springer. https://doi.org/10.1007/978-3-642-16373-9_37

Delot, T., Cenerario, N., & Ilarri, S. (2010). Vehicular event sharing with a mobile peer-to-peer architecture. *Transportation Research Part C: Emerging Technologies, 18*(4), 584–598. https://doi.org/10.1016/j.trc.2009.12.003

Dietzel, S., Kargl, F., Heijenk, G., & Schaub, F. (2011). Modeling in-network aggregation in VANETs. *IEEE Communications Magazine, 49*(11), 142–148. https://doi.org/10.1109/MCOM.2011.6069721

Dietzel, S., Petit, J., Heijenk, G., & Kargl, F. (2013). Graph-based metrics for insider attack detection in VANET multihop data dissemination protocols. *IEEE Transactions on Vehicular Technology, 62*(4), 1505–1518.

Dietzel, S., Schoch, E., Konings, B., Weber, M., & Kargl, F. (2010). Resilient secure aggregation for vehicular networks. *IEEE Network, 24*(1), 26–31.

Dzemydienė, D., Burinskienė, A., & Miliauskas, A. (2021). Integration of multi-criteria decision support with infrastructure of smart services for sustainable multi-modal transportation of freights. *Sustainability, 13*(9), 4675. https://doi.org/10.3390/su13094675

de Oliveira, K. M., Bacha, F., Mnasser, H., & Abed, M. (2013). Transportation ontology definition and application for the content personalization of user interfaces. *Expert Systems with Applications, 40*(8), 3145–3159. https://doi.org/10.1016/j.eswa.2012.12.028

do Prado Leite, J., Hadad, G., Doorn, J., & Kaplan, G. N. (2000). A scenario construction process. *Requirements Engineering, 5*(1), 38–61. https://doi.org/10.1007/PL00010342

Freschi, V., Delpriori, S., Klopfenstein, L. C., Lattanzi, E., Luchetti, G., & Bogliolo, A. (2014). Geospatial data aggregation and reduction in vehicular sensing applications: The case of road surface monitoring. In *Connected Vehicles and Expo (ICCVE), 2014 International Conference on*. https://doi.org/10.1109/ICCVE.2014.7297643

Galán-Jiménez, J., Berrocal, J., Garcia-Alonso, J., & Azabal, M. J. (2019). A novel routing scheme for creating opportunistic context-virtual networks in IoT scenarios. *Sensors, 19*(8), 1875.

Gricius, G., Drungilas, D., Andziulis, A., Dzemydiene, D., Voznak, M., Kurmis, M., & Jakovlev, S. (2015). Advanced approach of multiagent based buoy communication. *The Scientific World Journal, 2015*, 1–6. https://doi.org/10.1155/2015/569841

Holbrook, H., III. (1990). A scenario-based methodology for conducting requirements elicitation. *ACM SIGSOFT Software Engineering Notes, 15*(1), 95–104.

IEEE Std 1609.2-2006. (2006). *IEEE trial-use standard for wireless access in vehicular environments – Security services for applications and management messages* (pp. 0_1–105). https://doi.org/10.1109/IEEESTD.2006.243731

IEEE Std 1609.1-2006. (2006). *Trial-use standard for wireless access in vehicular environments (WAVE) – Resource manager* (pp. 1–71). https://doi.org/10.1109/IEEESTD.2006.246485

IEEE Std 802.11p-2010. (2010). IEEE standard for local and metropolitan area networks – Specific requirements Part 11: Wireless LAN medium access control (MAC) and physical layer (PHY) specifications amendment 6: Wireless access in vehicular environments, 51.

IEEE Std WAVE. (2010). Standard for wireless access in vehicular environments (WAVE) – Networking services – Redline.

IEEE Std 2857-2021. (2021). *IEEE standard for wireless smart utility network field area network (FAN)* (pp. 1–182). https://doi.org/10.1109/IEEESTD.2021.9464888

Jacobson, I. (1995). The use-case construct in object-oriented software engineering. In *Scenario-based design: Envisioning work and technology in system development* (pp. 309–336).

Kakkasageri, M. S., & Manvi, S. S. (2011). Safety information aggregation in VANETs using vehicle beliefs. In *Advanced networks and telecommunication systems (ANTS), 2011 IEEE 5th International Conference on*. https://doi.org/10.1109/ANTS.2011.6163641

Kakkasageri, M. S., & Manvi, S. S. (2013). Roadside networks for vehicular communications. In R. Daher & A. Vinel (Eds.). IGI Global. https://doi.org/10.4018/978-1-4666-2223-4

Kakkasageri, M. S., & Manvi, S. S. (2014a). Information management in vehicular ad hoc networks: A review. *Journal of Network and Computer Applications, 39*, 334–350. https://doi.org/10.1016/j.jnca.2013.05.015

Kakkasageri, M. S., & Manvi, S. S. (2014b). Regression based critical information aggregation and dissemination in VANETs: A cognitive agent approach. *Vehicular Communications, 1*(4), 168–180. https://doi.org/10.1016/j.vehcom.2014.07.001

Karculias, P. N. (1997). *A taxonomy of scenarios in software development*. The University of Texas at Arlington.

Koyamparambil Mammu, A. S., Jiru, J., & Hernandez Jayo, U. (2015). Cluster based semantic data aggregation in VANETs. In *Advanced information networking and applications (AINA), 2015 IEEE 29th International Conference on*. https://doi.org/10.1109/AINA.2015.263

Kyng, M. (1995). Creating contexts for design. In *Scenario-based design: Envisioning work and technology in system development* (pp. 85–107).

Padovitz, A. (2006). Context management and reasoning about situations in pervasive computing. Retrieved from https://books.google.lt/books/about/Context_Management_and_Reasoning_about_S.html?id=iIdiPAAACAAJ&pgis=1

Pascoe, J., Ryan, N. S., Morse, D. R. (1998). Human computer giraffe interaction: HCI in the field. Workshop on human computer interaction with mobile devices. Retrieved from internet. http://kar.kent.ac.uk/21665/1/Human_Computer_Giraffe_Interaction_HCI_in_the_Field.doc

Pohl, K., & Haumer, P. (1997). Modelling contextual information about scenarios. In *Third international workshop on requirements engineering: Foundation for software quality RESFQ', June 16–17, 1997, Barcelona, Spain.*

Rodden, T., Chervest, K., Davies, N., & Dix, A. (1998). Exploiting context in HCI design for mobile systems. In *Workshop on human computer interaction with mobile devices*. Retrieved from https://citeulike-article-id:3466581

Schilit, B. N., & Theimer, M. M. (1994). Disseminating active map information to mobile hosts. *IEEE Network, 8*(5), 22–32. https://doi.org/10.1109/65.313011

Sheikh, K., Wegdam, M., & van Sinderen, M. (2008). Quality-of-context and its use for protecting privacy in context aware systems. *Journal of Software, 3*(3), 83–93. Retrieved from http://www.academypublisher.com/jsw/vol03/no03/jsw03038393.html

Ucar, S., Ergen, S. C., & Ozkasap, O. (2014). VeSCA: Vehicular stable cluster-based data aggregation. In *Connected Vehicles and Expo (ICCVE), 2014 International Conference on*. https://doi.org/10.1109/ICCVE.2014.7297517

Wibisono, W., Zaslavsky, A., & Ling, S. (2009). Improving situation awareness for intelligent on-board vehicle management system using context middleware. In *2009 IEEE intelligent vehicles symposium* (pp. 1109–1114). IEEE. https://doi.org/10.1109/IVS.2009.5164437

Ye, J., Dobson, S., & McKeever, S. (2012). Situation identification techniques in pervasive computing: A review. *Pervasive and Mobile Computing, 8*(1), 36–66. https://doi.org/10.1016/j.pmcj.2011.01.004

Yu, B., Xu, C.-Z., & Guo, M. (2012). Adaptive forwarding delay control for VANET data aggregation. *Parallel and Distributed Systems, IEEE Transactions on*. https://doi.org/10.1109/TPDS.2011.102

Zhang, D., Huang, H., Lai, C.-F., Liang, X., Zou, Q., & Guo, M. (2011). Survey on context-awareness in ubiquitous media. *Multimedia Tools and Applications, 33*, 179. https://doi.org/10.1007/s11042-011-0940-9

Zhu, Y., Zhao, Q., & Zhang, Q. (2015). Delay-constrained data aggregation in VANETs. *Vehicular Technology, IEEE Transactions on*. https://doi.org/10.1109/TVT.2014.2335232

Chapter 3
Wireless Communication Network Infrastructure Analysis for Smart Service System Functioning

Dalė Dzemydienė

Abstract The communication infrastructure, as the basis for technical capabilities, provides the preconditions for interoperability between wireless computer networks and ground vehicles. Such infrastructure supports and enables the selection of mobility scenarios and helps to manage real transport situations. The communication infrastructure between wireless networks and terrestrial vehicles is described using an analysis of the methods proposed in the literature and the results obtained in the communication professional field. The aim is to identify the needs and capabilities of existing wireless networks to provide the necessary e-services on road infrastructure. The field of communication infrastructure is changing rapidly and is an area being developed by communication professionals, but the equipment is also involved in the operations of application systems (smart services). An overview of some important aspects of the applications of wireless communication networks is provided in this chapter by describing the main needs of wireless network infrastructure, the specifics of their characteristics, the results of the analysis of the characteristics of communication protocols, and their capabilities.

Keywords Wireless networks · Smart services · Service delivery system · Mobile connection · Automotive communication

3.1 Issues of Infrastructure Development for Delivery of Smart Services for Transportation

The development of a smart service delivery system is being addressed with the aim of helping to manage freight transport processes more efficiently, avoiding gaps in freight delivery processes and improving sustainable development. The process of

D. Dzemydienė (✉)
Vilnius Gediminas Technical University, Vilnius, Lithuania

Vilnius University, Vilnius, Lithuania
e-mail: dale.dzemydiene@mif.vu.lt

D. Dzemydienė et al., *Development of Smart Context-Aware Services for Cargo Transportation*, International Series in Operations Research & Management Science 330, https://doi.org/10.1007/978-3-031-07199-7_3

developing a system with intelligent components involves the use of best practices and helps to increase the efficiency of freight transport and the selection of new types of "green" freight corridors. Achieving the goals of more sustainable transport processes requires new efforts to develop infrastructure, reduce the burden on road transport, and increase freight transport by multimodal transport.

The EU expects that by 2030, 30% of freight transported by road over 300 km will be shifted to rail, water, or other modes. The EU has also faced new challenges, obligating Member States to take actions to achieve the 2050 target and increase autonomous transport capacity by more than 50%. These objectives will have an impact on the new development of transport corridor infrastructure with new transport options. Concrete seeks to focus and focus the efforts of all stakeholders on developing higher levels of transport process management efficiency. Such activities are mentioned in the Digital Agenda 2030. The Sustainable Development Agenda has led to lower emissions and inspires significant reductions in freight transport pollution by 2030 (COM, 2011; UN, 2021).

New insights are needed to assess multi-level transport infrastructure. Particular attention is paid to the development of smart services and the integration of IS and DW into the integrated management process to provide the necessary support.

Such activities are mentioned in the Digital Agenda until the 2030 year. The Sustainable Development Agenda activated achievements of lower levels of emissions and inspires significantly to reduce pollution in freight transport until the 2030 year (COM, 2011; UN, 2021).

By 2050, total freight traffic was expected to increase by about 50%. But in the conditions of Russia's unexpected war against Ukraine, unpredictable changes will occur. The digitalization of the freight transport sector will also become a major challenge in the coming years.

New insights are necessary for assessing of multi-layered infrastructure of transportation. Special attention is paid to providing necessary support for the development of smart services and the integration of ISs and DWs into a complex management process.

Examples of provision of semi-automatic services in organizing of trips are described in Dzemydienė et al. (2016); COOPERS (2010), and the results of the assessment of infrastructure of heterogenic service provision in freight transport are described in Dzemydienė et al. (2020).

The aims of the Digital Strategy of EC for achieving the results during 2021–2030 years will affect the efforts for the development of multi-layer infrastructure of ICT and will help in assisting in the development of smart services for transportation, standards and communication protocols, safety means and usage of open information structures (EC, 2021). Digital strategy largely integrates the policy of instruments of the Digital Agenda for Europe 2021–2030, forwarding reflections into the new EU-Central and Asia directions and influencing the activities of regrouping efforts according to the original objectives.

Innovation will be possible in transportation area, if the relevant types of information and communication technologies (ICT) are developed and their multiple functionality is enabled:

- Focusing research investigations for artificial intelligence;
- Implementation of cloud computing services;
- Application of the blockchain technology;
- Use of supercomputers and quantum technologies;
- Development of platforms for the Internet of Things (IoT).

The functionality of smart service provision systems can support under the background of environmental infrastructure that cannot be destroyed. The risks of such unforeseen threats arise with Russia's unforeseen and aggressive war against Ukraine and the world which started on 22 February 2022.

The aims of the new research projects are in line with new requirements. The development of processes focuses on the establishment of new kinds of ICT, system applications, and innovations. EU initiatives are related to significant investments in direct programmes for infrastructure development which are important tasks for the activation of strategies at the national level of EU Member States in particular and integration of activities with countries of the EU-Asia region.

A wide spectrum of investigations related to ICT development and innovative technologies are implemented in management processes of multimodal transport. The EU also aims to achieve more than 50% of innovative technologies, which are planned for implementation until 2050 year (C-ROADS, 2020). A more innovative and larger infrastructure has been planned in the development of corridor networks, including efforts to achieve high efficiency and low emission of freight transport (C-ROADS, 2020).

New ICT development directions cover new actions and tasks:

- To develop Single Digital Gateway-based complex services;
- To enable platforms for the free movement of data and protection of personal data;
- Efforts were taken to create more complex means for cybersecurity.
- Appeal to consumers, creation of a reliable technological platform for well-organized provision processes, and data collection have become important requirements for ensuring effective provision of digital services.
- The development of smart transportation services has been ongoing, particularly for priority functions, including:
- The development of smart, user-oriented, invisible, proactive, paperless services accessible to all;
- More intense use of data analytics tools, especially in the development of smart services;
- Encouragement of the development of a new generation of communications (5G and 100 Mb/s Internet access), which will primarily enable wider implementation of various innovations to meet the needs of businesses and population (e.g. IoT).

Horizontal assumptions that enable digital transformation are particularly important in the development of decision support systems, which can work online and enable the monitoring of transportation processes in real time.

It is important to develop and upgrade mass public digital services, including the implementation and improvement of interoperable solutions and key enablers. Solutions for the analysis of grant data used on artificial intelligence technologies play an important role. Subsidies for the creation, development, and renewal of complex public services are also important. The process of development of such services includes digitization of internal and external processes of institutions required for the provision of e-services.

Development and implementation of big data analytics tools (using artificial intelligence (AI) methods and technologies) and subsidies for the performance and improvement of basic infrastructure (e.g. Information Resource Interoperability Platform for service accessibility technologies) affect the possibilities of development of more effective smart transport services.

The research area is concerned with transportation and all modes of transport in multimodal transportation infrastructure. The approach for providing drivers with e-services, including the processes of estimating data flows, is necessary. Our research aims to assess smart services in freight delivery processes following network types. The development of the proposed methodology is based on methods of evaluation of the transport network and the structure of organizational and managerial procedures to evaluate data flow conditions and overloads and to minimize time gaps in data reporting during freight transport processes under multimodal transportation conditions.

The development of a new type of transportation support service is related to the development of intelligent transport systems (ITS) and the spread of the implementation of the advancing functions of ICT. New kinds of wireless sensor systems on road sites and vehicles have been used in the management and control of transport means (C-ITS platform, 2016; C-ROADS, 2020; C-mobILE, 2017). However, adaptable interaction between means of transport and infrastructure of road site units (RSU) requires new studies, especially those of smart and autonomous subsystems.

Achievements in the communication area named Vehicle Ad Hoc Networks (VANET) have an important influence on the development of new directions of ICT support for transport communication. Results of such achievements are described in various research works and C2C Communication Consortium reports (Chen & Tsao, 2015; Kakkasageri & Manvi, 2014; CAR 2 CAR Communication Consortium, 2011, 2014). There are several ways of use of VANET with long-distance and short-distance communication and some examples of them are proposed by (IEEE 2010; IEEE 802.11p-2010). Common ICT means, and the Internet of Things (IoT) in particular, are based on mobile protocols, such as 4G, 5G, or upcoming 6G networks (Campolo et al., 2015; Du et al., 2020; Kranz, 2016; Zagrouba & Kardi, 2021), also using short-range technologies, such as Wi-Fi and/or DSRC, and/or RFID (Lee et al., 2014).

Various types of sensors can be integrated into vehicles for monitoring engines, human health, and the environment, providing monitoring data and allowing to recognize context information more accurately. Sensors are becoming the sources for collecting row data about monitoring of transport means, goods, surroundings, and people and retranslating such data into needful information for using to sense the context of freight transport and make operative decisions in management processes. As well the capabilities of wireless sensors networks (WSN) are important too.

3.2 Possibilities of Integration of Different Services in the Changing Topology of Transportation and Communication Networks

Various sensors for monitoring cars, humans, and the environment and other information sources can be used to explain data for description of the context. The car's sensor system, as a source of information, generates large amounts of different information, including physical sensors from the car and the environment: GPS, speed, acceleration, temperature, radar, video, and virtual sensors. The content information can include:

- Traffic status information,
- Hazard warnings,
- Information about interactions with other vehicles,
- Information from systems like e-calls, monitoring of car's engine and working regime of vehicles and monitoring of surroundings, and similar.

Our previous research works in the development of intelligent transport systems (ITS) are proposed several solutions for the development of systems with providing possibilities of adaptive services (Dzemydienė & Burinskienė, 2021; Dzemydienė et al., 2019; Bielskis et al., 2012; Drungilas & Bielskis, 2012; Dzemydienė et al., 2010; Dzemydiene & Dzindzalieta, 2010; Gricius et al., 2015). The further development of adaptive systems for providing intelligent services requires the integration of direct identification of the surrounding environment and data transmission management methods.

Recently, the field of mobile communication networks with a changing topology has received a lot of attention, where vehicle drivers are facing an ever-increasing flow of information, making it important to retrieve useful knowledge from a variety of data sources and present it properly to augmented reality service systems. An important question is how to provide and adapt (adapt according to the needs of the user) various services in real time, according to the knowledge extracted from the context. Providing services to the driver through augmented reality systems raises several questions:

- What information is relevant or more relevant to the subject area (questions of prioritization)?
- What services are of interest to the consumer (issue of service provision and adaptation to needs)?
- The importance of security services should be provided automatically.

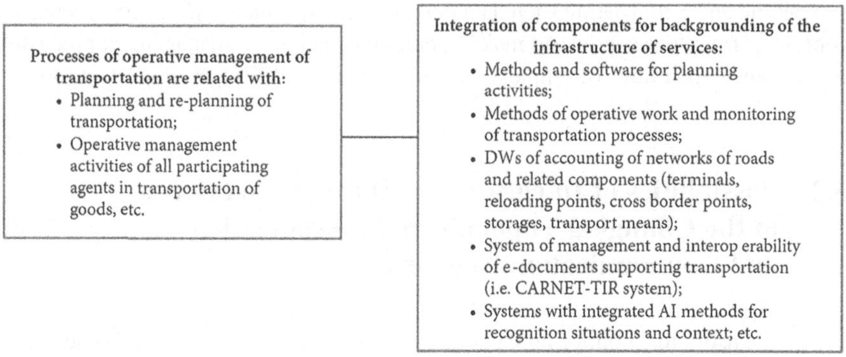

Fig. 3.1 Illustration of the multi-components integrated for management of transportation

The problem of efficient collection and dissemination of environmental information arises when developing a system for adapting services into dynamic environments (Dzemydienė et al., 2021; Kakkasageri & Manvi, 2014). The perception of contextual information and its proper interpretation in a computer system requires methods that allow the collection of relevant information, its understanding (or aggregation) and transmission, and its direct connection to the extraction of knowledge at a higher level of abstraction, i.e. the concept of environmental situations. Under these situations, the process of selecting, adapting, and improving the quality of relevant services would be automated.

In the processes of data collection, aggregation, and dissemination from different types of sensors, the specific problems are caused by the high mobility of the network, resulting in a rapidly changing topology and complex interpretation of information that requires a lot of network resources. To have a better understanding of the environment, it is necessary to exchange the available data with other nodes in the network, which in this case are cars, thus forming a cooperative system. When network nodes flood the network, transmitting all the collected context data to all surrounding nodes without any information about the network topology, a network load problem arises, leading to slow down or even complete downtime of the infrastructure in a given area. Data filtering and aggregation techniques can be used to reduce network load, but in this case, the question arises: how to choose which data to transmit to the surrounding nodes without losing the quality of the context data and without reducing the overall efficiency of the network. Solving these problems requires more adaptive intelligent methods to evaluate internally (location, time, environment, user status, and vehicle dynamics) and external context information (information received from other cars) as well as network mobility conditions and load based on data packets: formation and transfer processes.

The connection between reviewed processes of management of transportation and multi-components of integrated methods and systems is presented in Fig. 3.1.

Some specific and important components are included in the subsystems for operative management for supporting the processes, which are participating in the scheduling and planning of traffics of transportation of goods:

- Information about real road network conditions (in real time);
- Plans for reconstruction, renovation of roads and road parts (stages of plans at real time).

The plans for reconstruction of road, the original plan can be rescheduled for some reasons and can be incompatible, thus there is the same likelihood for the original plan to be changed or cancelled. Having made a decision, there will be a component modification, rescheduling or leaving without any changes. If the original plan cannot be shortened or rescheduled, then its duration may be reduced.

For development of software subsystems some layers of AI methods and ICT infrastructure are needful:

- Layer of AI methods for forecasting and assessment of arising situations (i.e. recognition of arising delays in transport corridors, delays in different road stretch and crossroad points);
- Layers of information interchange and communication network support;
- Layer of forecasting of traffics and scheduling of transportation (including the reloading points, terminals, crossroad nodes, etc.);
- Layer for monitoring environment and AI methods for recognition of abnormities.

A layer of AI systems for management of operations and automatic or semi-automatic support of services include machine learning, decision trees, genetic programming methods, multi-criteria decision-making methods, and ontology of the domain.

The layers of information interchange and communication network support describe the possibilities of implementation of the whole infrastructure of equipment. This infrastructure includes recording data in DWs, scenarios and protocols for wireless communication. The heterogeneous structure of all possible road site units (RSUs), outside board units (OBUs), and application units (AUs; DSRC, etc.) have to be included as well. We found recommendations on how to implement the protocols of communication in the real infrastructure (like IEEE 802.11 (p)/1609, etc.). But the communication standards have changed and new mobile technologies are appearing.

Analysis of services for helping in the areas of vehicle-to-vehicle (V2V) and vehicle-to-infrastructure (V2I) communication is related to decision support processes concerning some levels of the hierarchy of the decision-making in cargo transportation management, by trying to achieve the requirements of sustainable development.

3.3 Requirements of Smart Service Delivery System

The technical, functional, and non-functional requirements are described for the development of the smart service delivery system (Smart-SDS). The scope will cover many ground transportation processes. For development of such type Smart-SDS several new methods are needful. Suggested methods can integrate the contextual knowledge with the recognition process, obtaining raw data in a real-time regime, extract and transmit data from the environment and cooperating devices, etc. Integration of several methods for multi-criteria evaluation in the system demands communication with the whole infrastructure of information gathering. For assessing the usefulness of obtained information and for deciding—how to disseminate the data, information, and knowledge among other nodes, we need to examine the methods of collecting, aggregating, disseminating, and transforming contextual data between communication networks.

The system has to comply with different network mobility scenarios.

- Security messages should not be targeted at a specific recipient.
- Some parts of the collected data become irrelevant after some time or distance from the object.
- Data collection and dissemination must be ensured, even when some nodes are disconnected.
- It should be noted that not all vehicles will be equipped with communication equipment.
- The methods must work in both dense and sparse networks.
- The scatter must be resistant to fluctuations in the transmitted signal.

The methods will have the ability to identify the destination of the messages. The general statements of multi-criteria utility theory are used to develop a prototype system. The experiments are performed under different traffic and mobility conditions by using different scenarios. The results of the experiments were evaluated using statistical methods.

The special components are needed for the right communication implementation. The adaptation of the Dedicated Short-Range Communication standards (DSRC) and IEEE 802.11p and upper versions of standards help in the development of required infrastructure (IEEE, 2010; IEEE WAVE, 2010).

The use of various modelling, emulation and analysis software environments and platforms allows us to implement the different contextual data management scenarios. The estimation of the data storage capacity in the local DB is required for imitation of the exchanges with other vehicles and the exchanges with the hybrid VANET cloud.

Assessment of divisions of application areas of VANET helps in the development of the taxonomy of communication (Fig. 3.2). The communication structure involves such application processes, as field planning, operative management with control of online operations, etc. The connections between the vehicle and roadside units (RSU) have heterogeneous characteristics. Such connections are

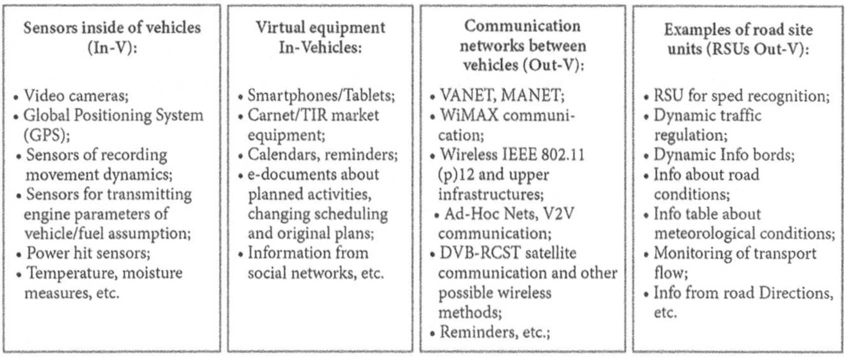

Fig. 3.2 Examples of heterogeneous sensors for transportation data recording and disseminating

communicative and create needful communication on different platforms. Vehicles can be equipped with several types of onboarding units (OBUs)—such units correspond for a type of communication with implemented in-vehicle (In-V) sensors. As well other types of units can be included for the application of outside units (AU)—they are implemented by the recommendations of C2C communication platforms (CAR 2 CAR Communication Consortium, 2014; IEEE, 802.11p-2010).

When designing a smart system, we have to take into account the fact that transportation takes place according to different scenarios in urban areas, rural areas, highways, crossing the border points of different countries and other places. Still, other scenarios are carried out at the locations of transshipment terminal locations for intermodal transport.

The algorithms for recognition of different situations are developed by the requirements of assessment of the importance of information and data and under making the decisions about their use of information. The messages are classified by usefulness for this type of transportation management situation. The classification algorithms influence the decision on provision/or not of them under online channels. The process of service provision to drivers has a significant influence and is based on the developed methods.

For designing the Smart-SDS, we choose some types of sensors for monitoring of surroundings, environment, and recording of needful raw data. The sensors are structured by different types, which were described in previous works (Dzemydienė & Burinskienė, 2021).

All possible types of communication have been taken into account: Ad hoc networks, vehicles equipped with OBUs, structures of exchange of information directly for drivers, which are in the near area and can make up MANET. MANET enables the distribution of information between vehicles by creating the precise coordination of actions. RSUs perform a fairly wide range of functions: they provide connection to infrastructure via a network, enable connectivity of OBUs, and provide access points for connections with the Internet and other access to network infrastructure (Chen et al., 2014; Chen & Tsao, 2015; CAR 2 CAR Communication Consortium, 2011).

The types of services for comfort and entertainment are considered less important as they are not related to safety. Such services may be provided in normal driving situations and may improve driver (and/or passenger) comfort. Comfort services can provide weather and traffic information, location information, prices at nearby gas stations, hotels, etc.

Important issues for providing information to drivers and passengers are collected and row data are recorded into the data warehouses (DW). The processes for delivery of the necessary information for drivers and for deriving of needful recommendations in the form of decisions requires the AI methods in the software systems.

An effective information collection mechanism can have the following characteristics according to the results of Kakkasageri and Manvi (2014). According to authors the provided information are as follows:

- Not to disturb the driver and passengers with information about the processes which are running autonomously without asking users to enter data or initialize other processes;
- Different mobility scenarios are taken into account;
- Databases are collected and transmitted in a highly dynamic changing topology environment;
- Security messages are not intended for a specific recipient;
- Data from the environment must be collected accurately and on time;
- After some time and distance, some of the collected data becomes insignificant;
- Data collection and dissemination algorithms must be fault-tolerant;
- Data collection and dissemination must be ensured, even when some nodes are disconnected;
- Assessment of fact that not all sets of cars are equipped with communication equipment;

Such facts have been taken into consideration for development of the Smart-SDS.

3.4 Methods for Collection of Contextual Data in Automotive Communication Systems

The methods for contextual data collection and recognition are reviewed here. Some issues are analysed by relating the collection and dissemination processes of contextual information processing in V2V communication systems. Some of these issues of system development are discussed in works of Hung and Peng (2010), Lee et al. (2012), Alam et al. (2015), Nzouonta et al. (2009), Kurmis et al. (2014) and Kurmis et al. (2015).

3.4.1 Sensor-Based Data Collection

Correcting operations of traffic safety programmes in-car communication environments require the accurate determination of the geographical location (position) of the vehicle. The most widely used object location technology for cars today is the Global Positioning System (GPS).

Traffic data collection platforms have been used to collect GPS data from cars for traffic monitoring purposes. On such platforms, each car is equipped with GPS modules and wireless network adapters that transmit GPS data to the server. Problems with these systems occur when a large number of cars try to transmit their GPS data to the server at the same time, because the wireless network cannot provide enough resources for such parallel connections.

High-rise buildings in cities do not allow this system to guarantee high positioning accuracy. The Radio Frequency Identification System allows to set up a DGPS (Differential Global Positioning System), which partially solves these problems. In RFID-enriched positioning systems, cars receive two types of information—the GPS coordinate from the GPS receiver and the physical coordinate via RFID. The RFID technology is discussed in greater detail. The GPS error is then calculated and shared with the surrounding cars so that they can correct inaccurate GPS coordinates (Lee et al., 2012).

This type of data exchange when each car reports its coordinates and its errors generates abundant data traffic. This problem is solved by offering a framework called Model-based Data Collection (MDC) to reduce the amount of GPS data transitions' (Hung & Peng, 2010). The programmes can run in a distributed manner both on the server and in each car. Having received a series of GPS points, on the side of the car the model features display GPS points. Each car reports certain coefficients that describe the movement rather than all the position information. The authors use linear regression and kernel regression algorithms to describe car motion.

Another important area of sensor-based data collection is adaptive cruise control systems. The possibilities to use an adaptive cruise control system installed in cars are presented in Milanés and Shladover (2014). Using information from system sensors, attempts are being made to derive dynamic models to predict vehicle dynamics. The controller proposed by the authors has been tested by performing experiments in different scenarios. The results showed that the controller works well when following another car, but has insufficient response time and poor stability if the traffic is more dynamic. This problem has been partially solved by integrating data exchange between cars in the system.

Another system of this type proposed for freight vehicles is based on information collected by GPS and radar, which was analysed by Alam et al. (2015). The authors explore engine control capabilities using speed, dynamics, and track change information from the front members of the group. The research was performed in the application of numerical methods, simulation modelling, and real system tests. The results showed that the system works with sufficient accuracy, although there are delays in some critical cases. More efficient engine models are needed to make the system more efficient, allowing control signals to be set more effectively for the engine control computer based on information received from other cars.

An interesting cooperative car positioning solution based on optical camera communication is proposed in Ifthekhar et al. (2015). Using this method, cars cooperate by exchanging position information over an optical wireless connection. This information is processed using computer vision and artificial neural network (ANN) methods to determine the position of the vehicle being tracked. The results of the simulation are shown in Ifthekhar et al. (2015) work and describe the structure of the ANN method that allows achieving higher accuracy than the computer vision methods. The model proposed by the authors is not very accurate, as it assumes that the road is smooth, does not have any slopes, and the position of the rear LEDs is well known. Under other conditions, a large system error is likely.

3.4.2 Data Collection Based on Routing Protocols

The Road-based method implemented in Vehicular Traffic (RBVT) protocols is offered in Nzouonta et al. (2009) work. The protocols are used real-time traffic information to create road-based network routes. Network routers are created at the intersections with the highest probability of successful communication. Geographical transmission methods are used to transmit data packets, which reduces the sensitivity of the route to the movement of an individual node. The simulation results showed that RBVT performs up to 40% better in urban conditions, based on the average ratio of delivered data compared to other routing protocols, such as AODV, OLSR, GPSR, and GSR.

The hybrid routing protocol HyBR is described by work (Bitam et al., 2013). Two procedures are used to manage VANET networks of different densities:

<*Condition*> when the network density is low → the geographic routing method is used;

<*Otherwise*> the topology-based routing method is used.

If necessary, both methods can be used simultaneously. The protocol divides the network. The results showed that this protocol outperforms the known AODV (5–61% in terms of average latency) and GPSR (from 69 to 112% in terms of average latency and from 58 to 92% in terms of packet delivery ratio) protocols.

One of the proposed solutions is the vaguely constrained Fuzzy Constraint Q-Learning (Wu et al., 2013). In this routing protocol, the system learns the routing path by using the Fuzzy Constraint Q-Learning algorithm that is based on AODV routing. The protocol uses fuzzy logic for the determination of the quality of communication channels by using some intervals in estimating the intervals of the assigned values for the context parameters and allowing their overlaps. For assessment of the quality of connections, such parameters are chosen like available data bandwidth, connection quality, and relative vehicle movement. The estimated connection quality helps the protocol to find out the route using the route request packet

RREQ and greeting messages. The protocol can obtain the trajectory of the car based on the information from adjacent cars if the position information of the car itself is not available. PFQ-AODV is independent of the lower OSI layers. The authors show that the protocol is more efficient than other widespread protocols in both simulation and real-world experiments.

An intelligent geographic routing protocol is analysed in Ghafoor et al. (2013). This protocol does not use the packets of signalling and operates according to the modes of usage of intersections and between intersections. Both modes depend on the selection of a distributed next node based on modified 802.11 RTS/CTS frames. The proposed protocol transmits data packets based on actual traffic. The results show higher efficiency in terms of the ratio of successfully delivered packets and the average latency compared to GPSR and Contention Based Forwarding (CBF) protocols.

3.5 Communication Infrastructure of Vehicles and Road Site Units

The communication between vehicles and roadside units (RSUs) for supporting all needful infrastructure is included in wide spectrum of possible communications. All such parts are forming the IoT technological platforms as well. The In-V area consists of the different types of onboard units (OBUs) and one or more application units (AUs). The Ad Hoc domain consists of cars equipped with OBUs and communicating directly with each other, thus forming a MANET (Mobile Ad Hoc Network) network that provides fully distributed, decentralized communication. In the area of infrastructure, the RSU can connect to infrastructure networks or the Internet, allowing the OBU to access infrastructure objects.

The onboard units are the WAVE devices, usually mounted in a car and used to exchange information between the RSU and other cars. It consists of a Resource Command Processor (RCP) and resources, including read/write memory used to store information, and a dedicated communication interface to connect to other OBUs or network devices.

One of the used types of communications is 802.11p radio technology. Additional devices can have other radio technologies, like as IEEE 802.11a/b/g/n/ac/ad, etc., 4G, 5G, LTE, etc.

The OBU can have a connection with the RSU or other OBU via a wireless connection by using the IEEE 802.11p radio frequency channel, which is responsible for communication with other OBUs and RSUs. Such types of communication channel help in the provision of communication services to the application devices. We can mention the key features of OBU, which can include: wireless access, Ad Hoc and geo-location routing, network load management, reliable messaging, data security, and IP mobility (Al-Sultan et al., 2014; CAR 2 CAR Communication

Consortium Manifesto, 2007, "IEEE Std 1609.1-2006," 2006; Olariu & Weigle, 2009).

The type of Application Devices (AUs) is set off the devices that concrete type of cars are equipped with. For example, the Service Provider can provide special services by using the capabilities of the OBU. An AU can be a device dedicated to specific security applications or a regular device, such as a smartphone or tablet, which provides Internet services. The distinctions between the functions of OBU and AU are often logical (CAR 2 CAR Communication Consortium Manifesto, 2007).

For the type of communication infrastructure that is named RSUs, i.e. roadside facilities that belong for devices, which can work on WAVE. RSUs can have a large spectrum of interactions and devices, which are installed near the road or in other specially designated areas. The areas of interaction can be crossroads, parking places, etc. These devices can be equipped with short-range wireless technology, such as IEEE 802.11p or others with wider ranges of connections, and enable interaction with the whole network infrastructure.

Several key functions and procedures related to RSU interoperability can be identified (CAR 2 CAR Communication Consortium Manifesto, 2007):

- Features that help expand the functionality of the Ad Hoc network by increasing the coverage distance when transmitting information to other OBUs or RSUs.
- Features of provision of safety-related functions and running safety programmes, like accident warnings, weather conditions, road surface information, etc. Such features are enabled by using the V2I communication mode and have properties of working as the sources of information.
- Functions of provision of the onboard computer with Internet access.

The directions of automotive communication development can be summarized as:

- The In-V link type, i.e. communication that can consist of the OBU and one or more AUs. It is possible to install both OBU and AU as separate devices and/or as a single device. In the automotive sector, contextual information is collected from installed sensors and is stored in a database.
- The V2V domain is a network of Ad Hoc created communication between vehicles equipped with OBU. Vehicles communicate with each other through the OBU, by using the formation of the MANET network, by enabling fully distributed and decentralized communication.

Vehicles communicate directly with each other, if a direct wireless connection is possible between them, by forming the Vehicle-to-Vehicle communication (V2V) type. If direct communication is not possible, the data is transferred to other cars as intermediaries until the addressee is reached. These communication forms can create a multi-hop connection infrastructure (Al-Sultan et al., 2014).

OBUs can also communicate with other nodes by providing services that are assessed as non-security services. Different heterogenic communication channels are implemented by using other mobile technologies, like GSM, GPRS, UMTS,

HSDPA, WiMAX, 4G, 5G, and others (CAR 2 CAR Communication Consortium Manifesto, 2007; IEEE Std 1609.3-2010 (Revision of IEEE Std 1609.3, 2010)).

The MANET communication infrastructure is used for ensuring distributed and decentralized communication between vehicles with special equipment. Based on MANET, RSUs can be connected to wireless networks and can connect OBUs to common communication for accessing other possible objects in infrastructure. RSUs, which are characterized as WAVE devices can be equipped near roads, in special designated places. These devices are equipped with short-range wireless communication technologies, working on the base of standards like IEEE 802.11p and others.

In situations, when direct communication is restricted, MANET enables data transfer to other cars, which are included in communication as intermediaries until the conditions of reachability of addresses will take effect.

In situations, when direct communication conditions are unable, the important data can be transferred to other cars (as intermediaries) until the addressee is reached. Al-Sultan et al. (2014) have described these forms of a multi-hop relationship.

Domain infrastructure relates to all these parts, especially the RSUs connecting network infrastructure or the Internet by providing OBUs with access to that network. OBU devices communicate with various nodes by providing non-security services using other mobile technologies like GPRS, GSM, 5G, HSDPA, UMTS, and WiMAX (IEEE, 802.11p-2010, IEEE, WAVE, 2011).

OBU devices can be classified as WAVE and are usually placed in the local car area and used for exchanging information between RSUs and other cars. It has Resource Command Processor (RCP) and can explore resources by its reading/ writing memory used for storing data in DW. OBU devices use interfaces dedicated to that type of communication that can connect a vehicle to network devices and other OBU devices by using radio technology in the application of the IEEE 802.11p standard.

Additional devices could be based on other radio technologies, like IEEE 802.11a/b/g/n/ac/ad, and others. Mobile access protocols, such as 4G, 5G, and the upcoming 6G, also LTE and others can be used.

Such functions are also within the competence of OBUs: Wi-Fi access, Ad Hoc, and geo-location-based routing, network load management, reliable messaging, security, and IP mobility (Bazzi et al., 2015; Cespedes et al., 2013; Campolo et al., 2015).

3.6 Areas of Automotive Communication

The domain that is named In-Vehicle explains the communication onboard units and application units, which can be equipped more than one in vehicle. OBUs and AUs can be implemented as a single piece of equipment. Automotive communication enables the collection of context data from sensors. Such data are stored in DW.

The V2V domain can be described as the type of Ad Hoc network. V2V can work on the base of MANET. Such communication we can characterize as fully distributed and decentralized, when a direct wireless connection is possible between vehicles. The infrastructure domain can be described as the RSUs, which can be connected to infrastructure and/or the Internet. The communication types through GSM, GPRS, UMTS, HSDPA, WiMAX, 4G, 5G, etc., are possible as well. The standards are described by IEEE Std 1609.3–2010 and are helpful in developments of such type communications (Revision of IEEE Std 1609.3, 2010).

Differences between these modern technologies can be expressed by types of dissemination of information. The Data Alliance (Data Alliance standards, 2022) provides a description of the LTE (Long Term Evolution) and a 4G and 5G communication standards that are designed to be 10× faster than standard 3G.

The ICT technology can provide the IP-Based communication of voice and multimedia and streaming at between 100 Mbit per sec and 1 Gbit per second. LTE has an algorithm that can send large chunks of data via IP. This approach streamlines the traffic and reduces latency.

GSM is a digital cellular technology that is used to transmit data and voice communication at a frequency range of 850–1900 MHz. GSM technology uses a Time Division Multiple Access (TDMA) technique to transmit data. The GSM system converts the data into a digital signal and sends it through two different time-stamped channels at a rate between 64 kbps and 120 kbps.

The Code Division Multiple Access (CDMA) uses a multiple access mode of communication. This is where several transmissions are made over the same channel simultaneously. Using a speed spectrum, each transmission is assigned a unique code that corresponds to the source and destination of the signal.

Industrial, scientific, and medical radio bands (ISM) are reserved for these purposes and not intended for telecommunication. Originally this band of radio frequencies was intended for use in industrial, scientific, and medical ISM machines that operate at this range in order not to interfere with the wider.

The standard 5G promises to boost cellular data transfer to speeds rivalling the fastest fibre optic networks. In general, 5G offers better broadband bandwidth, capacity, and reliability to levels far beyond 4G, and stands for fifth-generation wireless systems deployed from 2018 and beyond. The networks use 4G or new specified 5G frequency bands to operate. The ranges are promising of used frequencies of 5G diverse from 600 MHz until 6 GHz. Such ranges will enable the wider implementation of IoT and other smart service delivery systems.

3.6.1 Wireless Access in Car Communication Networks

There are currently some common wireless access standards and technologies for V2V, V2I, and direct in-car communication. Some of them are based on a centralized infrastructure for coordinating communication between nodes, while others operate through Ad Hoc distributed coordination. The most common wireless access technologies for wireless car communications are discussed below.

The standard for wireless access in automotive communication networks was introduced by the IEEE organization—IEEE 802.11p. This standard was developed following improvements to the IEEE 802.11 standard, taking into account the specifics of automotive communication networks. The main difference from the original IEEE 802.11 standard is that in the IEEE 802.11p standard is implemented the MAC (Medium Access Control) layer mechanism that is based on communication prioritization and the competition is based on EDCA (Enhanced Distributed Channel Access) channel access scheme Std 1609.4-2010 (Revision of IEEE Std 1609.4-, 2011). The standard uses a multi-channel WAVE) system mechanism (IEEE Std 1609.3-, 2010), under such a mechanism vehicles can operate through communication channels by periodically switching between a common control channel (CCH) for monitoring alarm and alert messages. As one of the examples of the information and multimedia service channels can be mentioned SCH and used for transmitting less high priority information. The DSRC (Dedicated Short-Range Communications) operates at 5.9 GHz and supports both public security and private I2V and V 2 V communication tools (Moustafa & Zhang, 2009).

In a dynamic vehicle communication environment, reducing network connection time is a challenge, making IEEE 802.11a device channel scanning and security solutions unsuitable for fast-moving vehicle security. The necessary changes were at both the basic levels of channel assignment and packet routing and decision-making. In this way, three ideas for the development and application of standards emerged. First, changes to the IEEE 802.11 wireless LAN standards were required to describe the DSRC spectrum and bands and allow rapid interaction to form the IEEE 802.11p physical layer. For changing the IEEE 802.11 layer, new security and channel selection services need to be offered at a higher level, which is attempted in the IEEE 1609 standards and WAVE—for wireless access in an automotive environment consisting of DSRC standards and a communication stack (Amadeo et al., 2012):

- Resource manager;
- Security services and management;
- Network services;
- Multi-channel operations.

3.6.2 Sort Review of Infrastructure of V2V Communication

The communication type of V2V can appear when a direct wireless connection is possible between vehicles by forming the type of -car-to-car connections. DSRC standards allow the development of systems in which cars can connect directly to a common channel in the DSRC range for safety, while other channels are available for less important communications. At the highest level of applications were SAE

J2735 means, which were developed by the Society of Automotive Engineers and DSRC Technical Committee.

The IEEE 802.11p standard (also called WAVE) has brought improvements to the IEEE 802.11 standard, without which the developments of automotive communication systems would not be possible. It includes data exchange between moving cars, as well as between cars and the RSU in the licensed 5.9 GHz frequency band up to 1000 m. WAVE includes new application classes related to road safety (e.g. collision avoidance) and emergency services (police, ambulance, fire, etc., special services). In some critical cases, the total time from signal detection to multi-frame switching should not exceed 100 ms. WAVE is used for control of several service channels to perform operations. The extension of physical layer is based on OFDM system. The OFDM system provides WAVE communications at the speed of such ranges like 3, 4.5, 6, 9, 12, 18, 24, and 27 Mbps on 10 MHz channels. Also, WAVE can operate on 20 MHz channels, in which case speeds of 6, 9, 12, 18, 24, 36, 48, and 54 Mbps are supported.

WAVE-based stations must be able to exchange messages between the RSU and vehicles travelling at speeds up to 140 km/h with a PER (packet error rate) of less than 10% when the PSDU is 1000 bytes in length; when the car is moving at a speed of up to 200 km/h—with less than 10% PER when the PSDU length is 64 bytes. At the V2V communication a PER can be less than 10% between cars running at 283 km/h with a PSDU length of 200 bytes that shall be ensured. These technical limitations of the standards ensure the support of security services, which is why WAVE is attracting a great deal of interest from the automotive industry (Weil, 2009).

3.6.3 Possibilities of Mobile Connections

Although mobile networks are mainly used for the provision of telephony services, in recent times it is also a key technology for data transmission in a mobile environment. First, to increase the speed transmitted over the GSM network (9.6 kb/s), GPRS (General Packet Radio Service) technology was proposed, which provided 0.47 Mb/s bandwidth. It was followed by EDGE technology, providing speeds of up to 1.9 Mbps. Later, this technology gave way to the third generation mobile. UMTS, the most advanced third generation mobile technology has been proposed. UMTS enhancement HSPA+ allows achieving 42 Mb/s reception and 22 Mb/s transmission speeds (Cheng et al., 2011).

The application of mobile networks to automotive systems began several years ago when GSM and GPRS systems were introduced for traffic information and alarm transmission (Masini et al., 2004). However, until the advent of 3G technology, low data rates led to low cellular usage in ITS systems. In the scientific literature, some authors advocate the use of UMTS for direct V2V communications, a technology that also applies to surveillance systems, but its application to V2V communications is still a major challenge due to the significant latency. The widespread adoption of

Table 3.1 Differences of 5G wireless frequency bands in different countries. Source: Data Alliance (2022)

Countries	5G wireless network frequency bands: Range of lower level of wireless frequency bands	Range of higher level of wireless frequency bands
Europe	3400–3800 MHz	
United States of America	3100–3550 MHz	3700–4200 MHz
The USA began 5 GMMW for pre-commercial deployment	27.5–28.35GHz	37–40GHz
China	3300–3600 MHz	4400–4500 MHz 4800–4990 MHz
Asian region including Japan	3600–4200 MHz	4400–4900 MHz

4G LTE communication technologies in recent years is being analysed as an alternative to the IEEE 802.11p protocol for automotive communications. The main advantage of this technology is the standardization of 3GPP organization's content delivery services, which allows efficient transmission of data packets to a large number of nodes. LTE is well suited for VANET because it provides relatively high data transmission and reception rates in both high-mobility and low-mobility traffic conditions (Atat et al., 2014; Vinel, 2012; Ucar et al., 2015). A comparison of the application of wireless access technologies to automotive communication networks is provided in Annex 1.

The main reason for discouraging the use of these technologies in intelligent transport systems (ITS) is the additional cost of using the cellular infrastructure of the operators, as well as the often unsatisfactory delays in providing critical services. Despite all the shortcomings, some researchers believe that common ITS communication technology and mobile communication can offer very wide coverage and support high car mobility (Bazzi et al., 2015; Cheng et al., 2014; Deruyck et al., 2011).

The fundamental differences between these modern technologies are related to special ways of transmitting and receiving information (Data Alliance, 2022). The stages of development of wireless communications from mentioning the Long Term Evolution (LTE) a 4G communication standard designed to be 10× faster than standard 3G. The technology is going forward to 5G with higher capabilities (Table 3.1). There are some differences of 5G wireless frequency bands applied in different countries, which are available for use (Table 3.1).

3.6.4 Key Features of Automotive Communication Networks

Car communication networks have special characteristics and features that distinguish them from other types of wireless networks. According to Lee and Gerla (2010) and Moustafa and Zhang (2009), the following unique features of a car as a mobile network node have been identified:

- higher energy reserve,
- movement according to templates.

Cars have a large energy reserve compared to conventional mobile devices. Energy can be obtained from batteries and charged with petrol, diesel, or alternative fuel engine. Vehicles can be equipped with physically larger and heavier computing (sensor) components (i.e. more powerful) compared to traditional wireless customers. Car computers can be larger, faster, and equipped with ultra-high-capacity storage devices, as well as powerful wireless interfaces that can ensure high communication speeds. Vehicles can travel at high speeds (160 km/h or more), making it difficult to maintain continuous, consistent V2V communication. However, the collection of statistics on transport movements, such as movement together according to certain patterns, can help to maintain the link between mobile car groups. Cars on the network can be out of range at any time (Wi-Fi, mobile, satellite, etc.), so network protocols need to be designed to reconnect to the communications network very quickly once the connection is re-established. This topic has been studied in more detail in the author's publications (Kurmis et al., 2011, 2013).

Despite its many unique positive features, the development of automotive networks also faces specific challenges, the main ones being:

- large-scale networks,
- high level of mobility,
- network fragmentation,
- changing topology,
- difficult to ensure quality communication.

Unlike the Ad Hoc networks described in the literature, which are relatively limited in size, automotive communication networks can, in principle, expand across the road network and include a large number of network devices (cars). The environment in which car networks operate is extremely dynamic and in some cases can be particularly different, e.g. on motorways, the speed can reach up to 300 km/h, on low-load roads the traffic can be only 1–2 cars per km. In cities, on the other hand, car speeds reach 50–60 km/h and car traffic is quite high, especially during peak hours. Car communication networks can often be fragmented. The dynamic nature of traffic can lead to large gaps between cars in sparsely populated areas, and several clusters of isolated network nodes can be created. The scenarios of car communication networks are very different from the classic Ad Hoc networks, as cars move and change positions constantly, the scenarios are very dynamic. In this way, the network topology changes very often due to the extremely frequent connections and disconnections between network nodes. The degree to which a network is connected depends on two factors: the distance between the wireless connections and the number of cars that can connect to the network.

The main purpose of general information and multimedia services is to offer services to the driver and passengers that ensure comfort and convenience in the car. An intelligent parking system structure for large parking lots providing real-time parking navigation, theft protection, and convenient dissemination of parking information has been proposed (Lu et al., 2009). The system is based on car communication. Digital billboards for advertising have been presented (Nandan et al., 2005) in a paper examining the possibilities of advertising distribution in automotive communication networks. An integrated AdTorrent system for content evaluation, retrieval, and delivery in this architecture is offered.

Using V2I communications based on the services of mobile service providers, it is possible to perform certain business management tasks in the car based on the idea of mobile office. The proposed CarTorrent P2P architecture allows the use of audio and video transmission over car networks, making long journeys more exciting (Lee & Gerla, 2010). A solution for the transmission of multimedia material over VANET networks in cities is proposed (Soldo et al., 2008). These technologies can also be used for commercial audio and video broadcasting purposes.

Road safety information services are always a priority to significantly reduce the number of road accidents. Car-to-car communication can be used to develop lane-change assistance systems, adaptive cruise control, and other systems to increase road safety and assist drivers in a variety of critical situations. It is proposed to use VANET networks to create a network of security surveillance cameras, which would be installed in buses, taxis, other public transport vehicles, and this collected surveillance data would be transmitted to data processing centres by wireless communication (Sirichai et al., 2011). In this way, it would be possible to help ensure public order, identify violators, and automatically respond to road accidents by calling special services. Using car communication networks, drivers are informed about traffic conditions and road load, which reduces travel time and fuel consumption (Dornbush & Joshi, 2007). The same problems are solved (Sommer et al., 2010) using UMTS technology. Creating in-car virtual trains is another way to increase traffic safety. Many authors offer their ideas for creating virtual vehicle trains using VANET communication technologies. By reducing the need to change lanes, increase or decreasing speed, this technology can significantly increase traffic safety and help increase fuel economy. Methods for building large-scale road trains in automated highway systems have been investigated (Jovanović et al., 2008). Also, combining an adaptive cruise control system with V2V communications can prevent many disasters due to human factors.

Traffic monitoring and management are essential to increase traffic flow and reduce congestion. In some cases, crossing intersections is difficult and dangerous. Proper and efficient traffic light control can facilitate intersections. Smooth traffic flow can significantly increase street throughput and reduce travel time. Self-distributed traffic regulation using VANET communications was analysed (Gibaud et al., 2011). The tool developed by the authors allows the modelling of innovative traffic management and confirms the promising scope of VANET. An intersection management system is proposed in which drivers and intersections are treated as autonomous agents of a multi-agent system (Dresner, 2008). New intersection management using the introduced reservation system has been proposed. Modelling has confirmed that this system has the potential to perform better than current intersection management systems.

The main purpose of general information and multimedia services is to offer services to the driver and passengers that ensure comfort and convenience in the car. An intelligent parking system structure for large parking lots providing real-time parking navigation, theft protection, and convenient dissemination of parking information has been proposed (Lu et al., 2009). The system is based on car communication. Digital billboards for advertising have been presented (Nandan et al., 2005) in a paper examining the possibilities of advertising distribution in automotive communication networks. An integrated AdTorrent system for content evaluation, retrieval, and delivery in this architecture is offered.

Using V2I communications based on the services of mobile service providers, it is possible to perform certain business management tasks in the car based on the idea of a mobile office. The proposed CarTorrent P2P architecture allows the use of audio and video transmission over car networks, making long journeys more exciting (Lee & Gerla, 2010). A solution for the transmission of multimedia material over VANET networks in cities is proposed (Soldo et al., 2008). These technologies can also be used for commercial audio and video broadcasting purposes.

Road safety information services are always a priority to significantly reduce the number of road accidents. Car-to-car communication can be used to develop lane-change assistance systems, adaptive cruise control, and other systems to increase road safety and assist drivers in a variety of critical situations. It is proposed to use VANET networks to create a network of security surveillance cameras, which would be installed in buses, taxis, other public transport vehicles, and this collected surveillance data would be transmitted to data processing centres by wireless communication (Sirichai et al., 2011). In this way, it would be possible to help ensure public order, identify violators, and automatically respond to road accidents by calling special services. Using car communication networks, drivers are informed about traffic conditions and road load, which reduces travel time and fuel consumption (Dornbush & Joshi, 2007). The same problems are solved (Sommer et al., 2010) using UMTS technology. Creating in-car virtual trains is another way to increase traffic safety. Many authors offer their ideas for creating virtual vehicle trains using VANET communication technologies. By reducing the need to change lanes, increase or decreasing speed, this technology can significantly increase traffic safety and help increase fuel economy. Methods for building large-scale road trains in automated highway systems have been investigated (Jovanović et al., 2008). Also, combining an adaptive cruise control system with V2V communications can prevent many disasters due to human factors.

Traffic monitoring and management are essential to increase traffic flow and reduce congestion. In some cases, crossing intersections is difficult and dangerous. Proper and efficient traffic light control can facilitate intersections. Smooth traffic flow can significantly increase street throughput and reduce travel time. Self-distributed traffic regulation using VANET communications was analysed (Gibaud et al., 2011). The tool developed by the authors allows the modelling of innovative traffic management and confirms the promising scope of VANET. An intersection management system is proposed in which drivers and intersections are treated as autonomous agents of a multi-agent system (Dresner, 2008). New intersection management using the introduced reservation system has been proposed. Modelling has confirmed that this system has the potential to perform better than current intersection management systems.

Real-time services (such as voice transmission) are sensitive to delays, where packets have been transmitted with a low packet rejection rate, but can tolerate high bit error rates. On the other hand, data transfer applications (e.g. point-to-point data exchange) are insensitive to delays but require extremely high data transfer accuracy.

One of the minimum MAC requirements for supporting multimedia services on VANET networks is to support best-effort services such as Internet browsing, file sharing, and advertising (Cuomo et al., 2014). This type of protocol is for cheap VANET with homogeneous information services. The provision of heterogeneous information services requires the prioritization of services according to the nature of the services (Cheng et al., 2011).

3.7 Control of Channel Resources for the Best Bandwidth

Car communication networks have higher energy reserve that can be achieved and these nodes are moving according to the some types of templates (Andziulis et al., 2012).

Cars have a large energy reserve compared to conventional mobile devices. Energy can be obtained from batteries and charged with petrol, diesel, or alternative fuel engine. Vehicles can be equipped with physically larger and heavier computing (sensor) components (i.e. more powerful) compared to traditional wireless customers. Car computers can be larger, faster, and equipped with ultra-high-capacity storage devices, as well as powerful wireless interfaces that can ensure high communication speeds. Vehicles can travel at high speeds (160 km/h or more), making it difficult to maintain continuous, consistent V2V communication. However, the collection of statistics on transport movements, such as movement together according to certain patterns, can help to maintain the link between mobile car groups. Cars on the network can be out of range at any time (Wi-Fi, mobile, satellite, etc.), so network protocols need to be designed to reconnect to the communications network very quickly once the connection is re-established. This topic has been studied in more detail in the author's publications (Kurmis et al., 2011, 2013; Bulbenkiene et al., 2011).

Despite its many unique positive features that enable the development of automotive networks it also faces specific challenges. The main of them are:

- large-scale networks;
- high level of mobility,
- network fragmentation,
- dynamically changing topology,
- difficult to ensure quality communication.

Unlike the Ad Hoc networks described in the literature, which are relatively limited in size, automotive communication networks can, in principle, expand across the road network and include a large number of network devices (and cars) (Amadeo et al., 2015). The environment in which car networks operate is extremely dynamic and in some cases can be particularly different, e.g. on motorways, the speed can reach up to 300 km/h, on low-load roads the traffic can be only 1–2 cars per kilometre. In cities, on the other hand, car speeds reach 50–60 km/h and car traffic is quite high, especially during peak hours. Car communication networks can often be fragmented. The dynamic nature of traffic can lead to large gaps between cars in sparsely populated areas, and several clusters of isolated network nodes can be created. The scenarios of car communication networks are very different from the classic Ad Hoc networks, as cars move and change positions constantly, the scenarios are very dynamic. In this way, the network topology changes very often due to the extremely frequent connections and disconnections between network nodes. In fact, the degree to which a network is connected depends on two factors: the distance between the wireless connections and the number of cars that can connect to the network.

The main purpose of general information and multimedia services is to offer services to the driver and passengers that ensure comfort and convenience in the car. An intelligent parking system structure for large parking lots providing real-time parking navigation, theft protection, and convenient dissemination of parking information has been proposed (Lu et al., 2009). The system is based on car communication. Digital billboards for advertising have been presented in Nandan et al. (2005) by examining the possibilities of advertising distribution in automotive communication networks. An integrated AdTorrent system for content evaluation, retrieval, and delivery in this architecture is offered.

Using V2I communications based on the services of mobile service providers, it is possible to perform certain business management tasks in the car based on the idea of a mobile office. The proposed CarTorrent P2P architecture allows the use of audio and video transmission over car networks, making long journeys more exciting (Lee & Gerla, 2010). A solution for the transmission of multimedia material over VANET networks in cities is proposed (Soldo et al., 2008). These technologies can also be used for commercial audio and video broadcasting purposes.

Road safety information services are always a priority to significantly reduce the number of road accidents. C2C communication can be used to develop lane-change assistance systems, adaptive cruise control, and other systems to increase road safety and assist drivers in a variety of critical situations. It is proposed to use VANET networks to create a network of security surveillance cameras, which would be installed in buses, taxis, other public transport vehicles, and this collected surveillance data would be transmitted to data processing centres by wireless communication (Sirichai et al., 2011). In this way, it would be possible to help ensure public order, identify violators, and automatically respond to road accidents by calling special services. Using car communication networks, drivers are informed about traffic conditions and road load, which reduces travel time and fuel consumption (Dornbush & Joshi, 2007). The same problems are solved (Sommer et al., 2010) using UMTS technology. Creating in-car virtual trains is another way to increase traffic safety.

Many authors offer their ideas for creating virtual vehicle trains using VANET communication technologies. By reducing the need to change lanes, increase or decreasing speed, the technology can significantly increase traffic safety and help increase fuel economy. Methods for building large-scale road trains in automated highway systems have been investigated (Jovanović et al., 2008). Also, combining an adaptive cruise control system with V2V communications can prevent many disasters due to human factors.

Traffic monitoring and management are essential to increase traffic flow and reduce congestion. In some cases, crossing intersections is difficult and dangerous. Proper and efficient traffic light control can facilitate intersections. Smooth traffic flow can significantly increase street throughput and reduce travel time. Self-distributed traffic regulation using VANET communications was analysed (Gibaud et al., 2011). The tool developed by the authors allows the modelling of innovative traffic management and confirms the promising scope of VANET. An intersection management system is proposed in which drivers and intersections are treated as autonomous agents of a multi-agent system (Dresner, 2008). New intersection management using the introduced reservation system has been proposed. Modelling has confirmed that this system has the potential to perform better than current intersection management systems.

The main purpose of general information and multimedia services is to offer services to the driver and passengers that ensure comfort and convenience in the car. An intelligent parking system structure for large parking lots providing real-time parking navigation, theft protection, and convenient dissemination of parking information has been proposed (Lu et al., 2009). The system is based on car communication. Digital billboards for advertising have been presented (Nandan et al., 2005) in a paper examining the possibilities of advertising distribution in automotive communication networks. An integrated AdTorrent system for content evaluation, retrieval and delivery in this architecture is offered.

Using V2I communications based on the services of mobile service providers, it is possible to perform certain business management tasks in the car based on the idea of a mobile office. The proposed CarTorrent P2P architecture allows the use of audio and video transmission over car networks, making long journeys more exciting (Lee and Gerla, 2010). A solution for the transmission of multimedia material over VANET networks in cities is proposed (Soldo et al., 2008). These technologies can also be used for commercial audio and video broadcasting purposes.

Road safety information services are always a priority to significantly reduce the number of road accidents. Car-to-car communication can be used to develop lane-change assistance systems, adaptive cruise control, and other systems to increase road safety and assist drivers in a variety of critical situations. It is proposed to use VANET networks to create a network of security surveillance cameras, which would be installed in buses, taxis, other public transport vehicles, and this collected surveillance data would be transmitted to data processing centres by wireless communication (Sirichai et al., 2011). In this way, it would be possible to help ensure public order, identify violators, and automatically respond to road accidents by calling special services. Using car communication networks, drivers are informed about traffic conditions and road load, which reduces travel time and fuel consumption (Dornbush & Joshi, 2007). The same problems are solved (Sommer et al., 2010) using UMTS technology. Creating in-car virtual trains is another way to increase traffic safety. Many authors offer their ideas for creating virtual vehicle trains using VANET communication technologies. VANET technology can significantly increase traffic safety and help in increasing fuel economy by including the means for reducing the need to change lanes, increase or decrease speed. Methods for building large-scale road trains in automated highway systems have been investigated (Jovanović et al., 2008). Also, combining an adaptive cruise control system with V2V communications can prevent many disasters due to human factors.

Traffic monitoring and management are essential to increase traffic flow and reduce congestion. In some cases, crossing intersections is difficult and dangerous. Proper and efficient traffic light control can facilitate intersections. Smooth traffic flow can significantly increase street throughput and reduce travel time. Self-distributed traffic regulation using VANET communications was analysed (Gibaud et al., 2011). The tool developed by the authors allows the modelling of innovative traffic management and confirms the promising scope of VANET. An intersection management system is proposed in which drivers and intersections are treated as autonomous agents of a multi-agent system (Dresner, 2008). New intersection management using the introduced reservation system has been proposed. Modelling has confirmed that this system has the potential to perform better than current intersection management systems.

Real-time services (such as voice transmission) are sensitive to delays, where packets have been transmitted with a low packet rejection rate, but can tolerate high bit error rates. On the other hand, data transfer applications (e.g., point-to-point data exchange) are insensitive to delays but require extremely high data transfer accuracy.

One of the minimum MAC requirements for supporting multimedia services on VANET networks is to support best-effort services such as Internet browsing, file sharing, and advertising. This type of protocol is for cheap VANET with homogeneous information services. The provision of heterogeneous information services requires the prioritization of services according to the nature of the services (Cheng et al., 2011).

The development of channel resource management methods to achieve the best bandwidth is designed to maximize the bandwidth of the communication channel while minimizing the number of collisions. The requirements for IEEE 802.11a, 802.11b, and 802.11g channels are the most widely used to support services. Channel access methods can be based on CSMA (Carrier Sense Multiple Access) technologies described in Alasmary and Zhuang (2012). The essence of the technology is such evaluation of conditions: if the channel is found to be free, the car sends packets, otherwise—the transmission is stopped.

The application of CSMA in large-scale multicast networks is problematic due to hidden terminals, which cause transmission conflicts, and exposed terminals, which unnecessarily suppress the radio signal, thus causing a decrease in system speed

(Alasmary & Zhuang, 2012). A request-to-send (RTS) and clear-to-send (CTS) mechanism can be used to address these issues and increase bandwidth. The sender first sends an RTS frame to the recipient. If the RTS is received successfully, the receiver responds with a CTS frame. All surrounding cars that hear RTS or CTS frame transmission delay their packet transmission.

Another effective method to avoid packet collisions in VANET networks is signal fluctuation. Using this method, the receiver sends a busy signal during packet reception, which performs two functions—confirming the possibility of data transmission to the sender and stopping the possible transmission of hidden terminals. This method is not as popular as CSMA-based methods because it consumes a lot of electricity. As mentioned earlier, power consumption is important in traditional MANETs, but VANET does not have this limitation, so protocols based on signal fluctuation may be valuable in the practical implementation of VANET (Cheng et al., 2011; Crichigno et al., 2008; Jarupan & Ekici, 2011).

These examined protocols can be applied to VANET networks with low to medium car speeds, but their application to high-mobility cars is inefficient. One of the main shortcomings is that the protocols do not take into account service priorities.

3.8 Conclusions

The main aspects of the development of new communication systems are observed. New service provision systems can be created by integrating the video spectrum of communication technologies, networks of wireless sensor equipment. The new adaptive algorithms are required for the collection, processing, and exchange of contextual information in transportation processes.

Systems can enable more efficient use of network infrastructure and configurable system resources. Therefore, new design studies are needed for updating adaptive approaches, which will develop environmentally aware and interoperable autonomous systems. These systems can be adaptable for new conditions of upcoming wireless communication channels, high mobility of moving objects, capabilities of VANET networks. Smart service provision systems are required in specific characteristics, which can support a variety of multi-criteria decisions in different traffic and network mobility conditions.

References

Alam, A., Mårtensson, J., & Johansson, K. H. (2015). Experimental evaluation of decentralized cooperative cruise control for heavy-duty vehicle platooning. *Control Engineering Practice, 38*, 11–25. https://doi.org/10.1016/j.conengprac.2014.12.009

Alasmary, W., & Zhuang, W. (2012). Mobility impact in IEEE 802.11p infrastructureless vehicular networks. *Ad Hoc Networks, 10*(2), 222–230. https://doi.org/10.1016/j.adhoc.2010.06.006

Al-Sultan, S., Al-Doori, M. M., Al-Bayatti, A. H., & Zedan, H. (2014). A comprehensive survey on vehicular Ad Hoc network. *Journal of Network and Computer Applications, 37*, 380–392. https://doi.org/10.1016/j.jnca.2013.02.036

Amadeo, M., Campolo, C., & Molinaro, A. (2012). Enhancing IEEE 802.11p/WAVE to provide infotainment applications in VANETs. *Ad Hoc Networks, 10*(2), 253–269. https://doi.org/10.1016/j.adhoc.2010.09.013

Amadeo, M., Campolo, C., & Molinaro, A. (2015). Forwarding strategies in named data wireless ad hoc networks: Design and evaluation. *Journal of Network and Computer Applications, 50*, 148–158. https://doi.org/10.1016/j.jnca.2014.06.007

Andziulis, A., Plėštys, R., Jakovlev, S., Adomaitis, D., Gerasimov, K., Kurmis, M., & Pareigis, V. (2012). Priority based tag authentication and routing algorithm for intermodal containers RFID sensor network. *Transport, 27*(4), 373–382. https://doi.org/10.3846/16484142.2012.750622

Atat, R., Yaacoub, E., Alouini, M.-S., Filali, F., & Abu-Dayya, A. (2014). Delay-sensitive content distribution via peer-to-peer collaboration in public safety vehicular ad-hoc networks. *Ad Hoc Networks, 16*, 182–196. https://doi.org/10.1016/j.adhoc.2013.12.007

Bazzi, A., Masini, B. M., Zanella, A., & Pasolini, G. (2015). IEEE 802.11p for cellular offloading in vehicular sensor networks. *Computer Communications, 60*, 97–108. https://doi.org/10.1016/j.comcom.2015.01.012

Bielskis, A. A., Guseinoviene, E., Dzemydiene, D., Drungilas, D., & Gricius, G. (2012). Ambient lighting controller based on reinforcement learning components of multi-agents. *Electronics and Electrical Engineering, 121*(5), 79–84. https://doi.org/10.5755/j01.eee.121.5.1656

Bitam, S., Mellouk, A., & Zeadally, S. (2013). HyBR: A hybrid bio-inspired Bee swarm routing protocol for safety applications in Vehicular Ad hoc NETworks (VANETs). *Journal of Systems Architecture, 59*(10), 953–967. https://doi.org/10.1016/j.sysarc.2013.04.004

Bulbenkiene, V., Pareigis, V., Andziulis, A., Kurmis, M., & Jakovlev, S. (2011). Simulation of IEEE 802.16j mobile WiMAX relay network to determine the most efficient zone to deploy relay station. *Elektronika Ir Elektrotechnika, 112*(6), 81–84. https://doi.org/10.5755/j01.eee.112.6.451

Campolo, C., Molinaro, A., & Scopigno, R. (2015). From today's VANETs to tomorrow's planning and the bets for the day after. *Vehicular Communications, 2*(3), 158–171. https://doi.org/10.1016/j.vehcom.2015.06.002

CAR 2 CAR Communication Consortium Manifesto. (2007). Retrieved from https://www.car-2-car.org/index.php?id=31

CAR 2 CAR Communication Consortium. (2011). Memorandum of understanding for OEMs within the CAR 2 CAR Communication Consortium on Deployment Strategy for cooperative ITS in Europe. Retrieved from https://www.car-2-car.org/index.php?id=231

CAR 2 CAR Communication Consortium. (2014). CAR 2 CAR related projects. Retrieved from https://www.car-2-car.org/index.php?id=6

Cespedes, S., Lu, N., & Shen, X. (2013). VIP-WAVE: On the feasibility of IP communications in 802.11p vehicular networks. *Intelligent Transportation Systems, IEEE Transactions on*. https://doi.org/10.1109/TITS.2012.2206387

Cheng, H. T., Shan, H., & Zhuang, W. (2011). Infotainment and road safety service support in vehicular networking: From a communication perspective. *Mechanical Systems and Signal Processing, 25*(6), 2020–2038. https://doi.org/10.1016/j.ymssp.2010.11.009

Cheng, C. M., & Tsao, S. L. (2015). Adaptive lookup protocol for two-tier VANET/P2P information retrieval services. *Vehicular Technology, IEEE Transactions on*. https://doi.org/10.1109/TVT.2014.2329015

Cheng, N., Lu, N., Zhang, N., Shen, X. (Sherman), & Mark, J. W. (2014). Vehicular WiFi offloading: Challenges and solutions. *Vehicular Communications, 1*(1), 13–21. https://doi.org/10.1016/j.vehcom.2013.11.002

Chen, M., Ong Mau, D., Zhang, Y., Taleb, T., & Leung, V. C. M. (2014). VENDNET: VEhicular Named Data NETwork. *Vehicular Communications, 1*(4), 208–213. https://doi.org/10.1016/j.vehcom.2014.09.002

C-ITS Platform. (2016). C-ITS platform, final report. Retrieved from internet: https://www.polisnetwork.eu/wp-content/uploads/2019/09/c-its-platform-final-report-january-2016.pdf

C-MobILE. (2017). *C-MobILE accelerating C-ITS mobility innovation and deployment in Europe-description of work*. C-MobILE Consortium.

COM. (2021). European Commission. New transport proposals target greater efficiency and more sustainable travel. Strasbourg, 14 December 2021.

COM (2021). 813 final 2021/0419 (COD). European Commission. Proposal for a Directive of the European Parliament and of the Council Amending Directive 2010/40/EU on the Framework for the Deployment of Intelligent Transport Systems in the Field of Road Transport and for Interfaces with Other Modes of Transport. Strasbourg. Retrieved from internet: https://ec.europa.eu/commission/presscorner/detail/en/ip_21_6776

COM. (2011). 144 Roadmap to a Single European Transport Area – Towards a competitive and resource efficient transport system. http://eur-lex.europa.eu/LexUriServ/LexUriServ.do?uri=COM:2011:0144:FIN:en:PDF

COOPERS. (2010). COOPERS co-operative networks for intelligent road safety, final report on demonstration.

Crichigno, J., Wu, M.-Y., & Shu, W. (2008). Protocols and architectures for channel assignment in wireless mesh networks. *Ad Hoc Networks, 6*(7), 1051–1077. https://doi.org/10.1016/j.adhoc.2007.10.002

C-ROADS. (2020). C-ROADS – The platform of harmonized C-ITS deployment in Europe. Retrieved October 15, 2020, from https://www.c-roads.eu/platform.html

Cuomo, F., Rubin, I., Baiocchi, A., & Salvo, P. (2014). Enhanced VANET broadcast throughput capacity via a dynamic backbone architecture. *Ad Hoc Networks, 21*, 42–59. https://doi.org/10.1016/j.adhoc.2014.04.008

Data Alliance. (2022). Data Alliance information about wireless communication standards. https://www.data-alliance.net/blog/5g-4g-3g-standards-lte-gsm-cdma-ism-wcdma-hspa/

Deruyck, M., Tanghe, E., Joseph, W., & Martens, L. (2011). Modelling and optimization of power consumption in wireless access networks. *Computer Communications, 34*(17), 2036–2046. https://doi.org/10.1016/j.comcom.2011.03.008

Drungilas, D., & Bielskis, A. A. (2012). Cloud interconnected affect reward based automation ambient comfort controller. *Electronics and Electrical Engineering, 18*(10), 49–52. https://doi.org/10.5755/j01.eee.18.10.3060

Dornbush, S., & Joshi, A. (2007). StreetSmart traffic: Discovering and disseminating automobile congestion using VANET's. In *2007 IEEE 65th Vehicular Technology Conference - VTC2007-Spring* (pp. 11–15). IEEE. https://doi.org/10.1109/VETECS.2007.15

Dresner, K. (2008). A multiagent approach to autonomous intersection management. *Journal of Artificial Intelligence Research, 31*, 591–656. Retrieved from http://www.aaai.org/Papers/JAIR/Vol31/JAIR-3117.pdf

Dzemydienė, D., & Burinskienė, A. (2021). Integration of context awareness in smart service provision system based on wireless sensor networks for sustainable cargo transportation. *Sensors (Basel), 21*(15), 5140. https://doi.org/10.3390/s21155140

Dzemydienė, D., Maskeliūnas, S., Dzemydaitė, G., & Miliauskas, A. (2016). Semi-automatic service provision based on interaction of data warehouses for evaluation of water resources. *Informatica, 27*(4), 709–722.

Dzemydienė, D., Kurmis, M., Baikštienė, V., & Dzindzalieta, R. (2019). Possibilities to support heterogeneous mobile services in vehicle communication networks. *Informacijos Mokslai, 85*, 176–192. https://doi.org/10.15388/Im.2019.85.22

Dzemydienė, D., Burinskienė, A., & Miliauskas, A. (2020). An assessment of provision of heterogeneous services for sustainable cargo transportation process management by roads. *Sustainability, 12*(20), 1–20.

Dzemydienė, D., Burinskienė, A., & Miliauskas, A. (2021). Integration of multi-criteria decision support with infrastructure of smart services for sustainable multi-modal transportation of freights. *Sustainability, 13*(9), 4675.

Dzemydienė, D., Bielskis, A. A., Andziulis, A., Drungilas, D., & Gricius, G. (2010). Recognition of human emotions in reasoning algorithms of wheelchair type robots. *Informatica, 21*(4), 521–532. Retrieved from http://dl.acm.org/citation.cfm?id=1923922.1923926

Dzemydiene, D., & Dzindzalieta, R. (2010). Development of architecture of embedded decision support systems for risk evaluation of transportation of dangerous goods. *Technological and Economic Development of Economy, 16*(4), 654–671. https://doi.org/10.3846/tede.2010.40

Ghafoor, K., Abu Bakar, K., Lloret, J., Khokhar, R., & Lee, K. (2013). Intelligent beaconless geographical forwarding for urban vehicular environments. *Wireless Networks, 19*(3), 345–362. https://doi.org/10.1007/s11276-012-0470-z

Gibaud, A., Thomin, P., & Sallez, Y. (2011). Foresee, a fully distributed self-organized approach for improving traffic flows. *Simulation Modelling Practice and Theory, 19*(4), 1096–1117. https://doi.org/10.1016/j.simpat.2011.01.003

IEEE Std 1609.1-2006. (2006). *Trial-use standard for wireless access in vehicular environments (WAVE) – Resource manager* (pp. 1–71). https://doi.org/10.1109/IEEESTD.2006.246485

IEEE Std 802.11p-2010. (2010). IEEE standard for local and metropolitan area networks – Specific requirements Part 11: Wireless LAN medium access control (MAC) and physical layer (PHY) specifications amendment 6: Wireless access in vehicular environments, 51.

IEEE Std WAVE. (2010). Standard for wireless access in vehicular environments (WAVE) – Networking services – Redline.

IEEE Standard for Wireless Access in Vehicular Environments (WAVE) – Multi-channel Operation. (2011). https://doi.org/10.1109/IEEESTD.2011.5712769

IEEE Std 1609.3-2016. IEEE standard for wireless access in vehicular environments (WAVE) – Networking services, revision of IEEE Std 1609.3-2010 (pp. 1–160). https://doi.org/10.1109/IEEESTD.2016.7458115

IEEE. 1609.4-2016/Cor 1-2019. IEEE standard for wireless access in vehicular environments (WAVE)—Multi-channel operation – Corrigendum 1: Miscellaneous corrections. https://standards.ieee.org/ieee/1609.4-2016_Cor_1/7465/

Ifthekhar, M. S., Saha, N., & Jang, Y. M. (2015). Stereo-vision-based cooperative-vehicle positioning using OCC and neural networks. *Optics Communications, 352*, 166–180. https://doi.org/10.1016/j.optcom.2015.04.067

Hung, C.-C., & Peng, W.-C. (2010). Model-driven traffic data acquisition in vehicular sensor networks. In *2010 39th International Conference on Parallel Processing* (pp. 424–432). IEEE. https://doi.org/10.1109/ICPP.2010.50

Gricius, G., Drungilas, D., Andziulis, A., Dzemydiene, D., Voznak, M., Kurmis, M., & Jakovlev, S. (2015). Advanced approach of multiagent based buoy communication. *The Scientific World Journal, 2015*, 1–6. https://doi.org/10.1155/2015/569841

Jarupan, B., & Ekici, E. (2011). A survey of cross-layer design for VANETs. *Ad Hoc Networks, 9*(5), 966–983. https://doi.org/10.1016/j.adhoc.2010.11.007

Jovanović, M. R., Fowler, J. M., Bamieh, B., & D'Andrea, R. (2008). On the peaking phenomenon in the control of vehicular platoons. *Systems and Control Letters, 57*(7), 528–537. https://doi.org/10.1016/j.sysconle.2007.12.004

Kakkasageri, M. S., & Manvi, S. S. (2014). Regression based critical information aggregation and dissemination in VANETs: A cognitive agent approach. *Vehicular Communications, 1*(4), 168–180. https://doi.org/10.1016/j.vehcom.2014.07.001

Kranz, M. (2016). *Building the Internet of Things: Implement new business models, disrupt competitors, transform your industry*. Wiley.

Kurmis, M., Andziulis, A., Dzemydienė, D., Jakovlev, S., Voznak, M., & Gricius, G. (2015). Cooperative context data acquisition and dissemination for situation identification in vehicular communication networks. *Wireless Personal Communications*. [New York]: Springer LLC, *85* (1), 49–62. https://doi.org/10.1007/s11277-015-2727-1

Kurmis, M., Dzemydienė, D., & Andziulis, A. (2011). Daugialypių paslaugų duomenų perdavimo galimybių tyrimas kritinėse mobilių objektų komunikacijos situacijose. *Social Technologies, 1* (2), 427–438. https://repository.mruni.eu/handle/007/11322?show=full&locale-attribute=lt

Kurmis, M., Dzemydiene, D., Andziulis, A., Voznak, M., Jakovlev, S., Lukosius, Z., & Gricius, G. (2014). Prediction based context data dissemination and storage model for cooperative vehicular networks. In I. Zelinka, P. N. Suganthan, G. Chen, V. Snasel, A. Abraham, & O. Rössler (Eds.), *Nostradamus 2014: Prediction, modeling and analysis of complex systems SE - 3* (Vol. 289, pp. 21–30). Springer International Publishing. https://doi.org/10.1007/978-3-319-07401-6_3

Kurmis, M., Dzemydiene, D., Didziokas, R., & Trokss, J. (2013). Modeling and simulation of cloud computing solution for distributed space data storage and access in mobile communication networks. *Latvian Journal of Physics and Technical Sciences, 50*(5), 20–29. https://doi.org/10.2478/lpts-2013-0030

Lee, U., & Gerla, M. (2010). A survey of urban vehicular sensing platforms. *Computer Networks, 54*(4), 527–544. https://doi.org/10.1016/j.comnet.2009.07.011

Lee, E.-K., Oh, S. Y., & Gerla, M. (2012). RFID assisted vehicle positioning in VANETs. *Pervasive and Mobile Computing, 8*(2), 167–179. https://doi.org/10.1016/j.pmcj.2011.06.001

Lee, E., Lee, E.-K., Gerla, M., & Oh, S. Y. (2014). Vehicular cloud networking: Architecture and design principles. *Communications Magazine, IEEE, 52*(2), 148–155. https://doi.org/10.1109/MCOM.2014.6736756

Lu, R., Lin, X., Zhu, H., & Shen, X. (2009). SPARK: A new VANET-based smart parking scheme for large parking lots. In *IEEE INFOCOM 2009 – The 28th Conference on Computer Communications* (pp. 1413–1421). https://doi.org/10.1109/INFCOM.2009.5062057

Masini, B. M., Fontana, C., & Verdone, R. (2004). Provision of an emergency warning service through GPRS: Performance evaluation. In *Proceedings. The 7th International IEEE Conference on Intelligent Transportation Systems (IEEE Cat. No. 04TH8749)* (pp. 1098–1102). IEEE. https://doi.org/10.1109/ITSC.2004.1399060

Milanés, V., & Shladover, S. E. (2014). Modeling cooperative and autonomous adaptive cruise control dynamic responses using experimental data. *Transportation Research Part C: Emerging Technologies, 48*, 285–300. https://doi.org/10.1016/j.trc.2014.09.001

Moustafa, H., & Zhang, Y. (2009). *Vehicular networks: Techniques, standards, and applications*. CRC Press.

Nandan, A., Das, S., Pau, G., Gerla, M., & Sanadidi, M. Y. (2005). Co-operative downloading in vehicular ad-hoc wireless networks. In *Second Annual Conference on Wireless On-demand Network Systems and Services* (pp. 32–41). IEEE. https://doi.org/10.1109/WONS.2005.7

Nzouonta, J., Rajgure, N., Wang, G., & Borcea, C. (2009). VANET routing on city roads using real-time vehicular traffic information. *Vehicular Technology, IEEE Transactions on*. https://doi.org/10.1109/TVT.2009.2014455

Olariu, S., & Weigle, M. C. (2009). Vehicular networks: From theory to practice. Retrieved from http://dl.acm.org/citation.cfm?id=1530872

Sirichai, P., Kaviya, S., Fujii, Y., & Yupapin, P. P. (2011). Smart car with security camera for road accidence monitoring. *Procedia Engineering, 8*, 308–312. https://doi.org/10.1016/j.proeng.2011.03.057

Soldo, F., Casetti, C., Chiasserini, C.-F., & Chaparro, P. (2008). Streaming media distribution in VANETs. In *IEEE GLOBECOM 2008 IEEE Global Telecommunications Conference* (pp. 1–6). https://doi.org/10.1109/GLOCOM.2008.ECP.126

Sommer, C., Schmidt, A., Chen, Y., German, R., Koch, W., & Dressler, F. (2010). On the feasibility of UMTS-based traffic information systems. *Ad Hoc Networks, 8*(5), 506–517. https://doi.org/10.1016/j.adhoc.2009.12.003

Ucar, S., Coleri Ergen, S., & Ozkasap, O. (2015). Multi-hop cluster based IEEE 802.11p and LTE hybrid architecture for VANET safety message dissemination. *Vehicular Technology, IEEE Transactions on*. https://doi.org/10.1109/TVT.2015.2421277

UN. (2021, October). United Nations Department of Economic and Social Affairs. The importance of sustainable transport for the 2030 Agenda and the Paris Agreement. In *Sustainable transport, sustainable development* (pp. 2–13). https://doi.org/10.18356/9789210010788c003

Vinel, A. (2012). 3GPP LTE Versus IEEE 802.11p/WAVE: Which technology is able to support cooperative vehicular safety applications? *Wireless Communications Letters, IEEE*. https://doi.org/10.1109/WCL.2012.022012.120073

Weil, T. (2009). Service management for ITS using WAVE (1609.3) networking. In *2009 IEEE Globecom Workshops* (pp. 1–6). IEEE. https://doi.org/10.1109/GLOCOMW.2009.5360714

Wu, C., Ohzahata, S., & Kato, T. (2013). Flexible, portable, and practicable solution for routing in VANETs: A fuzzy constraint Q-learning approach. *Vehicular Technology, IEEE Transactions on*. https://doi.org/10.1109/TVT.2013.2273945

Zagrouba, R., & Kardi, A. (2021). Comparative study of energy efficient routing techniques in wireless sensor networks. *Information, 12*(1), 42. https://doi.org/10.3390/info12010042

Chapter 4
Structure of Smart Service Provision System for Operative Management of Transportation

Dalė Dzemydienė

Abstract The aim of this part of the research is to examine the system architecture required for Smart-SDS in freight transportation processes. Smart-SDS is designed to help prevent accidents, warn of other accidents or adverse situations on the road, and avoid traffic jams. The requirements for developing this type of system are related to the context recognition functionality. The infrastructure supporting such a system integrates networks, equipment to record the required data in data warehouses (DWs), analyse DWs, and requires AI methods to develop intelligent service performance algorithms. The system will be able to support multi-criteria solutions in the stages of operational transportation management processes. All system components will be able to work in compatibility and wireless style. Interoperability of multi-dimensional data warehouses (DWs) using cloud computing technology will be required. The services integrated into the system must be assessed in the light of their prioritization at various stages of transport. The system infrastructure will be able to monitor the transport process and enable the changing topology of transport networks and communication networks.

Keywords Smart service delivery system · Ontology of freight transportation · Transmission of context data · Assessment algorithms

4.1 Componential View of Smart-SDS Architecture

The analysis of intelligent transport systems (ITS) and their development methods allowed to evaluate innovations in the transport management sector (Lim & Maglio, 2018; Mo et al., 2010; Osmundsen et al., 2018; Du et al., 2021; Al-Sultan et al., 2013; Cespedes et al., 2013). A design methodology of the Smart-SDS is analysed in

D. Dzemydienė (✉)
Vilnius Gediminas Technical University, Vilnius, Lithuania

Vilnius University, Vilnius, Lithuania
e-mail: dale.dzemydiene@mif.vu.lt

D. Dzemydienė et al., *Development of Smart Context-Aware Services for Cargo Transportation*, International Series in Operations Research & Management Science 330, https://doi.org/10.1007/978-3-031-07199-7_4

this subsection. This methodology enables the evaluation of information, which is extracted from the environment (surroundings) by implementing all possible types of onboard units (OBUs) and road side units (RSUs), as well as heterogenic communication channels. The context of cooperating facilities is implemented in the communication infrastructure working at real-time conditions by following the requirements of ICT standards (IEEE, 2010a, b) and recommendations of developing wireless sensor networks (WSNs) following (Bazzi et al., 2014; Chen et al., 2014). The methods for multi-criteria decision-making are developed for evaluation of usefulness of obtained information and context information following the results of works (Chang et al., 2017; Padovitz, 2006; Dzemydienė et al., 2022; Le Dinh et al., 2021). The algorithms for limitation and selection of context information are developed by following results of works (Golestan et al., 2016; Al Sultan et al., 2013, Boyraz et al., 2012; Chang et al., 2017). The structures and algorithms for the dissemination of knowledge among other vehicle nodes have been integrated (Osmundsen et al., 2018; Kurmis et al., 2013, 2015; Dzemydienė & Burinskienė, 2021).

The development of service delivery system for transportation of goods have follow to the requirements of Sustainable Development Agenda (Colglazier, 2015) and green transport (COM, 2011) by improving the pollution and safety of such processes. The requireents decribed by Andrea and Savigni (2001) help in requirements engineering stages of context aware services development. The infrastructure equipment of roads and communication chanels and standartds enable us for such kind software development of the system (C-ROADS, 2020; IEEE Std, 2006). Results of previous works influence the development of structures of the system following the results of works (Fan et al., 2019; Holbrook, 1990; Kakkasageri & Manvi, 2014; Karciulias, 1997; Le Dinh & Pham Thi, 2012; Pereira et al., 2014; Schilit & Theimer, 1994). Results of our previous works influence our development methodology as well (Dzemydienė et al., 2016, 2021; Kurmis et al., 2013, 2015).

The architecture of a prototype of the Smart-SDS system is developed and includes the description of the topology of automotive communication networks, management components for the provision of various services (Fig. 4.1). The structure of the system integrates the contextual data storage and exchange models and algorithms for reduction of amount of transmitted data. The system cooperatively use network channel resources in the automotive communication processes following the works (Cheng et al., 2011; De Oliveira et al., 2013; Yang et al., 2013) and by implementing ideas from innovative initiatives (C- MobILE, 2017; COOPERS, 2010; C-ITS Platform, 2016).

The separate modules of the system are included in the whole structure of the system:

- The subsystem for context data acquisition (S-CDA), i.e. for obtaining the data from sensors in vehicle environment (In-V) and from sensors in outside of vehicle environment (Out-V);
- The data warehouses (DWs) are used as big data stories for recording (monitoring) data from sensors;

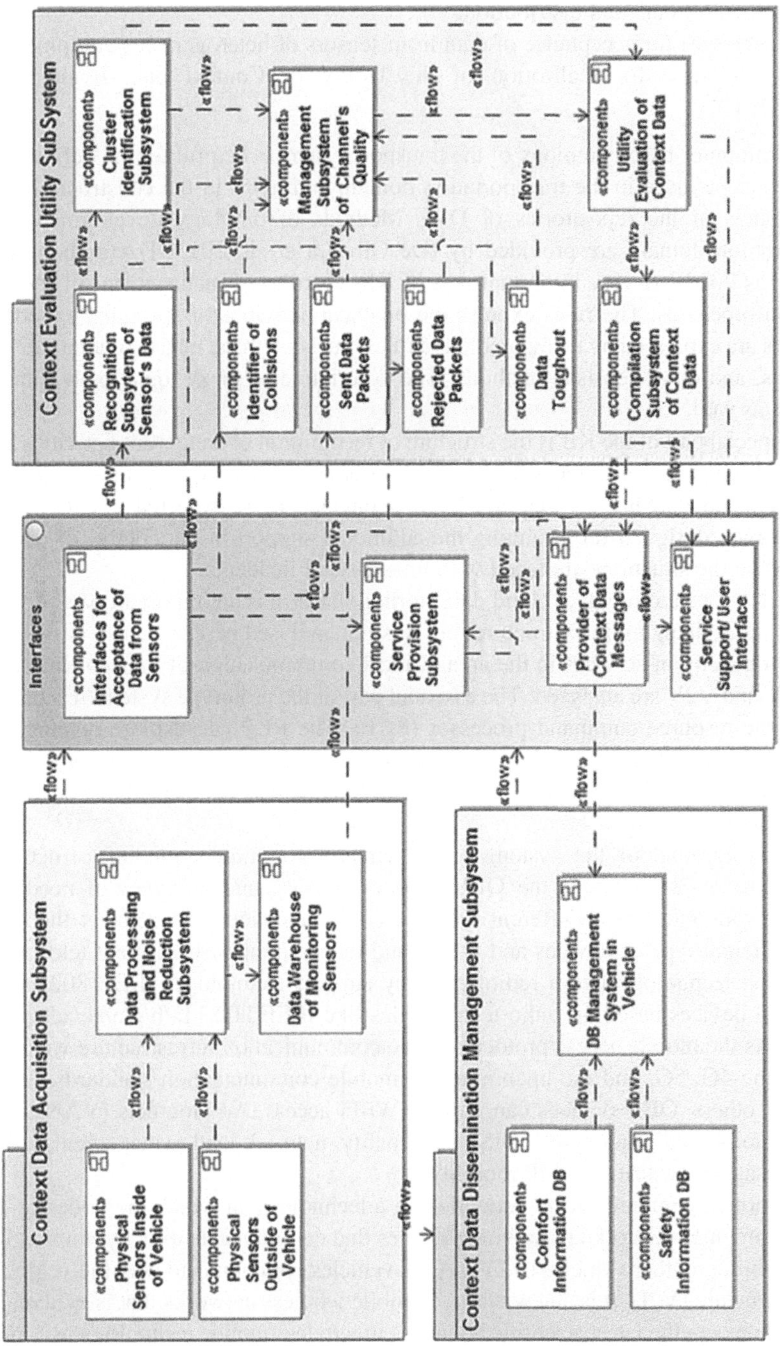

Fig. 4.1 The structure of subsystems integrated into the prototype of the Smart-SDS for the recording of monitoring data of transportation. Sources: (Kurmis et al., 2015; Dzemydienė & Burinskienė, 2021)

- The subsystem for context data dissemination (S-CDD) is designed for the assessment of data and distribution;
- The interfaces for acceptance of data from sensors of heterogeneous equipment;
- The subsystem for realization of the Utility of Context data Evaluation (S-CDEU).

The computer-based ontology of the transport system is helpful for the definition of the concepts used in the transportation domain and helps in the construction of meta-models of the repositories of DWs Methods of ontology development in transportation domain are provided by (De Oliveira et al., 2013; Dzemydienė & Miliauskas 2013). A very important part of KB are rules of management of transportation processes. The rules express the practical activities of specialist-experts. The rules are expressed by using predictive rules, decision trees, matrix of alternative decisions, and other heuristic methods. The descriptive or fuzzy logic expressions are used as well.

The special part of the KB is the structure of recognition of dangerous activities in transportation situations, identification of important indications of the alarming situations, and enabling the control with avoidance of unexpected events. The interfaces are designed for arranging the additional support in the control of decisions, when the situations are faced with unsuspected incidents.

The cloud technology like cloud data storing platform is an important part of the whole data exchanging infrastructure, but it is not analysed here.

Heterogeneity of services in the area of local communication of transport means like In-V and V2V are analysed. The essential part of the prototype system of Smart-SDS is the resource command processor (RCP). The RCP can explore resources, which are used for data recording by implementing reading/writing memory. The RCP is responding for storage of data in distributed DWs. The DWs have support interoperability.

The background of the system is the wireless communication infrastructure (WCI). The WCI integrates the OBU devices, WSNs, and all types of needful heterogeneous interfaces. Different types of communications can connect the car to the different type of devices and OBUs and they can employ different telecommunication technology (from radiofrequency support technology, IEEE 802.11p, additional devices based on radio technologies like IEEE 802.11a/b/g/n/ac/ad, and others). As the mobile access protocols in the communication infrastructure we can include the 4G, 5G, and the upcoming 6G mobile communication standards, also LTE and others. OBU devices can include Wi-Fi access, Ad Hoc nets (VANET), geo-location-based routing with GIS functionality, network load management, reliable messaging, security, and IP mobility.

Automotive communication networks are a technology in which the nodes of a mobile computer network are moving vehicles that communicate directly with each other or through other vehicles (V2V). Also, vehicles communicate with the vehicle to infrastructure (V2I). It is a new class of mobile wireless networks that is evolving rapidly thanks to the latest scientific advances in wireless mobile technology and the automotive industry. Car communication networks can be formed between moving

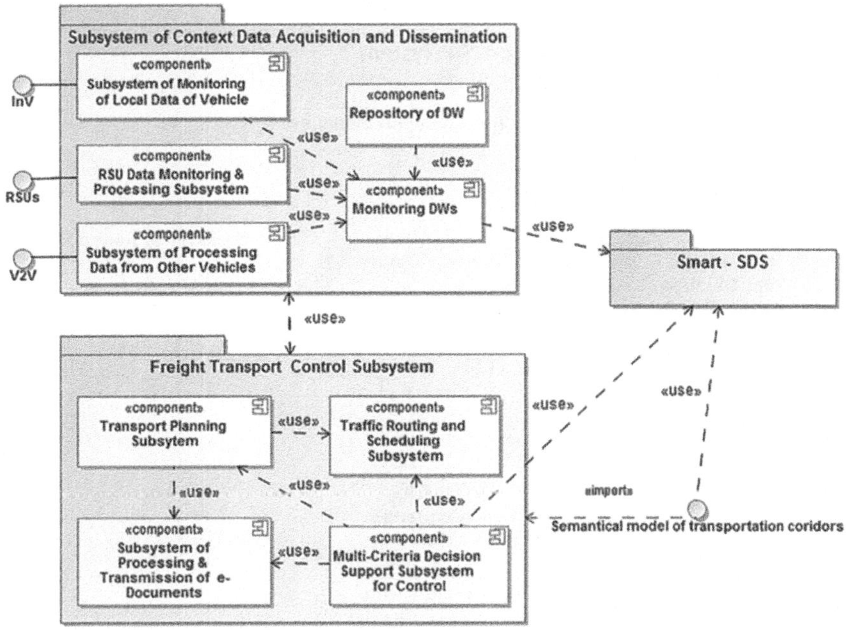

Fig. 4.2 Structure of packages in relation with the Smart-SDS

vehicles, which are equipped with homogeneous or heterogeneous wireless interfaces (802.11a/b/g/n/p, WiMAX, 4–5 G, LTE, etc.).

The structure of the subsystems of transport management and scheduling is presented as the service delivery process. Such part of the system is interacting with other subsystems of the Smart-SDS (Fig. 4.2).

The main components of the subsystem for freight transport control are:

- The subsystem of processing and transmission of accompanying e-documents;
- Traffic routing and scheduling subsystem;
- The context data acquisition and dissemination subsystem consist with:
- The subsystem of monitoring of local data in vehicle (InV);
- RSU data monitoring and processing subsystem;
- The subsystem of processing data from other vehicles.

All of such three subsystems have interfaces with required sensor equipment. The detailed description of components of the Smart-SDS is designed as package of the Service Support Subsystem presented in Fig. 4.3.

The classification of services is provided by the component of "Adaptation of Preferences" that uses opinions of drivers. The cloud service performance possibilities are used as well.

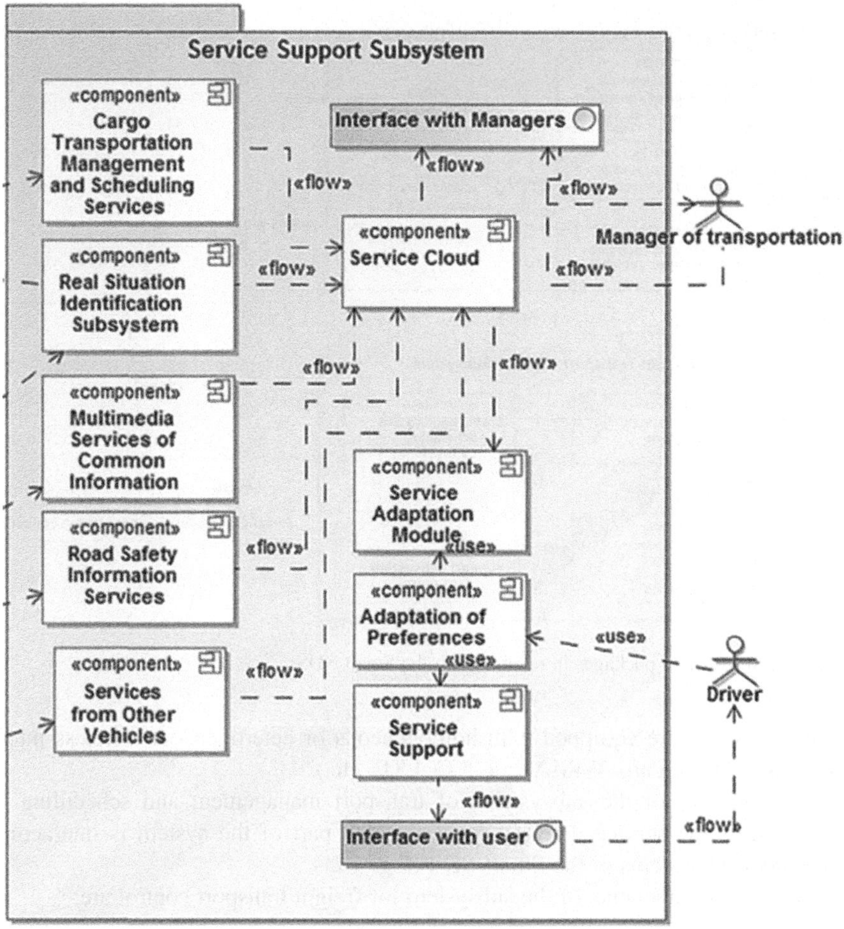

Fig. 4.3 Detailization of componential structure of the package of subsystem for service support. Source: following (Kurmis et al. 2015; Dzemydienė & Burinskienė, 2021)

4.2 Classification of Heterogeneous Services Provided for Transportation

This subsection reviews some studies in the area of e-service development for the vehicle-to-infrastructure (V2I) communication. The EU focuses on the development of Cooperative Intelligent Transport Systems (C-ITS) (C-ICT Platform, 2016). The systems, like C-ITS, are recognized as a way to improve traffic safety, efficiency, and comfort (Bazzi et al., 2014; Campolo et al., 2015; Ohn-Bar et al., 2015; Cheng et al., 2011; Du et al., 2021; Sathe & Deshmukh, 2016; Sun et al., 2010). Multiple EU financial mechanisms show the support efforts for the development of different components of C-ITS. Research works are directed towards the development of necessary infrastructure (Chen et al., 2014; IEEE, 2010a; Kranz, 2016; Sternberg et al., 2014). Various communication technologies are selected for these purposes from the digital audio broadcasting (DAB) technology, Infrared light, General Packet Radio Service (GPRS) to the Global Standard for Mobile Communications (GSMC), and others.

There are several types of services, which are important for the development of the C-ITS Platform, and we would like to considerate them in detail:

- Safety services—accident warning, incident warning, wrong-way driver warning, weather condition warning, roadwork information, lane crossing, lane-keeping, auxiliary lane accessibility, legal speed limit, traffic congestion warning, recommended speed limit;
- Convenience services—international service handover, road pricing, estimated travel time, recommended next link, map updates.

The implementation covers various aspects from the building of an in-vehicle platform to the whole safety system architecture and the needed infrastructure, including business and legal aspects [5]–[7]. Services provision by vehicle was grouped into clusters:

- Side collision—road junction safety, lane-change manoeuvres, safe overtaking;
- Longitudinal collision—head impact warning, rear impact, speed limit and safe distance, frontal collision warning;
- Exit from the road—road condition—slippery road, curve warning;
- Vulnerable road users—identification of vulnerable road users and revision.

The development of new kinds of services and technological platforms, which can support the relation and communication of vehicles with information infrastructure (V2I) and between vehicles (V2V) can help in increasing road safety and management efficiency (C-ICT Platform, 2016). Service platforms are based on wireless networks, new kinds of communication protocols and services provided for vehicles and infrastructure.

This part of services can be grouped into the following parts:

- *Cooperative urban services* are services of such type which are related to traffic management and route planning in urban scenarios.
- *Cooperative inter-urban services* support driver awareness of the road.
- *Cooperative freight and fleet services* are components for professional drivers and fleet managers, like rest area parking, loading space, booking, monitoring the carriage of dangerous goods, and others.

When services come to vehicle communication networks, they are classified as of the type of safety or the type of comfort and entertainment application. The services are characterized as safety programmes that enhance the protection of drivers and passengers by sending and receiving information related to vehicle safety. The services of the safety category usually include collision alerts, lane departure alerts, roadside alerts or even video streaming, other incident management. Safety category services are sent directly to drivers or are received through an automatic active safety system. Services that fall into the category of comfort and entertainment are often characterized as unsafe programmes. They are only needed to improve the level of comfort for drivers and passengers, e.g. choosing a place to rest on a trip, a stopover, and so on.

Comfort and entertainment services can provide information to drivers or passengers on weather and traffic information and data retrieved from the GIS, i.e. the location of the nearest restaurant, gas station, or hotel and their prices. Passengers can play online games, connect to the Internet, and send or receive instant messages when the vehicle is connected to the infrastructure network. Investments by vehicle manufacturers show that these programmes are becoming extremely popular.

The services can be characterized by more than 4 categories following the proposal of Crainic & Montreuil (2016): services which can indicate the safety conditions of the road and/or unsafety conditions of the road; services that provide traffic monitoring data; services that provide general information (data on weather conditions and prognosis); multimedia services; services for management needs; services which provide vehicle engine monitoring data.

The class diagram of the main components of the computer-based ontology of the freight transportation process is designed (Fig. 4.4) by using UML class diagram notation and express main interrelated concepts. The road segments are characterized by the parameters, which are important for DW structures. The ontology is applied for service construction as semantic model. The parameters of road class description (i.e. road ID, road type, number of rows in the road segment; max speed in the road) are important for the service delivery process.

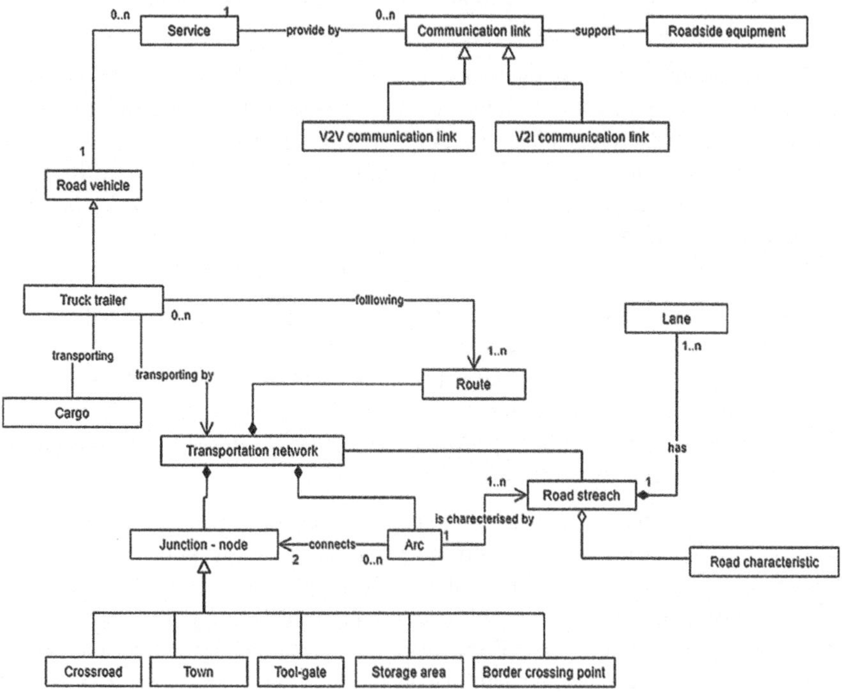

Fig. 4.4 Computer-based ontology of freight transportation process with the assessment of infrastructure elements

The road segments are connecting with stationary nodes (i.e. inter-cross points, crossroads, traffic rules, etc.). Risks are related to the accident scenarios which were in history and are affected by types of dangerous goods carried, and the type of surroundings.

Different parameters are designed for road segment description, like speed limits of vehicles and statistical averages of the density of nodes during different journey periods. The values of such parameters differ in highways, cities, and/or rural areas. The speed usually differs in different year periods and depends on the road pavement characteristics. In the summer period, the speed of vehicles has average value in rural areas, low or very low in towns, and very high on highways. For example, inference possibilities are low in rural areas, average in towns, very high in cities, and low in highways.

Vehicles can support heavier capacities of computing, radio frequencies, the large spectrum of sensors In-V and wireless interfaces with high-speed communication. The analysis of historical, statistical data can help in the recognition of certain patterns on peaks, links between mobile automotive groups.

Development of service delivery in different networks of vehicles is faced with specific issues on how to control the networks with large scale and high level of mobility, including fragmentation of networks, changing topology, and differentiation of assurance quality (C-MobILE, 2017; COOPERS, 2010). The description of fragmentation of road segments with different characteristics has to be used in data structures of service provision systems. The description of different characteristics of road segments must entangle different dynamicity properties of the surroundings (i.e. on the highways the speed can reach up to 300 km/h, in urban areas 50–60 km/h). On low-traffic density roads, vehicle density may be 1–2 cars per km.

The service classification structure is designed by analysing the types of the services (Fig. 4.5).

The Smart-SDS is an automatic, active, road safety system that informs drivers by providing warnings generated by sensors, which primary data are obtained in many subsystems of the system. In the system are included GIS-enabled cameras and GPRS subsystems. The embedded subsystems with built-in functions (such as vehicle collision alert, lane departure warning, emergency video broadcast, prediction and adverse event warning) are included in the infrastructure supporting all intelligent services. The wireless communication network and subsystems based on it provide the information needed by users. Primary raw data sets are provided from many heterogeneous network nodes and are processed, adapted, and presented as useful information needed to meet user needs. This information is adapted for managers. The algorithms of filtering of the services are provided and properly manage the required contextual data. Services that are of interest to users at a given point in time, depending on the specifics of the environment, the type of environment, the capabilities of users, and refer to data in this context. Context data can be diverse and dynamically changing, making user-initiating service discovery impractical and of little use. In a dynamic environment that controls transport processes, consumers should be constantly informed about the services that are important to them, and the information should be adapted by choosing between synchronous and asynchronous, i.e. the data management mode must be triggered

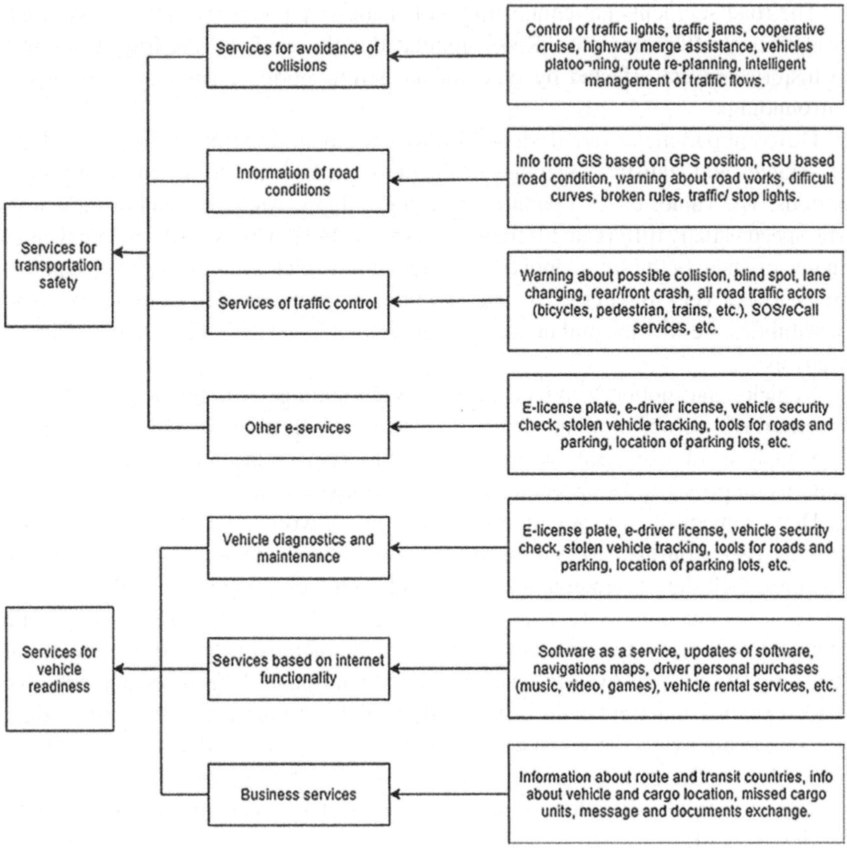

Fig. 4.5 Illustration of classification of services for freight transport

and organized according to changes in context data and their importance (Rasch et al., 2011).

Service-Oriented Architecture (SOA) has been used to create and implement services. Typically, service discovery methods have been used in traditional SOA techniques. User queries provide the main service discovery factors. In the automotive communication environment, the user's context and choices become essential factors in determining consumers' choice of the services of the greatest interest to them in a given situation. The user context is constantly changing, and in the transport process, the following key decisive parameters determine the context data:

- The location of a vehicle user;
- The situation of management time;
- Environmental information;
- User status;
- Information from the surroundings of cars;
- Information from monitoring devices.

These ever-changing aspects pose new challenges in the development of state-of-the-art methods for the development, delivery, discovery, and selection of services for automotive communication systems. The system must find services based on changes in the context data, even if the user sometimes does not initiate a service search engine. The ability to respond to these predicted service request scenarios allows for a significant improvement in the user experience using the system (Rasch et al., 2011).

Understanding the context in a mobile environment requires an interdisciplinary approach that combines programming paradigms, operating systems, embedded systems, computer networks, mathematical analysis, and other areas of science. This creates a huge gap between a high level of requirements for mobile system services and the complexity of operations in dealing with contexts derived from the environment. Application requirements cover various aspects such as adaptability, flexibility, and intelligence. Due to the heterogeneity of devices, the high level of mobility, and the changing topology, managing contextual information is extremely complex and error-prone.

Applications are exposed to the raw context and its comprehension modules. The context preprocessing and reasoning is needful to express in algorithms for the service provision systems with smart components. Context recognition corresponds to AI methods because the management in a rapidly changing topology of mobile objects and dynamical conditions in an environment changing require additional methods for knowledge management.

The physical location of moving objects and situations in surroundings are needful for the assessment of the importance of contextual information. In one type of condition, the contextual information can become very important, in another insignificant, or useless at all. Due to the frequent disconnection of nodes from the context source, maintaining up-to-date context information may be difficult.

The modern meaning of the context is manifested through dynamic changes, through the mobility of clients or context providers. The temporal importance of the context can change dynamically depending on the vehicle's location. This problem can cause the accompanying problems of contextual ambiguity. Lack of coordination in a V2V communication infrastructure can lead to ambiguity and redundancy since a similar context can be offered for multiple nodes with different values and attributes.

As well the safety and reliability of the information sources became the problem that needful to solve. The information can be of varying and influence the quality and reliability of transportation processes.

4.3 Description of Algorithms for Assessment of Importance of Transmission of Messages and Context Data

For ensuring the supporting process of delivery of necessary services for the management of freight transport processes and enabling the right process of transmission data via communication channels the algorithms are developed. They are included in the Smart-SDS as part of ensuring selectivity of obtained context data by avoiding the disturbance of drivers with transmitting of non-important information and with the ability to avoid flooding of communication channels. The development of the algorithms is aimed at defining the mathematical expression for the construction of schemas on how to express writing mechanisms and how to explain the prioritization and selection process of transmission of messages (Kurmis et al., 2013; Kaffash et al., 2021; Kurmis et al., 2015; Dzemydienė et al., 2020). The developed algorithms are included in the recognition process of conditions of provision of the necessary and context-aware services, which have importance for the safety of cargo transportation operative management. Different movement scenarios affect the situations in transportation. Differences are observed, when vehicles are moved on highways, in cities, rural areas, cross-border points, reloading terminals, or standing parks.

The analysis of the subject covers the following aspects of classification of the importance of obtained information from sensors (Dzemydienė & Radzevičius, 2020; Wibisono et al., 2009; Sun & Yu, 2014; Sun et al., 2010).

The algorithm is constructed for the classification of information by accessing the importance of monitoring data. Some of the incoming data will be rejected, another part will be left for further processing and another part will be accepted promptly and expeditiously as critical for transport security. The main parameters, which are important for the identification of all sensor data, which are stored in data warehouses are:

- Vehicle identification (ID),
- Location of vehicle (longitude and latitude coordinates);
- Time of provided data;
- Environment status is important to understand from what conditions the obtained data are delivered;
- User status;
- Vehicle dynamics at this time moment.

The network conditions, available resources for data packet formation and transmission as important parameters but they are about communication capabilities.

The primary (row data) matrix is constructed. All data obtained are forwarded for assessment of their usefulness. The arrived sensing data are obtained at the moment ti and at the location $l\{la,\ lo\}$ and are transferred from the sensor $s_{j,m}$ of type m, indicating the vehicle v_k.

$$M_L = \begin{pmatrix} d_{L_{11}} & d_{L_{12}} & \dots & d_{L_{1n}} \\ d_{L_{21}} & d_{L_{22}} & \dots & d_{L_{2n}} \\ \dots & \dots & \dots & \dots \\ d_{L_{l1}} & d_{L_{l2}} & \dots & d_{L_{ln}} \end{pmatrix} \tag{4.1}$$

All data are forwarded for assessment of their usefulness by following the formula:

$$d_{L_{ij}} = \left(Ty_j + H_j + Ex_j\right) m_i cr_i Pr_i, i = 1, \ldots, l, j = 1, \ldots, n \tag{4.2}$$

where Ty_j indicate the importance of the concrete data (message) type. The messages from comfort areas are assigned to value 1; messages from mixed type areas (they have properties of comfort and safety) are assigned to value 2; and the messages, which have the character of safety are assigned to value 3.

Another set of parameters *{H}* is used for indicating the importance of messages for future decisions. The assessment value is assigned for a concrete message from [0;1] and indicates—whether a message should be stored for a longer time or not.

The parameter *Exj* expresses the data exchange area, and digital values from [1–4] were assigned to the message/data, indicating the following cases: (1) for messages of type Vehicle to Manager (V2M) communication, (2) for messages in the local environment of vehicle (InV), (3) for messages of type of Vehicle to Infrastructure (V2I), and (4) for messages of type Vehicle to Vehicle (V2V).

Crj indicated the coordinates of the location of obtained data/message and time t_i, expressing such values as $cr_j\ t_i = \{crj\ t_i,\ latitude;\ crj\ ti,\ longitude\}$.

The rule for priority assignment is calculated by using the formula: $\mathrm{Pr}_i = 1 + \frac{I_j}{A_j}$ and is introduced in the algorithm. The normalized values of $\mathrm{Pr}_i = 1 + \frac{I_j}{A_j}$ are assigned according to satisfying the appropriate conditions. The pseudocode of this algorithm is

```
 <condition> <if information of messagej is critical and must be sent
immediatelly=True>,
     <when> <Pri:=3>; („it means that messagej is very important for
safety and the information about this sitution is stored in the
appropriate status DW");
 <condition> <if information of messagej is so much significant=True>,
     <when> <Pri:=2>; („it means medium importce of messagej for safety
conditions and the informations is stored for appropriate DW");
 <condition> <if information of messagej is not significant=True>,
    <when> <Pri:=1>; („it means that messagej is not important for safety
conditions and information will be rejected").
```

Also, by reducing the amount of irrelevant information, the relative age of the received message data is calculated as I_j, and the rule for such reductions is expressed as

$$A = \begin{cases} 1, & \text{if } T_M > 5s \\ 2, & \text{if } 1 < T_M < 5s\,. \\ 3, & \text{if } T_M < 1s \end{cases} \tag{4.3}$$

$T_M$is the relative age of storage of the received messages.

For reducing the number of forwarded messages, another algorithm is entered. For these purposes the matrix MO is constructed, which values express the importance of data provided for other moving objects:

$$M_O = \begin{pmatrix} d_{O_{11}} & d_{O_{12}} & \dots & d_{O_{1n}} \\ d_{O_{21}} & d_{O_{22}} & \dots & d_{O_{2n}} \\ \dots & \dots & \dots & \dots \\ d_{O_{l1}} & d_{O_{l2}} & \dots & d_{O_{ln}} \end{pmatrix}. \tag{4.4}$$

Each item of matrix M_O is expressed by predictive usefulness (the weight) of the message mi, which is transmitted to the other vehicle (car) v_j. The items of matrix M_O are assessed by the formula:

$$d_{O_{ij}} = (Ty_j + \text{Exc}_j + Z_j)m_i cr_i \text{Pr}_i n_i, \quad i = 1, \dots, l, j = 1, \dots, n. \tag{4.5}$$

The parameter Exc_j is included in the weight expressions and has different assignments:

1. for communication of type from V2M;
2. for communication In-V;
3. for communication of type V2I;
4. for communication of type of V2V.

n_j_indicates the number of interacting cars in cluster j. The parameter Z_j is defined as a predictive indicator that helps to express the quality of the communication channel, which can be calculated using the formula:

$$Z_t = \frac{1 + \left(\frac{C_t + D_t}{2}\right)}{Tr}. \tag{4.6}$$

Because the collisions have a probability to appear in different types of communications, we need to include the collision assessment parameter C_t into the algorithm, which is expressed as

$$C = 1 - \left(\frac{1}{1 + c_{t-1}}\right). \tag{4.7}$$

As well as for more adaptable solution of the assessment procedure, the possibility of rejection of packages during message transfer sessions can appear, the parameter of package rejection is introduced: $D = 1 - \left(\frac{1}{1+d_{t-1}}\right)$. In the formula of evaluation of the predictive indicator of the quality of the communication channel must be included the parameter Tr—for expression of the bandwidth of the communication channel:

$$Tr = 1 + \left(\frac{tr_{t-1}}{100}\right). \tag{4.8}$$

For expression of context utility values, which are important, when the Smart-SDS system transmits messages via communication channel from hybrid VANET to the cloud system, the new type of matric $M_C$is needful to apply. The items of matrix express the context utility values of message m_j which are provided for recipient entity r_i:

$$M_C = \begin{pmatrix} d_{C_{11}} & d_{C_{12}} & \cdots & d_{C_{1n}} \\ d_{C_{21}} & d_{C_{22}} & \cdots & d_{C_{2n}} \\ \cdots & \cdots & \cdots & \cdots \\ d_{C_{l1}} & d_{C_{l2}} & \cdots & d_{C_{ln}} \end{pmatrix}. \tag{4.9}$$

The function for assigning weights for assessment of the usefulness of each transmitting data, when the message is passed to recipient entity is expressed as

$$d_{C_{ij}} = \left(Ty_j + Hx_j + Exc_j + Z_j\right)m_i cr_i \mathrm{Pr}_i, \quad i = 1, \ldots, l, \quad j = 1, \ldots, n. \tag{4.10}$$

The parameter Ty_j is applied for the expression of the importance of the message for safety. When the value 2 is assigned for this matrix item, it means that the message data are very important for the safety process of transportation and 1 is assigned, then importance is less. The parameter Hx_j is introduced for expression of the importance of storing data for the future and has obtained values from the discrete range [0;1]. Other parameters have meanings, which are the same as in previous formulas.

4.4 Illustration of Experimental Results

Simulation modelling tools and certain numerical methods were used to evaluate the performance of prototype of Smart-SDS system. The package of National Chiao Tung University Network Simulator (NCTUns) was chosen as the simulation modelling environment according to the recommendations (Wang & Chou, 2009; Wang & Huang, 2012).

The package NCTU is the environment for simulation and emulation of links between moving objects and helps to make experiments with networks and object's mobility. The package includes a modelling tool and an emulator to model the various protocols which can be used in both wired and wireless networks. It is based on a modern kernel re-introduction method that gives NCTUns many advantages that are lacking in other tools such as ns-2 with OPNET (Wang et al., 2008, 2009, 2011).

When developing the prototype system of Smart-SDS we are designed the imitational models, which are helpful in assessing of working conditions of the wireless communication networks. Special methods were included for multi-criteria

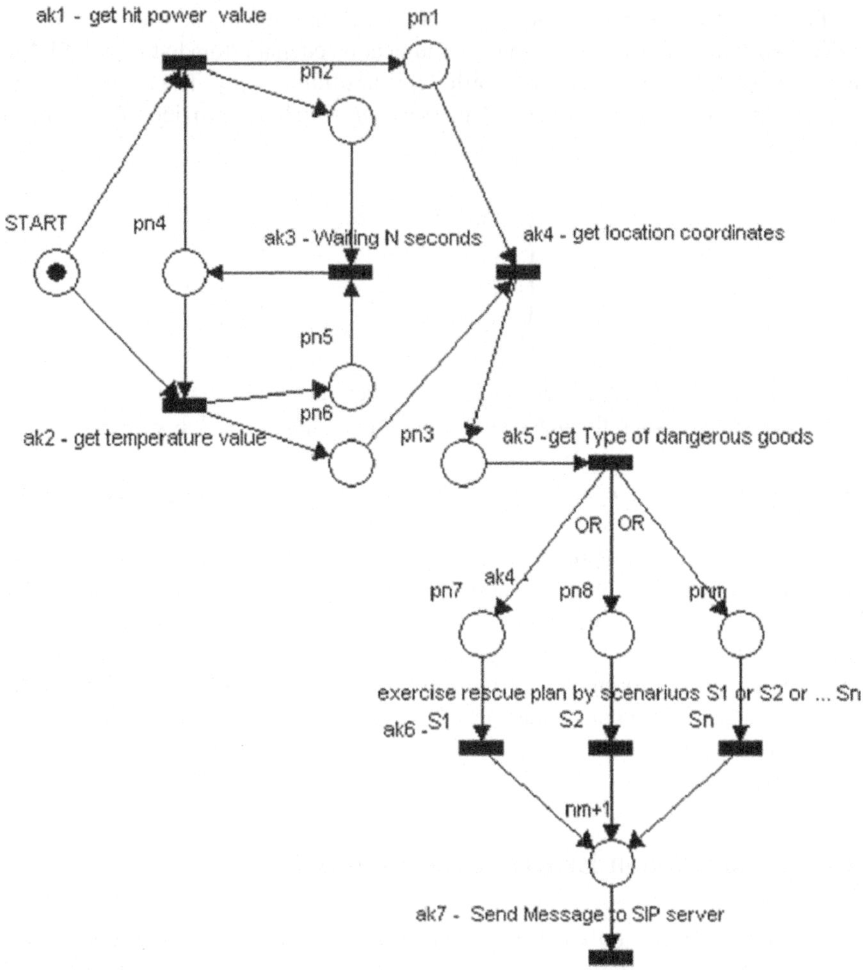

Fig. 4.6 Modelling of monitoring scenario of wireless sensors in vehicles (using notation of Petri nets)

evaluation, machine learning, and training. Also, stages of contextual data collection and dissemination are examined (Fig. 4.6).

Using different scenarios, experiments were performed under different traffic and mobility conditions. The examples of scenarios' for assessment of conditions of wireless communication possibilities are constructed by using modules of simulation software. The examples of created scenarios enable us to imitate the working conditions of wireless communication between nodes in-car networks and OBU, RSU, and other sensors by using the base of LabView modules and other components for sensing and proceeding of row data (Fig. 4.7), designed using notation of the sequence diagrams of UML.

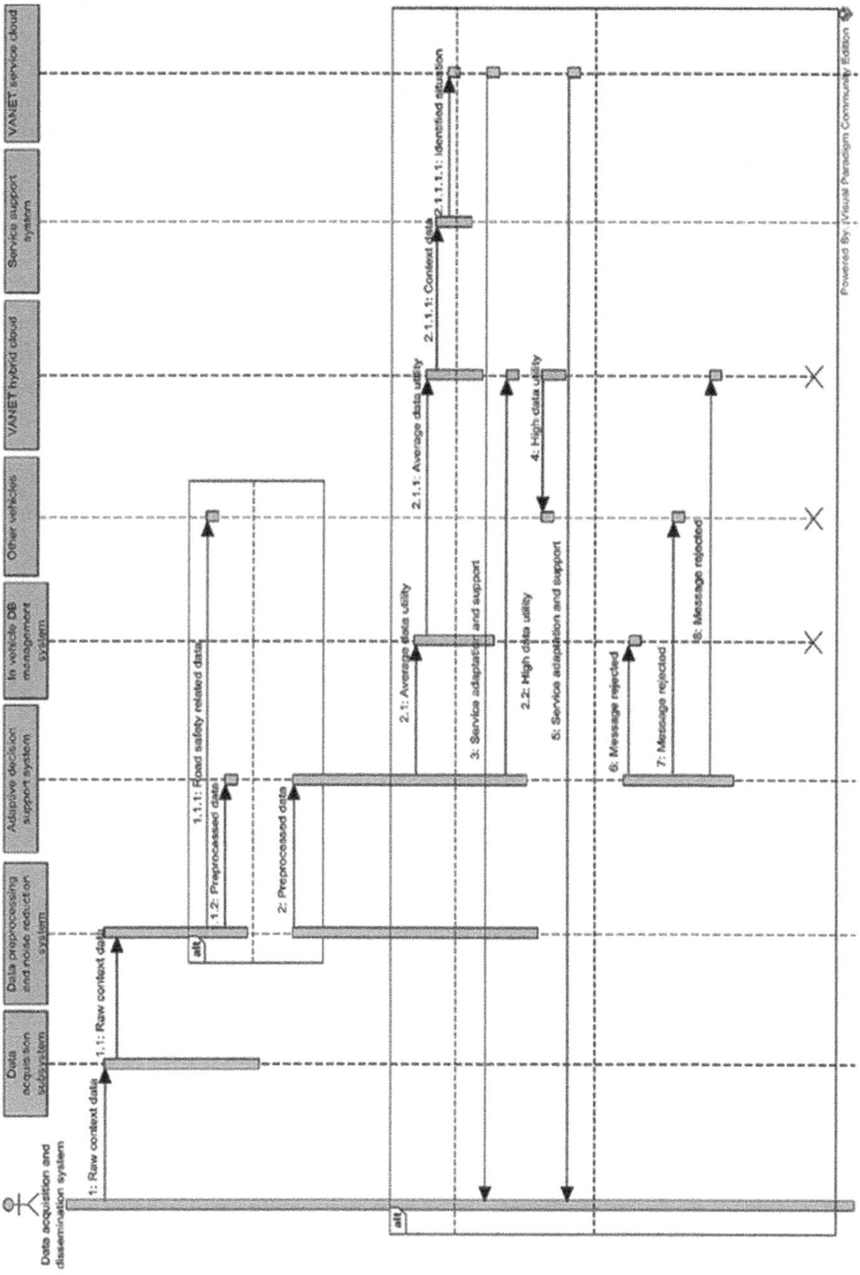

Fig. 4.7 Illustration of description of the scenario for communication between components of Smart-SDS (Source: Kurmis et al., 2013, 2015)

The prototype system is modelled for a specific purpose and has the functionality to understand the environment. The new versions of NCTUns package enable intuitive user interface applications (Wang et al., 2011) and simplify the modelling process and eliminate the need to model emerging communication-compliant scenarios with complex scripts.

Key benefits offered by the NCTUns 6.0 version and the later version of Estinet 9.0 are possibilities to support package transferring and imitate the basic patterns of driver behaviour. The main road network elements and wireless access methods are included in the imitational model (i.e. RSUs, OBUs and IEEE 802.11 (b) infrastructure mode, GPRS, Ad Hoc mode, and WiMAX communication). As well, satellite communication (like DVB-RCST) was included in the package.

The NCTUns version 5.0 allowed to simulate important improvements in VANET network modelling, i.e., efficient node mobility management, high-volume automotive communication networks, automatic road network creation from SHAPE format map files, and enabled IEEE 802.11 (p)/1609 standards for automotive communication network organization.

Different types of communication scenarios were simulated during the stages of the modelling process. One of them was designed to evaluate the capabilities of communication processes when generating messages and transmitting data from vehicles to a hybrid VANET cloud server. Computer ontologies have been developed for appropriate structures of DW. The conditions of the communication network were measured when the network enabled the storage of data in the DW server, some types of 802.11p communication of RSUs were installed in the modelling laboratory. In this case scenarios of communication allow modelling from 1 to 100 units of 802.11p type OBU. To run this scenario, statistical data about the parameters of the messaging requirements was obtained (Table 4.1).

Table 4.1 Evaluation of data transmission parameters for heterogeneous service support in automotive communication networks

ID	Activities of transportation	Packet bandwidth (KBps)	Likelihood of data loss	Data Rate	Permissible delay (microseconds)
S1	Lane change	~100/1	Average	Event	~100
S2	Changing the position of the traffic light	~100/1	Average	Periodical	~100
S3	Hazard warnings	~100/1	High	Event	~100
C1	Multimedia transmission	~100/1	Average	Periodical	~100
C2	IPTV	~1300/500	Average	Periodical	<200
C3	VOIP	~100/64	Average	Periodical	<150
C4	Exchange of video/ audio packages	High	High	Periodical	–
C5	Leisure and games activities	High	High	Periodical	–

Table 4.2 Types of sensors used in data monitoring scenarios

Types of sensors	Frequency of data transmission	Classification of primary data sources	Types of application area
Speed	High	Vehicle operating data	In-V
Acceleration	High	Vehicle operating data	In-V
The temperature inside the vehicle	Low	Measurement of car thermometer	In-V
The temperature outside the vehicle	Average	Thermometer data outside machine and RSU	In-V
Fuel level	Low	Vehicle operating data	In-V
Number of passengers	Low	Sensors in car seats	In-V
Vision	High	Cameras	In-V
Transmission of voice commands	Average	Multimedia data	In-V
Data from millimetre wave radar system	High	Data from vehicles	In-V
GPS	High	GIS GPS	In-V, V2I
WSN quality	Average	Data from environment	V2I
Conditions of wireless interface	Low	Wireless interface equipment	In-V, V2I
Data from virtual additional types of equipment (calls, calendar, clock)	Low	Data from smartphones, clocks, timers	V2M
Notification of reminders	Low	Data from smartphones	V2M
User preferences	Low	Data from smartphones	V2M
Data about road conditions	High	Data from environment	V2I, V2M, V2V
Warning notification	High	Data from another vehicle, governmental institutions, and the environment	V2I, V2M, V2V
V2V communication	Average	Data from environment	V2I, V2M, V2V

In determining the parameters of communication capacity, the data flows in two directions were simulated, i.e. from the vehicle to the DW cloud server and from the cloud DW server to the vehicle. The types of sensors are presented in Table 4.2, which are included in the experimental scenario.

The imitational duration of data transmission cycles was chosen—60 s. The data transfer rate was chosen—27 MB/s, and the packet size was approximately 1000 Mb. The experiment used the algorithm for determining the priority assignments for the provision of services. The trends of parameters (i.e. accumulated value, predicted utility *Excj*, normalized *Zj*, and reduced *Tyj*) were obtained during experimental cycles (Fig. 4.8).

The results of the experimental study show that the changes in the accumulated value over time depend on the quality parameters of the communication channels. Various data overlaps, collisions, channel bandwidth, and the number of dropped

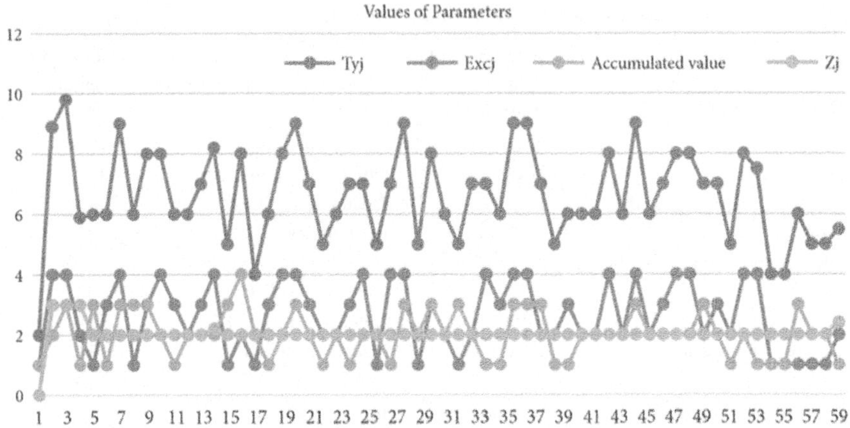

Fig. 4.8 Trends of parameters during imitational cycles

packets can also be estimated. Algorithms deal with the description of the main components of infrastructure for the provision of services.

4.5 Conclusions

The proposed architecture of the smart service delivery system enables the identification of transport situations, understanding their context and managing these situations. The proposed method allows the use of environmental context information, to evaluate vehicle dynamics and information transmission processes in the conditions of wireless networks. Various means of communication are being considered to enable the monitoring of the transport environment and the receipt of relevant information from other vehicles. The capacity of the wireless network shall be assessed taking into account the network bandwidth usage of the data messages in ensuring transport security conditions.

A prototype of the Smart-SDS was developed, and simulation experiments were performed, which help to assess certain messaging bandwidth conditions. The proposed algorithms were integrated into the architecture of the Smart-SDS system.

The WSNs classes and their application areas are used for assessment of identifying situations and collecting data, and for determination of applied methods. The results illustrate the operations of the Smart-SDS. Some limitations of such research were in restriction of our research areas. The typology of unplanned events has not been presented. The authors do not demonstrate several road accidents involving heavy vehicles. The research object is very complex and we do not describe the analysis of incidents, such as damage of tires, unexpected need for vehicle repairs, the policy stopping the vehicle, traffic jams at customs, and route mistakes, also occur, which we are planned to include in our future research works.

References

Al-Sultan, S., Al-Bayatti, A.-H., & Zedan, H. (2013). Context-aware driver behaviour detection system in intelligent transportation systems. *IEEE Transactions on Vehicular Technology, 62*(9), 4264–4275.

Andrea, A. F., & Savigni, A. (2001). A framework for requirements engineering for context-aware services. In *Proceeding of 1st international workshop from software requirements to architectures (STRAW 01)* (pp. 200–201).

Bazzi, A., Masini, B. M., Zanella, A., & Pasolini, G. (2014). IEEE 802.11p for cellular offloading in vehicular sensor networks. *Computer Communications, 60*, 97–108. https://doi.org/10.1016/j.comcom.2015.01.012

Boyraz, P., Yang, X., & Hansen, J.-H. (2012). Computer vision systems for "context-aware" active vehicle safety and driver assistance. In J. Hansen, P. Boyraz, & K. Takeda (Eds.), *Digital signal processing for in-vehicle systems and safety* (pp. 217–227). Springer.

Campolo, C., Molinaro, A., & Scopigno, R. (2015). From today's VANETs to tomorrow's planning and the bets for the day after. *Vehicular Communications, 2*(3), 158–171. https://doi.org/10.1016/j.vehcom.2015.06.002

Cespedes, S., Lu, N., & Shen, X. V. I. P.-W. A. V. E. (2013). On the feasibility of IP communications in 802.11p vehicular networks. *IEEE Transactions on Intelligent Transportation Systems*. https://doi.org/10.1109/TITS.2012.2206387

Chang, J., Yao, W., & Li, X. (2017). The design of a context-aware service system in intelligent transportation system. *International Journal of Distributed Sensor Networks, 13*(10), 1–8. https://doi.org/10.1177/1550147717738165

Chen, M., Ong Mau, D., Zhang, Y., Taleb, T., & Leung, V. C. M. (2014). VENDNET: VEhicular Named Data NETwork. *Vehicular Communications, 1*(4), 208–213. https://doi.org/10.1016/j.vehcom.2014.09.002

Cheng, H. T., Shan, H., & Zhuang, W. (2011). Infotainment and road safety service support in vehicular networking: From a communication perspective. *Mechanical Systems and Signal Processing, 25*(6), 2020–2038. https://doi.org/10.1016/j.ymssp.2010.11.009

C-ITS Platform, C ITS platform, final report, 2016.

C-MobILE. (2017). *C-MobILE (accelerating C-ITS mobility innovation and depLoyment in Europe) description of work*. C-MobILE Consortium.

Colglazier, W. (2015). Sustainable Development Agenda: 2030. *Science, 349*, 1048–1050. https://doi.org/10.1126/science.aad2333

COM. (2011). 144 Roadmap to a Single European Transport Area – Towards a competitive and resource efficient transport system. http://eur-lex.europa.eu/LexUriServ/LexUriServ.do?uri=COM:2011:0144:FIN:en:PDF

COOPERS (2010). COOPERS-co-operative networks for intelligent road safety, final report on demonstration.

Crainic, T. G., & Montreuil, B. (2016). Physical internet enabled hyperconnected city logistics. *Transportation Research Procedia, 12*, 383–398.

C-ROADS. (2020). C-ROADS – The platform of harmonized C-ITS deployment in Europe. Retrieved October 15, 2020, from https://www.c-roads.eu/platform.html

De Oliveira, K. M., Bacha, F., Mnasser, H., & Abed, M. (2013). Transportation ontology definition and application for the content personalization of user interfaces. *Expert Systems with Applications, 40*(8), 3145–3159. https://doi.org/10.1016/j.eswa.2012.12.028

Directive (EU) 2016/1148 of the European Parliament and of the Council of 6 July 2016 concerning measures for a high common level of security of network and information systems across the Union. 2016. OJ L 194:1–30.

Du, Y., Wang, J., Rizos, C., & El-Mowafy, A. (2021). Vulnerabilities and integrity of precise point positioning for intelligent transport systems: Overview and analysis. *Satellite Navigation, 2*(3), 1–2. https://doi.org/10.1186/s43020-020-00034-8

Dzemydienė, D., & Burinskienė, A. (2021, Jul). Integration of context awareness in smart service provision system based on wireless sensor networks for sustainable cargo transportation. *Sensors (Basel), 21*(15), 5140. https://doi.org/10.3390/s21155140

Dzemydienė, D., & Miliauskas, A. (2013). Possibilities of automatic and semi-automatic end-user driven web service composition. *Information Sciences (Informacijos mokslai), 65*, 34–44. https://doi.org/10.15388/Im.2013.0.2057

Dzemydienė, D., & Radzevičius, V. (2020). An approach for networking of wireless sensors and embedded systems applied for monitoring of environment data. In G. Dzemyda, J. Bernatavičienė, & J. Kacprzyk (Eds.), *Data science: New issues, challenges and applications. Studies in computational intelligence* (pp. 61–82). Springer. https://doi.org/10.1007/978-3-030-39250-5_4

Dzemydienė, D., Maskeliūnas, S., Dzemydaitė, G., & Miliauskas, A. (2016). Semi-automatic service provision based on interaction of data warehouses for evaluation of water resources. *Informatica, 27*(4), 709–722.

Dzemydienė, D., Burinskienė, A., & Miliauskas, A. (2020). An assessment of provision of heterogeneous services for sustainable cargo transportation process management by roads. *Sustainability, 12*(20), 8405. https://doi.org/10.3390/su12208405

Dzemydienė, D., Burinskienė, A., & Miliauskas, A. (2021). Integration of multi-criteria decision support with infrastructure of smart services for sustainable multi-modal transportation of freights. *Sustainability, 13*(9), 4675.

Dzemydienė, D., Dzemydaitė, G., & Gopisetti, D. (2022). Application of multicriteria decision aid for evaluation of ICT usage in business. *Central European Journal of Operations Research*, 1–21. https://doi.org/10.1007/s10100-020-00691-9

Fan, Y., Behdani, B., Bloemhof-Ruwaard, J., & Zuidwijk, R. (2019). Flow consolidation in hinterland container transport: An analysis for perishable and dry cargo. *Transportation Research Part E: Logistics and Transportation Review, 130*, 128–160.

Golestan, K., Soua, R., & Karray, F. (2016). Situation awareness within the context of connected cars: A comprehensive review and recent trends. *Inform Fusion, 29*, 68–83.

Holbrook, H., III. (1990). A scenario-based methodology for conducting requirements elicitation. *ACM SIGSOFT Software Engineering Notes, 15*(1), 95–104.

IEEE. (2010a). 1609.3-2016 - IEEE standard for wireless access in vehicular environments (WAVE) – Networking services – Redline.

IEEE. (2010b). 802.11p-2010 – IEEE standard for local and metropolitan area networks – Specific requirements Part 11: Wireless LAN medium access control (MAC) and physical layer (PHY) specifications amendment 6: Wireless access in vehicular environments, 51.

IEEE Std. 1609.1–2006. IEEE trial-use standard for wireless access in vehicular environments (WAVE)-resource manager. https://doi.org/10.1109/IEEESTD.2006.246485.

Kaffash, S., Nguyen, A. T., & Zhu, J. (2021). Big data algorithms and applications in intelligent transportation system: A review and bibliometric analysis. *International Journal of Production Economics., 231*, 107868. https://doi.org/10.1016/j.ijpe.2020.107868

Kakkasageri, M. S., & Manvi, S. S. (2014). Regression based critical information aggregation and dissemination in VANETs: A cognitive agent approach. *Vehicular Communications, 1*(4), 168–180. https://doi.org/10.1016/j.vehcom.2014.07.001

Karculias, P. N. (1997). *A taxonomy of scenarios in software development*. The University of Texas at Arlington.

Kranz, M. (2016). *Building the Internet of Things: Implement new business models, disrupt competitors, transform your industry*. Wiley.

Kurmis, M., Dzemydienė, D., Didžiokas, R., & Trokss, J. (2013). Modeling and simulation of cloud computing solution for distributed space data storage and access in mobile communication networks. *Latvian Journal of Physics and Technical Sciences, 50*(5), 20–28. https://doi.org/10.2478/lpts-2013-0030

Kurmis, M., Andziulis, A., Dzemydienė, D., Jakovlev, S., Voznak, M., & Gricius, G. (2015). Cooperative context data acquisition and dissemination for situation identification in vehicular communication networks. *Wireless Personal Communications, 85*(1), 49–62. https://doi.org/10.1007/s11277-015-2727-1

Le Dinh, T., & Pham Thi, T. T. (2012). Information-driven framework for collaborative business service modelling. *International Journal of Service Science, Management, Engineering, and Technology (IJSSMET), 3*(1), 1–18. https://doi.org/10.4018/jssmet.2012010101

Le Dinh, T., Thi, T. T. P., Pham-Nguyen, C., & Nam, L. N. H. (2021). A knowledge-based model for context-aware smart service systems. *Journal of Information and Telecommunication.* https://doi.org/10.1080/24751839.2021.1962105

Leite, J. C. S. d. P., Hadad, G. D. S., Doorn, J. H., & Kaplan, G. N. (2000). A scenario construction process. *Requirements Engineering, 5*(1), 38–61. https://springerlink.metapress.com/openurl.asp?genre=article&issn=1432-010X&volume=5&issue=1&spage=38

Lim, C., & Maglio, P. P. (2018). Data-driven understanding of smart service systems through text mining. *Service Science, 10*(2), 154–180. https://doi.org/10.1287/serv.2018.0208

Mo, T., Li, W., Chu, W., & Wu, Z. (2010). CABS3: Context-awareness based smart service system. In *2010 6th International Conference on Wireless Communications Networking and Mobile Computing (WiCOM), IEEE* (pp. 1–4).

Ohn-Bar, E., Tawari, A., & Martin, S. (2015). On surveillance for safety critical events: In-vehicle video networks for predictive driver assistance systems. *Computer Vision and Image Understanding, 134,* 130–140.

Osmundsen, K., Iden, J., & Bygstad, B. (2018). Digital transformation: Drivers, success factors, and implications. In *Proceedings of the Mediterranean conference on information systems, Corfu, Greece, 28–30 September 2018* (pp. 1–15).

Padovitz, A. (2006). Context management and reasoning about situations in pervasive computing. Retrieved from https://books.google.lt/books/about/Context_Management_and_Reasoning_about_S.html?id=iIdiPAAACAAJ&pgis=1

Pereira, C., Zaslavsky, A., Christen, P., & Georgakopoulos, D. (2014). Context aware computing for the Internet of Things: A survey. *IEEE Communications Surveys and Tutorials, 16*(1), 414–454.

Rasch, K., Li, F., Sehic, S., Ayani, R., & Dustdar, S. (2011). Context-driven personalized service discovery in pervasive environments. *World Wide Web, 14*(4), 295–319. https://doi.org/10.1007/s11280-011-0112-x

Sathe, A. D., & Deshmukh, V. D. (2016). Advance vehicle-road interaction and vehicle monitoring system using smart phone applications. In *2016 online international conference on green engineering and technologies, Coimbatore, India, 19 November 2016* (pp. 1–6). IEEE.

Schilit, B. N., & Theimer, M. M. (1994). Disseminating active map information to mobile hosts. *IEEE Network, 8*(5), 22–32. https://doi.org/10.1109/65.313011

Sharma, H., Haque, A., & Blaabjerg, F. (2021). Machine learning in wireless sensor networks for smart cities: A survey. *Electronics, 10,* 1012. https://doi.org/10.3390/electronics10091012

Sternberg, H., Prockl, G., & Holmström, J. (2014). The efficiency potential of ICT in haulier operations. *Computers in Industry, 65*(8), 1161–1168.

Sun, Y., & Yu, X. (2014). An innovative nonintrusive driver assistance system for vital signal monitoring. *IEEE J Biomed Health Information, 18*(6), 1932–1939.

Sun, J., Zhang, Y., & He, K. (2010). Providing context-awareness in the smart car environment. In *Proceedings of the IEEE 10th international conference on computer and information technology* (pp. 13–19). IEEE.

Wang, S. Y., & Chou, C. L. (2009). NCTUns tool for wireless vehicular communication network researches. *Simulation Modelling Practice and Theory, 17*(7), 1211–1226. https://doi.org/10.1016/j.simpat.2009.04.008

Wang, S., & Huang, Y. (2012). NCTUns distributed network emulator. *Internet Journal, 4*(2), 61–94.

Wang, S.-Y., & Lin, C.-C. (2008). NCTUns 5.0: A network simulator for IEEE 802.11(p) and 1609 wireless vehicular network research. In *Vehicular technology conference, 2008. VTC 2008-Fall. IEEE 68th*. https://doi.org/10.1109/VETECF.2008.464

Wang, S.-Y., Wang, P.-F., Li, Y.-W., & Lau, L.-C. (2011). Design and implementation of a more realistic radio propagation model for wireless vehicular networks over the NCTUns network simulator. In *Wireless communications and networking conference (WCNC)*. IEEE. https://doi.org/10.1109/WCNC.2011.5779455

Wibisono, W., Zaslavsky, A., & Ling, S. (2009). Improving situation awareness for intelligent on-board vehicle management system using context middleware. In *2009 IEEE intelligent vehicles symposium* (pp. 1109–1114). IEEE. https://doi.org/10.1109/IVS.2009.5164437

Yang, Y., Chen, B., & Suetal, L. (2013). Research and development of hybrid electric vehicles CAN-bus data monitor and diagnostic system through OBD-II and android-based smartphones. *Advances in Mechanical Engineering, 5*(3), 741240.

Chapter 5
Towards the Transport and Corresponding Sectors

Aurelija Burinskiene

Abstract This chapter focuses on the freight transport sector, which has been getting increasingly more attention. Most of the topics focus on freight delivery, technologies, and their application important for services oriented to context transfer and understanding. Progress in communication channels and technologies, environmental and system changes, and contemporary studies of freight transport has been of great interest.

Nowadays, the modelling and design of transport systems require application of modern tools offered by various theories. Nevertheless, there are decision-making algorithms that should be considered for building and expanding knowledge and documents digitization. Finally, the chapter ends with efficiency and sustainability topics and their evidence in transport studies.

Keywords Modelling · Digitalization · Context-aware service · Freight transport · Efficiency · Sustainability

5.1 Importance of Transport Sector

At the European Union (EU) level, the EU freight transport logistics sector contributes to the EU economy, mainly by creating new jobs and adding value. The EU's transport and storage sector employs around 11 million people, representing more than 5% of full-time employment and nearly 5% of EU gross domestic product (European Commission, 2018). Nevertheless, the road transport sector still lags in digitalization behind other sectors of our society and the economy.

The sector of transport creates 8% of gross domestic product and has 12,000 employees. This shows an important impact on the economics of Lithuania. In 2019, 84 % of freight was carried by inland transport modes (road and rail) (The Lithuania Department of Statistics, 2019).

A. Burinskiene (✉)
Vilnius Gediminas Technical University, Vilnius, Lithuania
e-mail: aurelija.burinskiene@vilniustech.lt

D. Dzemydienė et al., *Development of Smart Context-Aware Services for Cargo Transportation*, International Series in Operations Research & Management Science 330, https://doi.org/10.1007/978-3-031-07199-7_5

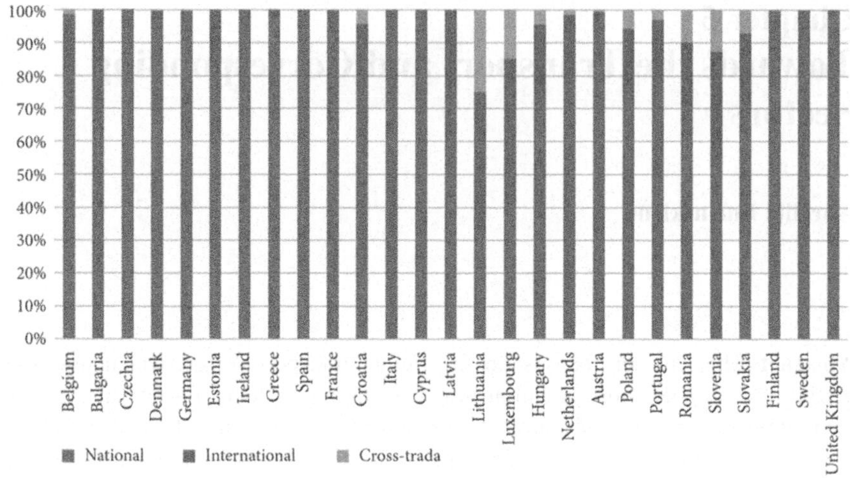

Fig. 5.1 Transportation of freights, 2019; Source: Eurostat (2021)

In most of EU countries, significant growth of national (3.5%), international (4.6%), and cross-trade (8.5%) flows has been observed. These values in Lithuania are higher than the EU-28 average (7.2%, 16.8%, and 31.1%, respectively) and impact the development of the sector in the country (Fig. 5.1).

The transport sector also depends on the flows of international trade. For further analysis, services provided by land transport have not been separated, including these data in the sector figures together with the data representing other transport services.

A study focusing on the changes of the sector and the value of equity of transport companies has been carried out and considered as two separate topics.

5.2 Modelling the Transport Sector

The transport sector could be analysed on different levels. On the analytical level, there are participants, processes, and the level of analysis, which could be linked to function, environment, and technology aspects. These important categories help to figure out the equity of the sector.

In further analysis, the efficiency of the sector could be measured, comparing service transport prices and evaluating financial indicators. The research of the transport sector has many directions.

Table 5.1 Modelling of the transport sector

Structural dimension	Analytical category	
Transport sector	Participants	Carriers, transport service providers, terminals, customers
	Level of analysis	Type of networks, application of smart services and infrastructure support, digitalization of processes
	Processes	Freight delivery, reloading, back haulage
Modelling	Purpose of model	Economic analysis, network analysis, decision support analysis, choice analysis, demand and supply analysis, growth evaluation, market-oriented service review
	Model type	Gordon's growth model, econometric models, performance evaluation, balanced scorecard (BSC), data envelopment analysis (DEA)
	Modelling approaches	Forecasting, benchmarking, econometrical estimation of productivity and efficiency

There are many models suitable for modelling various aspects of the transport sector. Some of those models have been used for evaluation, including benchmarking or performance evaluation (Table 5.1).

Gordon's growth model is a well-known evaluation method. This model reveals the expansion of the transport sector as well as operational conditions of transport companies that are evident on the sectoral level.

5.3 Gordon Growth Model and Technology Adoption

The Gordon Growth Model helps to compare the transport sector with other sectors and assumes the following conditions (Gordon, 1962):

- The sector's activity is stable and runs without any operational changes;
- The sector growth rate is constant;
- The sector has financial stability.

The application of the Gordon Growth Model could help to combine both the revision of the sector and the estimation of the value of equity in the sector. Such focus was placed on the matters of digitalization and smart service systems in this sector.

Gordon Growth Model is one of the evaluation methods, which follows the equity value estimation procedure. The application of the Gordons Growth Model renders companies with practical results which are also important for the sector analysis. It is suitable for the estimation of the equity value of the transport sector in the selected country and for benchmarking purposes. Because of the simplicity of the model, it is only usually used for the evaluation of sectors with stable growth rates.

In Lithuania, the increase in flows was higher than the EU-28 average (Eurostat, 2021). The absolute growth of sales in Lithuania during the period from 2001 to

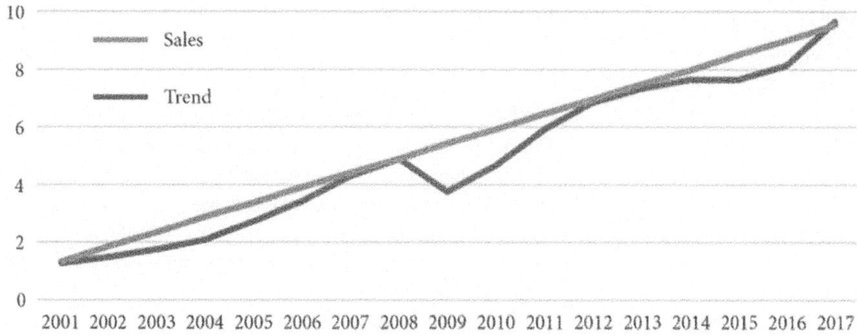

Fig. 5.2 Sales and their trend in the Lithuania transport sector, mln Eur (Burinskas & Burinskienė, 2020)

2017 was five times higher. The research results revealed that the average annual growth of earnings accounts for 11.20% in the transport sector of Lithuania.

Moreover, the revision of various growth rates in different periods has been incorporated. The Gordon Growth Model was extended for use with varying growth rates in the calculation. This model can be used during business cycle fluctuations and includes ongoing and unusual financial activities. The sector had two downtimes in 2008–2009 and 2014–2015 (Fig. 5.2).

The evolution of the sector is affected by macroeconomic indicators, which also show the macroeconomic implications of technology.

When it comes to technology, it can be described as introduced during the period under analysis; second, it was used in production activities (i.e. activities that contribute to GDP); and third, they were introduced in many countries.

The process of technological adaptation is important for the economic growth of the country and its development. Technology–adaptation interactions can explain cycles of different periods related to the spread of various innovations. Taking advantage of many innovative opportunities, the adaptation of technology in the business cycle implies development (Fusari, 1996).

5.4 Development of Transport Sector Towards Synchro-Modality and ICT

The sector of transport is very important to the national economy and supports the development of international trade. However, there are some environmental effects (emissions, pollution, noise, congestions), which have encouraged to investigate modal, greener system structure that allowing to explore the potential of the sector. Recently, higher attention is given to the effective harmonization of road transport modes, such as lorry, sea and rail transport, helping to avoid the change of mode of transport. This solution removes cargo mode changes, increases safety, minimizes freight damages and losses, and contributes to faster cargo transportation (Ho et al., 2015).

In recent years, studies focused more on freight delivery and green freight allowing lower CO_2 emissions in particular. The key topics related to electric trucks (Ehrler & Hebes, 2012; Quak et al., 2016), the assessment of freight regulations (Bhuiyan et al., 2015), and the estimation of ecological footprint (Mishra et al., 2017; Muñuzuri et al., 2010) in the area.

The use of ICT helps to build hyperconnected logistics networks (Crainic & Montreuil, 2016), allows planning stakeholders' collaboration (Anand et al., 2016; Awasthi et al., 2016; Khayyat & Awasthi, 2015; Lagorio et al., 2017), integrated freights delivery planning (Awasthi & Proth, 2006; Crainic et al., 2009) and improves logistics service quality (Awasthi et al., 2018) has also been taken into consideration.

Due to innovations in the integration of ICT and data systems, the door-to-door supply chain may see a greater leap forward in intermodality (Verweij, 2011). Digitization in such a sector could create real "synergies" among partners in logistics. The synchro-modality concept was developed to make optimize the usage of transport capacity (Mes & Iacob, 2016).

Reviewing its relationship to the transport paradigms that have formed since the 1980s, namely, multimodality, intermodality, combined transportation, is important. The author indicates that paradigms overlap but are ambiguous (Reis, 2015), emphasizing the idea of synchro-modality (Tsertou et al., 2016). The synchro-modality is on the way to becoming a new leading transport paradigm capable of coping with new supply chain trends (Putz et al., 2015).

Multimodal transport involves the application of different modes of transport used for the cargo carriage only. Intermodality involves the transportation of cargo from door to door in the supply chain (Zilliaskopoulos & Wardell, 2000). The concept of a unit of cargo, i.e., a physical object on to which pallets or containers are loaded, supports the change of transport mode and includes efficient management of particular operations. Combined transport is used as a branch of intermodality that focuses on the minimization of the negative effect of transport and the usage of sustainable transport modes (such as water and rail transport) and encourages less extensive use of lorries (Nabais et al., 2015). Finally, intermodality aims at the sustainable use of various transport modes jointly and separately (Chang et al., 2007).

Synchro-modality can be considered a further development of these transport paradigms. Synchro-modality logistics covers different tightly integrated modes of transport (Zhang & Pel, 2016). The aim is to create a sustainable and optimized supply chain that focuses on efficient usage of resources (Rivera & Mes, 2016). Synchronous modality inherits a high-level vision of flexible corridor management from intermodality to include supply chain partners' interests, not just parties involved in freight transportation (Nabais et al., 2016).

Synchro-modality as a concept seeks to give more significance to the role of inland waterway transport and rail, instead of offering a friendly transition from those modes to the mode of road transport. Freight transport is viewed from the perspective of the whole modal network for choosing optimal transport modes and routes (Tavasszy et al., 2010).

Synchronous modality is defined as the development of concepts of intermodal and intermodal transport, where the stakeholders of the transportation chain interact in a common network to allow flexible planning and the ability to change modes of transport adapted to existing vehicles in real time (Buiel et al., 2015). The consignor lays down solely the basic shipment requirements: cost, CO_2 emission, in advance. The optimized usage of transport modes could be revised as separate ones or as integrated ones (Hofman, 2014).

The following are the main driving forces that can speed up the interconnection of networks into asynchronous mode systems:

- High and volatile fuel prices, leading to savings in transport solutions,
- A significant increase in road congestion, raising environmental and public awareness of side effects of road traffic on local communities,
- Strict environmental rules, described in the Paris Agreement and other agreements signed by the members of the International Maritime Organization until 2050, provide for reducing emissions by at least 50% compared to 2008.

The application of the synchronized transport system in ports is complex (Batarliene & Sakalys, 2020). First, it includes networking and collaboration based on trust and deals with the concept of customer relationships. The creation of such a network helps to ensure mutual respect and trust, which is the key condition for synchronized processes. As many actors are reluctant to collaborate with their competitors, however, the synchronous network could be built then competitors trust each other and identify the advantages of cooperation. The next challenge is the complexity of operations and planning. The planning and modelling of transport routes are important for creating efficient synchronous operations. Elements, such as preferences of new customers, traffic flows in routes, and the available resources in a logistics node should be assessed and examined before planning. For optimizing transport performance monitoring and demand forecasting tools need to be developed. The third limitation—the interconnection of ICT and platforms for data-sharing (Sarraj et al., 2014). A high-performance platform for data-sharing is needed to have the exchange of data among various stakeholders: delivering companies, forwarders, and terminals.

To reach synchronous modality (Fig. 5.3) is the way to transmit from non-modal transport to a dynamic, smart network (Hofman, 2015). This system uses different modes of transport from intermodality to co-modality (Tadić et al., 2019). The implementation of an efficient synchronous mode follows the requirements (Tavasszy et al., 2015), where transport resources have to be used in an optimized and sustainable way (Fatnassi et al., 2015). The concept is largely based on synergies and systematic coordination between stakeholders in the transport network, where close integration is necessary for efficient and sustainable use of resources (Hofman et al., 2016).

Development in ICT has several streams. First, the stream is linked with embedded systems. The spread of radio frequency identification (RFID) technology made

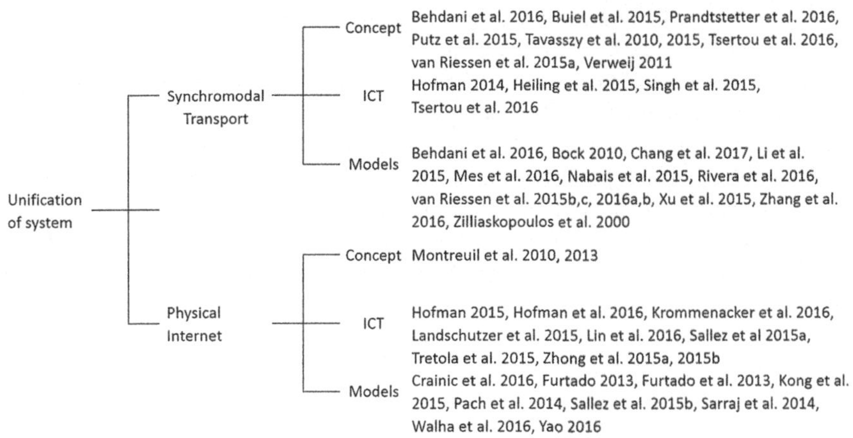

Fig. 5.3 Overview of developments in the area of cargo transportation

embedded systems the key technology now. Sensors and actuators have been installed in the embedded systems, which allows the systems to record, store, and process various data from the environment, which can later be used to enable them to influence their environment. These systems turn objects into intelligent objects and the environment—into smart environments (Zhong, Xu, et al., 2015a).

Second, simultaneous mobile connection and wireless local area network (WLAN) along with input Internet Protocol IPv6 in 2012 led to significant Internet development (Lin & Cheng, 2016), which means that embedded systems facilitate mutual communication, exchange of data and serve as online services. Here, the flexibility of digital functions is implemented into physical objects allowing for the transformation of embedded systems into the cyber-physical system (Tretola et al., 2015). Several cyber-physical systems collect huge amounts of data from the environment and provide digital information about processes (Sallez, Berger, et al., 2015b; Sallez, Montreuil, & Ballot, 2015a). Such shifts from manual to automatic transfer of data to media with all the errors inherent in this process can now occur automatically. In addition, constantly evolving sensor technology ensures an increasingly higher data resolution for fine-grained monitoring of the environment (Montreuil et al., 2010).

Third, cloud computing now provides affordable storage for an exponentially increasing amount of data created by smart objects. Large amounts of data stored in "data lakes" are often referred to as large data and can be extracted using intelligent algorithms based on correlations and probability calculations (Montreuil et al., 2013). A review of the data and the revision of models that generate information allow generating new knowledge. Thus, cloud computing offers advanced services, supported by digital infrastructure in respect to business needs, including service-oriented architecture (SOA) covering a variety of freight delivery processes (Awasthi et al., 2011; Heilig et al., 2015).

Fourth, any technological device can exchange information on things, data, and services online at high speed with any other device or a person anytime and anywhere. The information about the individual installations and their environment can be accessed at any time. Technology infrastructure can even be monitored and managed at a distance. Networks allow simultaneous management and optimal coordination of various processes that are technologically complex (Ko et al., 2006).

For managing urban traffic the information reached from intelligent navigation systems handled by drivers of various types of vehicles are important (Zhong, Peng, et al., 2015b). Connecting production systems and transport processes in real time and in various companies—from order to freight delivery—will also become possible (Burrough, 1992).

5.5 The Digitalization of Documents in the Transport Sector

The EC for transport sector emphasized that digitalization will be vital for the success and the survival of transport and logistics companies. Many freight operations in the EU still require the application of paper, so international truck drivers still have to carry a significant amount of paper copies in cabins. According to an impact assessment[1] prepared by DG MOVE in support of the forthcoming legislation, almost all international freight delivery operations in the EU in recent years still involve documentation at one stage or another. Slow uptake of digitalization has been associated with inefficiencies in transport and logistics market participants, cross-border transport, and unnecessary burdens on the environment. In addition, slow digitalization makes it more difficult for national authorities to enforce rules, as paper copies require inspections. Given that by 2050, the overall freight transport traffic will increase by 50%, the slow pace of digitization will be the bigger issue in the further periods. Against this background, the European Parliament has repeatedly called for a simplified, paperless, smooth, transparent, secure, and reliable flow of information between businesses, customers, and public authorities. The EC supports the recognition of the introduction of electronic consignment notes for freight transport. An e-waybill has been promoted in the Communication from the Commission on a European Digital Single Market Strategy (May 2015)[2] announced as an initiative that opens up new opportunities for transport and other sectors. In a broader context, the Commission aims to digitize documents for all modes of transport and to

[1]European Commission (2018). Commission Staff Working Document, Impact Assessment, Accompanying the document, Proposal for a Regulation of the European Parliament and of the Council on electronic freight transport information SWD/2018/183 final–2018/0140 (COD).

[2]European Commission (2015). Communication from the Commission to the European Parliament, the Council, the European Economic and Social Committee and the Committee of the Regions: A Digital Single Market Strategy for Europe, https://eur-lex.europa.eu/legal-content/EN/TXT/?uri=celex%3A52015DC0192.

encourage their acceptance by public authorities. The European Commission is making such a commitment to the EU's e. government action plan to be published in the 2016–2020 period. Priorities for ICT standardization have also been set out in a document for the Digital Single Market. An EU regulation is currently being prepared to encourage and implement the acceptance of digital transport-related data by all EU authorities. As a result, the European Commission has published a Proposal for an eFTI Regulation,[3] and negotiations between institutions are currently underway (European Commission, 2019). The Forum of Digital Transportation and Logistics (DTLF)[4] mainly focus on the development of solutions for the post-implementation period.

Freight processes involve both consignee requesting for freight delivery and the service provider that could provide such service (Awasthi, 2015). Such an agreement describes the application of roles in practice. To this end, the elements must be identified, including integration of the necessary data, transport capacity management, service provision, freight invoicing, service charges, backup, and privacy guidelines.

Messaging and electronic documents are important for freight transport. Internal needs include building and identifying a route crossing international countries, as well as route control (Van Riessen et al., 2016b). In freight transport, the companies control drivers by revising the routing information and travel time. This data is important to clarify driving time balance, follow dispatch instructions, obtain information on vehicle accidents or inspections, and identify freight lost in accidents and take actions after delivery (i.e. to archive data and provide statistics).

Information about the loading of a consignment and the formalities to be taken by the consignor, planning the arrival of the consignment at the consignee or the customs clearance office and the identification steps are among the obvious needs. When the carrier has full freight delivery, the sender and receiver are tracking the truck carrying consignment, and, in partial delivery case, the interested party could monitor only its freight (Batarliene & Sakalys, 2020).

For public purposes, the numbers of the vehicle can also be applied to identify situations of cabotage, assess customs activity, and determine waiting times for tolls paid. In addition, freight and vehicle are weighed, and their weights have to be linked to the vehicle's weight. This is necessary for authors to continue to respond to business needs (Age, 2011). Generally, a lot of different documents are used in the freight transport process. Part of the documents includes the content related to freight delivery (for example, bill of lading, where is mentioned the data about dangerous or live animal freights), other documents—about vehicles (registration and insurance of vehicle), and drivers (data about driving licences and driver's health insurance).

---

[3] Proposal for a Regulation of the European Parliament and of the Council on electronic freight transport information COM/2018/279 final—2018/014/COD, https://eur-lex.europa.eu/legal-content/EN/TXT/?uri=CELEX%3A52018PC0279.

[4] Digital Transport and Logistics Forum, www.dtlf.eu.

Table 5.2 Data and documents supporting freight delivery process

Objects	Additional conditions	Result	eDocuments	
			Reported in	Transferred to
Loading-unloading points	The network of available roads	The delivery route, including the lists of countries for transit, the successive carriers, and vehicle licence plate number	eCMR	eCMR
Instructions for shipping (temperature regime, danger level, etc.)	Space available in the vehicle	The assignment of the vehicle including vehicle registration number	eCMR	eTIR eCMR
Loading date Transit deadline	Tracking for vehicle and freight location Traffic jams Accidents	The delivery on-time following the instructions during shipping	eCMR eTIR eTIR eTIR	eTIR eCMR

Recipients of freight services and carriers use many documents defining the cargo and transportation requirements (i.e. instructions on temperature regime, weight, fragility, settings, amount of transport units), pick-up and drop-off location, time of planned activities, agreeing on the notification about the status of the carriage of goods within the time limits for the provision of transport services.

To meet the above needs, sustainable modelling of information flow is necessary. Stefaniuk et al. in Stefaniuk & Chomicka, 2019 analysed information flow modelling and determined that it is significant for the management of the freight delivery process (Table 5.2).

First, the carrier receives information of the locations of pick-up and drop-off, and following the network, indicates the potential routes and shipping instructions. After reviewing the fleet data, the carrier designates a vehicle that meets the specified requirements. Before the loading of freight, the date and time is planned. When the route starts on the day of loading, the vehicle and cargo are tracked, and the network conditions (such as road repairments, traffic spikes, accidents, etc.) are revised. In case the network functions under normal conditions, the driver will be able to meet the delivery deadline and bring the freight to the drop-off point on time (Van Riessen, Negenborn, Dekker, & Lodewijks, 2015b).

The important information is provided on the agreement of parties: consignor/ consignee and the shipper. Freight transport processes rely on two main documents: Transports Internationaux Routiers on International Road Transports (TIR) and the Agreement on an International Delivery of Freights by Roads (CMR) (Digital Transport and Logistics Forum, 2019). Such data is copied on the carrier's documentation paper or electronic form. The locations of pick-up and drop-off are placed on the eCMR including delivery instructions and pick-up date. The TIR indicates the transit duration and the date of drop-off (Van Riessen, Negenborn, & Dekker, 2015a). Information on sequential shippers is stated on the eCMR, indicating vehicle(s) number(s) in eTIR and eCMR. TIR contains information on the arrival at the customs office, while eCMR indicates the execution of dispatch instructions, where the carrier can also state its doubts about freights and/or the compliance with dispatch instructions. The accident situation is marked in the eTIR.

5.6 Analysis of TIR and CMR Statistics for the Evaluation of E-Freight Documents in European Countries

Freight transportation documents are used for several reasons: (a) to indicate the presence of a contract on freight delivery and liability corresponding to this activity, (b) to indicate the owner of the cargo, and (c) to provide information about freight and the delivery features. These indications inform about the freight transport activities:

- Status. The information is constantly updated on the main documents;
- Cargo details. Specifications of cargo: temperature regime, danger level, handling instructions, positions related to the handling of freight;
- Terms of delivery. Shipment conditions specified in the consignment note CMR and delivery conditions called Incoterms;
- Payment for delivery service. This provides information on which entity pays duties and taxes.

More detailed information will be presented in the next paragraphs.

Assessment of eTIR Usage

The computerization of freight delivery documents changes processes and requires some time for implementation. Computerization is slow mainly due to inadequate infrastructure that supports the process and complexity of cross-border procedures. Results of the application of the International Road Transport Union (IRU) carried in driver's cab in international routes are presented in Table 5.3.

Freight delivery operations in the EU28 still require truck drivers to carry paper copies of TIR between European countries. For the international freight movements, electronic versions are used rarely compared to paper documents. To simplify the process, it is necessary to integrate documents flows. The data exchange of information in electronic documents is organized via ICT means. All information helps to monitor the delivery of freight through borders.

The digitalization of TIR Carnets began in 1995. The inefficiency of transit procedures, long taking procedures in customs have been major challenges faced by developing countries strengthening their competitive position worldwide.

The evolution of the implementation of the electronic TIR version began after 2002. This evolution is supported by business because it creates value during the freight delivery process, by providing extended information enabling to monitor risks during cargo arrival.

Table 5.3 Number of TIR Carnets distributed in 2019 for international road transport trips in the EU 28 countries (UNECE, 2021)

States	Amount of TIRs	States	Amount of TIRs
Poland	81,000	Finland	500
Lithuania	41,000	Denmark	300
Bulgaria	13,200	Croatia	150
Romania	10,600	Spain	150
Latvia	9500	France	50
Hungary	3650	Belgium	0
Slovakia	3000	Cyprus	0
Czech Republic	2900	Ireland	0
Slovenia	2000	Luxembourg	0
Greece	1950	Malta	0
Germany	1900	The Netherlands	0
Estonia	1300	Portugal	0
Austria	1200	Sweden	0
Italy	600	United Kingdom	0
Total 174,950			

The studies distinguish the problem of assigning them to the international route. For that selection rule is necessary. The system may take into account transit route selection behaviour, involving sequential carriers acting on the territory. One of the key features is that the system detects the impact on the transport system, in case of congestions and capacity matters. At first, the departure time and transfer point of the changing carriers are identified. It is assumed that drivers reduce cargo transportation time when travelling between service points. Driving time is randomly assigned to the drivers.

Equilibrium is reached when none of the carriers thinks that a change of route can extend travel time. This is described as the condition of stochastic equilibrium (Suresh & Vasantha, 2018). Traffic assignment can be used for the procedure during which available time slots are used for the assignment of a transit route through the network. Having selected a solution, the task loads delivery journeys between points (especially the point of origin and destination) in search of the shortest international route. If there are several carriers, for the identification different route codes may be used. The carrier follows a route code and defines the information of the consignment up to the postal code. If the cooperation agreement is concluded between several carriers, the same route ID can be used for various indications for transportation and loading purposes.

The Provision of the Use of eCMR

In road haulage (CMR), paper consignments still account for about 88% of all cross-border transport operations in European Countries.

The IRU has initiated digitalization of the CMR protocol, which allows the CMR to be processed electronically and must contain all the information which are printed on the paper. The digital form of CMR protocol was promoted on 5 June 2011, and

20 out of the 28 European Union states have joined it to this day, including Bulgaria, Croatia, and other 18th European countries. Such allowed supply chain and transport partners to select which form of CMR they would like to use. For the first time, eCMR at EU level was launched in January 2017 between two countries: France and Spain.

Later on, additional countries launched pilot projects without a common eCMR data content system. The key challenge is to harmonize the content of exchanges between freight participants so that the digital form has a probative value.

An electronic CMR form must follow all procedures, namely, issuing and cancelling are used with the specified reason code, returning (change of consignee with return reason code) and closing of the CMR (when the consignee acknowledges the receipt indicating the number of transport units unloaded).

References to dangerous freight identifications (DG for short) are obligatory for the Directorate-General.

The delivery terms including Incoterms rules and instructions that may affect transportation are specified for the carrier. The carrier must make reservations and remarks during loading by indicating the number of transport units, the information on their packaging. Recording any damage to the cargo is necessary, indicating the quantity that is damaged and the reason for that. Additional information about net weight, TIR carnet number can also be present on CMR.

In general, the use of electronic CMR version allows faster procedures, giving up paper printing, processing and archiving data; the reduced time necessary to issue a paper version).

Development of a Maturity Model for the Usage of Electronic Transport Documents

Electronic data interchange is a joint solution developed by multiple companies (Suzuki & Williams, 1998), which requires collaboration and commitment between senders and receivers of electronic transport documents (Jun et al., 2000), which are taking any risk and payoff identified by making the uptaking of investments decision (Kim & Sanders, 2002). A four-step approach can be used for B2B services. It includes: (1) understanding the objectives of EDI (electronic data interchange); (2) automated data delivery through the technical environment; (3) integrating data throughout business partners; and (4) establishing long-term business-to-business (B2B) integration.

New ICTs allow companies to connect faster and to establish better cooperation. Electronic data interchanges (EDI) will reduce costs incurred by both senders and receivers of electronic transport documents.

Achrol and Stern (1988) propose that organizations interchange electronic data between partners (Table 5.4). Authors propose that data interchange in B2B relationships can occur in three ways (looking from the service provider level): (1) delivery of data about customers' needs (i.e. orders, forecasts, sales data), (2) receipt of data on upcoming shipments (i.e. advanced shipment notification, shipment manifest),

Table 5.4 Freight delivery outsourcing reference model

Buy	Ship		Pay
E-Order	Freight departure e-document	Delivery confirmation e-document	E-Payment
Cagliano et al. (2017), Hwang et al. (2008)	Civelek et al. (2017), Moon et al. (2011)	Civelek and Seckin (2017), De (2011)	Jetter et al. (2011), Ngeno and Kinoti (2017)

Table 5.5 Components detailed for each EDI network party

Level	Benefits		Costs	Investments		Costs of security loss
The application of EDI	Web EDI	Traditional EDI	All types of EDI	Web EDI	The application of EDI	Web EDI
Sender	Lower	Higher	Service fee	0	Sender	Lower
Receiver	Higher		Service fee	One-time		Losses
Service provider	Service sales	Maintenance	One-time	Multiple	Service provider	Service sales

and (3) delivery of information on shipments arrived (i.e. delivery confirmation, discrepancy reports).

Supplier and buyer behaviour focused on business integrations reducing expenses for each selling and buying transaction. The theory of transactional costs could be applied to describe the efficiency increase.

Research by Downing (2002) examines three categories of companies in terms of their use of EDI: (1) enterprises not using EDI; (2) enterprises using integration-based EDI solution; and (3) enterprises using online web EDI tool. The use of each EDI solution has been examined in clarifying process related costs; efficiency of transactions; responsibility share among EDI parties; and overall situation. The calculation shows that enterprises applying web EDI experience have better clarification on roles and responsibilities, while companies using traditional EDI have better internal efficiency and overall productivity. When it comes to costs and benefits, different specifics have to be analysed. For example, costs incurred by the document sender and receiver associated with service fees are benefits (service sales) for EDI service providers and so on. Table 5.5 lists components for each EDI network party, namely, benefits, costs, investments, and costs of the loss of security.

The model of maturity is constructed by revising different integration of process and performance stages (Salah et al., 2014). Business process transformation in the models of maturity (BPRMM) has been designed including the first part BPR (process transformation) and second part PML (continuous process improvement).

Capgemini (2012) developed a maturity model for electronic freight documentation which encompasses seven layers of the digitalization of freight documentation and indicates how data was collected. Under the first layer entry of freight documents by hand is foreseen; the second-level is scanning paper freight documents; the third-level is a portal solution for hosting freight documents online (Landschützer et al., 2015); the fourth level is the Electronic Data Interchange; the fifth layer is the platform of EDI providers; the sixth layer is the fourth party model, and the seventh level means self-authorization. Capgemini's maturity model focuses on identifying information and responsibilities related to the freight documentation process. These seven layers could be integrated into the maturity model of business process management (BPMMM). The EDI solution works among enterprises having direct or indirect transactions.

This author's approach covers different steps (direct and indirect) and phases that involve the transition from a single stand-alone solution to virtual community electronic freight documentation solutions (Table 5.6). Following the business process management (BPM) literature, they can be assigned to the maturity model of business processes (BPMMFisher) showing the transition handled in a primitive and smart way.

The transaction cost method (TCA) is used in the implementation of the maturity model.

Transaction costs include all expenses incurred in organizing economic activity when transactions are the object of analysis. Transaction cost theory explains the costs and effects of market and business transactions (Punnakitikashem et al., 2009). The TC theory identifies the essential features which distinguish operations, taking a static view of cost and economic efficiency only.

It can combine interdisciplinary insights on supply chain and organizational behaviour presented in the studies. According to the theory of transactions, their cost increases when an enterprise (calling a "transaction") provides the resource to the design of the solution.

In the first stage, an enterprise starts using the electronic version of freight documentation for operations among its branches. Such is called the one-step approach. The enterprise introduces edocuments including the initiation and development of IT solution's design.

The second stage includes the business-to-business activity and is the two-tier approach. Senders and recipients establish their relations by interacting when the supplier sells goods, and the recipient gets them together with an electronic freight document (Age, 2011). Adapting specific relationships is a traditional business practice. The solution of this request meets the possibilities to exchange documents (Table 5.6). First, single businesses with business integrations were popular but later on, many-to-many integrations are supported by EDI service companies.

Sender-to-recipient behaviour focused on B2B integration reduces some transaction costs. The theory of transaction costs could be followed to describe the gains associated with efficiency reached by such behaviour.

In the next stage, EDI solution provider is involved, also possibly involving him in this interaction.

First of all, customers understand that the service provider creates value for them. The creation of value using the service company is interesting for business, which means access to new markets, advanced technology, and real-time data. Herein,

Table 5.6 Maturity in electronic freight documentation (constructed by the author)

Maturity levels	Internal boundaries (B)	External boundaries (B2B)	Triangle (B2B & SP)	Multiple service providers (SP2SP)	E-platform (B2 (B2B))	Central position in multi-tier e-platform
Standardization	Stands on its format	Each connection stands in a different unique format	Mandatory to agree on the standard	Service providers agree on the standard	Multiple standards possible	The company selects the standard
Costs	Initial development	A process requiring employees for integration	SP takes technical control, monitoring, and data transfer insurance	Corporate investment into system performance security	Costs of joining e-platform	Main costs are constant
Benefits	Internal optimization	Number of transactions reached from 1–1 connection	Corporate information exchange, faster freight document processing	Minimization of losses due to security issues	Expansion of e-platform services for new users	Reach positive results after transactions are used for day-to-day cases

recipients identify the possibility to reach sustainability. The transfer of the service of electronic freight documentation in Europe providers reported an increase in 2015, a 27% growth rate and an additional € 1.25 billion in electronic freight transport document procedures (Academic Conferences and Publishing Limited, 2017).

The service provider invests money in capital investments, which also demonstrates that profits within counterparties are evident only when the companies are aiming to adopt solutions and invest to specific in a particular relationship. Finally, the created value means that it can be higher.

Under this model, the relationship between the recipient, the supplier, and the service provider is characterized by the following key elements: the investment that may be required to enter into the transfer of documents and the incremental adjustments, later on, the economic value generated due to the investment decisions reduces expenses, increases revenue, and improves electronic freight document management. Time-consuming problems often occur, including data entry errors, moving to e-service, or rearrangement of accounting procedures. Therefore, the procedure of checking electronic freight documents must be introduced.

While initially, investments by service providers might increase at the beginning, later on, the investments would be equal and compensatory and dedicated for maintenance cases.

In the further stage, operations are expanded to other geographical areas (Batty et al., 2012). If the consignor aims to deliver products cross border, involvement of abroad EDI providers for transferring documents is required. Such is required as the activities of EDI service are linked to a certain geographical area (Crooks & Heppenstall, 2012). An excellent example of this could be electronic freight documentation between South American and Western European countries. Typically, service providers join a foreign partner in support of cross-border transactions. Both the supplier and the recipient use an EDI service in the local country. Transactions in these cases follow the incentives from different countries and business behaviour.

Timely transfer of documents is necessary for higher value creation. For that, the functioning of EDI systems is very important. Data submitted by one party to the system is not provided to the other party on time due to technical system problems. Loss incurred by the service provider can be expressed into the sum of losses of all networks incurred due to system failures. To reduce losses of service providers, investments in the safe operation of the system are needed.

In the fifth stage, the "classic" B2B service is extended to the concept of multilevel B2 (B2B) services, requiring the implementation of EDI platforms that communicate with different actors. This allows multiple recipients and senders to transfer documents online. If international companies were interested in such a relationship the most, nowadays small businesses are also willing to join such platforms. Companies could use a particular platform actively, passively or intermediately. Other factors relate to shared behaviours and expectations, and adaptive behaviours depending on business changes.

The service is used by several companies (Walha et al., 2016). And no need for the recipient to maintain many connections with different vendors, having only one of them on the e-platform.

In the sixth stage, companies reach their central position in the e-platform. The proposed derivatives take a B2B-oriented perspective. This means that the synergy effect is achieved by the company that has acquired the central position, while new members are just starting to acquire. Following this logic, a company that accumulates all companies it has the relationships with on the e-platform benefits the most from this. Since the underlying costs are fixed, the higher the number of documents transfer the higher is the benefit. Nowadays, the major office furniture manufacturers of the United States have already adopted electronic freight documentation aiming to increase efficiency.

In general, an environmental service is called an environmental service having certain specifics. Public goods or shared resources are considered environmentally friendly services. However, electronic or paperless freight documents are also a part of these services, and such services have been growing rapidly.

Beckstrom's law provides many insights into different processes that take place in sectors and the efficiencies that can be achieved in different chain structures oriented to economic, ICT, and industrial behavioural aspects.

Under the Beckstrom rule, each buyer and seller get benefits which are linked with the transfers that the user performs on the network. This model could be applied to a network of any size if it has ten or billions of users. For the parties involved, a network means electronic services used by entities, which can be estimated by accounting for the cost and benefit analysis. Costs and benefits are revised from each user's perspective seeking to calculate them for every single position (Beckstrom, 2009).

The paperless freight documentation approach is beneficial for the companies involved (Asare et al., 2011), it could offer the following advantages:

1. Help to avoid manual typing, printing, posting, and archiving;
2. Track the movement of freight by using real-time documents transfer approach;
3. Involve companies providing other services: registration, declaration;
4. Provide a sustainable solution.

The implementation of eCMR has legal solutions ready, is tested with IT means, and has a positive effect on business.

5.7 The Role of ICT in Transport for Achieving Efficiency and Sustainability

The implementation of ICT solutions helps to reach sustainability by meaning safety, efficiency, and health effects of road freight transport. These problems are stated by the authors (McKinnon & Piecyk, 2009; Palsson & Kovacs, 2014; Perego et al., 2011; Wang et al., 2015). The externalities are different for freight delivery in cities, outside the cities, and between the cities (Hajduk, 2017). In Europe, for example, the sustainability of road freight transport has been provided for by a European Commission's (EC) priority, listing such initiatives as FIDEUS, KOMODA, MERCURIO, and ERTRAC, focusing on sustainable development of freight transportation. From an academic point of view, the authors (Ballantyne et al., 2013; McKinnon, 2007; Richardson, 2005) have investigated initiatives in freight delivery. Such sustainability initiatives have several different mechanisms to intervene and address the external consequences of road freight transport. For example, the development of the green freight delivery corridor has been researched and discussed in the literature (Clausen et al., 2012). There were studies by Wang et al. (2015), Button et al. (2001), Tob-Ogu et al. (2018) that explored the application of ICT means to support sustainability in freight delivery (Winebrake et al., 2012).

ICT has such components: software (programmes), hardware, and data transfer tools (Harris et al. 2015; Marchet et al., 2009; Wang et al., 2015). By solving the management issues of communication and relationships, business practices are highlighted by Sternberg et al. (2014), Crainic et al. (2009), and Ando and Taniguchi (2006). Information, communication technologies are generally presented as a positive thing affecting road haulage due to its operational efficiencies (Davies et al., 2007; Wang et al., 2015), providing social benefits (Button et al., 2001; Sternberg et al., 2014), costs minimization, and efficiency as mentioned by Marchet et al. (2012), reducing administration workload for drivers (Sternberg et al., 2014). The authors were ones who clearly stated in the documents the social aspects of the potential of ICT concerning the sustainability of road freight transport (Button et al., 2001; Sternberg et al., 2014). More targeted research may be needed to revise the ICT use and problems relating to freight delivery.

Many studies are devoted to the environmental and economic aspects linked with ICT application specified by Ando and Taniguchi (2006), Button et al. (2001), Davies et al. (2007), Marchet et al. (2012), and Walker and Manson (2014). They focus on the environment, emission reduction, and integration of ICT in the freight delivery process. The development of these initiatives could simplify decisions (DSSs) making during the planning and performing operations.

Efficiency approach have several efficient research directions, which are planned like:

- Emphasizing difference in costs;
- Development of a model that can be used to evaluate efficiency and provision of empirical evaluation.

Imagine several companies with balance sheets and income statements, but different environments: some of them have a high turnover, volume, and potential, and other ones face fierce competition. More importance will be given to the former in this case. Of course, various factors affect competitive advantage, but the author focuses on economic performance, which creates economic efficiency (Kalai et al., 1990).

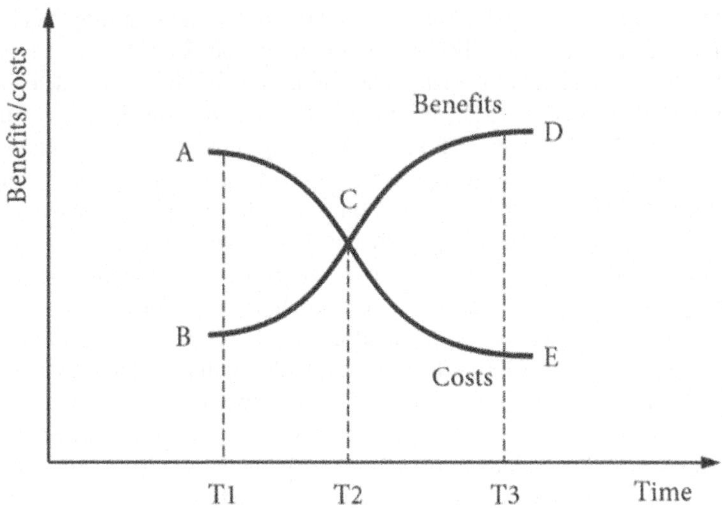

Fig. 5.4 Costs and benefits in the historical perspective

To illustrate this, curves that show the transformation of a company has been used. Initially, at T1 (Fig. 5.4) transaction costs at point A are quite high, and the benefits at point B are small due to the resource consumption process. When the application of such transactions expands, the cost per transaction decreases (to C and later to E), while the benefits of economies of scale shift in the opposite direction (to C and then to D). During this time, economic benefits are rapidly increasing, demonstrating growth of more than 100% over the period (between T1 and T3).

There are several ways to measure cost efficiency: real and expected efficiency. The indices that analyse the real-time capability of the technologies used have been analysed to describe the actual efficiency. Usual efficiency can be determined when it comes to assessing the benefits of upcoming efficient transport management (Christopher, 1992). Historical data and forecasts that reflect the experience of other companies have been used to estimate the expected performance. One way of measuring the likely efficiency is to evaluate the investments and benefits (increase in revenue or decrease in costs) of companies.

The authors analyse investments inefficient transport management, dividing them into short-term and long-term investments, basic (used to acquire and deploy technology) and permanent (used many times to develop efficient transport management).

Various methods of evaluation of efficiency include methods that analyse benefits and costs. Such techniques have been designed to evaluate the long-term costs and results of technological development. In practice, these techniques help to invest in ICT decisions. An example of this is the model proposed by Bergendahl (2005) to evaluate the benefit of efficient transport management from the time perspective.

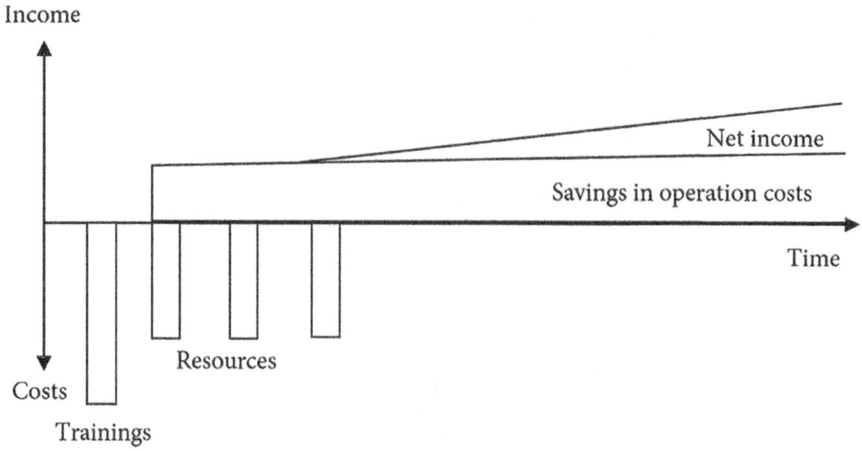

Fig. 5.5 The cash flows model which is generated by investments into efficient transport management (following Bergendahl, 2005)

In the Bergendahl model (2005), efficient expansion of transport management and sales growth are the key factors determining the profitability of efficient transport management investments. Therefore, the extension of efficient transport management must be assessed from a time perspective. Investment income is expected to be generated immediately after investing. Income (I_t) increases over time (5.1):

$$I_t = I_0(1 + q)^t. \tag{5.1}$$

Variables:

I_t—income accumulated over the period t,
I_0—present income,
q—accumulation coefficient,
t—time duration (in calendar years).

Income (I_t) may increase incrementally (q) and unevenly.

The Bergendahl model (2005) (Fig. 5.5) is useful in measuring efficient transport management's investment and the funds that companies allocate for training and resources.

While enterprises invest inefficient transport management, investments in labour and other resources (used for the implementation of efficient orientation) and training (or consultations) are made. Villarreal et al. (2017) find these constraints to be important.

Having implemented efficient transport management, the growth in sales and the reduction of costs are expected (Bergendahl, 2005) (Fig. 5.5).

At the start of the implementation of efficient transport management, average costs are evident but they decrease when a business reaches maturity in the

Table 5.7 Change in costs having achieved efficiency and sustainability

Costs	Direction	Authors
Operating costs	Reduction	Izadi et al. (2020)
Personnel costs	Reduction	Jakob et al. (2006)
Maintenance costs	Reduction	Ahuja and Khamba (2008); Purdy and Wiegmann (1987)
Losses	Reduction	Anholcer et al. (2019)
Sales	Increase	Ho et al. (2015)
Investments	Reduction	Cohen (2007)

application. Such application forms the right economy (Xu, Cao, et al., 2015a; Xu, Cheng, & Huang, 2015b).

Well-organized efficient transport management usually generates additional revenue, as its costs tend to decrease. Therefore, to identify the benefit of efficient transport management, Bergendahl (2005) proposes discounting (a way to revise the cash flows in future).

Efficiency could be evaluated in case of a one-time investment in transport management (one significant investment in training and resources). Methods used in sustainable freight transportation analysis are presented below.

Following the Sustainability Approach

A review of literature on sustainable transport management application allows stating that sustainable transport management funds can be allocated in multiple ways (several ongoing investments in audit, expertise, and maturity). As freight delivery minimizes waste, such improve cash flows.

The cash flow approach shows that an operating system includes further investments into audits. Cash flows, cash equivalents, and income are generated over time after making investments (Mackevičius & Senkus, 2006).

The advantage of the cash flow approach is the calculation of the current value of investments, turnover, and savings. The method depends on the risk premium that may be included in the interest rate. The author offers to use a variety of risks in implementing sustainable transport management, including organizational, technical, and financial risks. Although the method seems simple, justifying the interests level applied in the approach is difficult.

There are also some challenges in the application of the cash flow approach, in particular the discounting factors, such as varying inflation and solvency of the company, which is usually disregarded. While these factors may affect the size of cash flows, they remain underestimated.

The cash flow model is universal, which can be used to analyse historical (actual) and generated (probable) data. However, in all cases, positive cash flows were modelled. Belhadi et al. justify the assumption—they provide evidence of the benefits of sustainable transport management and changes in costs and direction (as specified in Table 5.7).

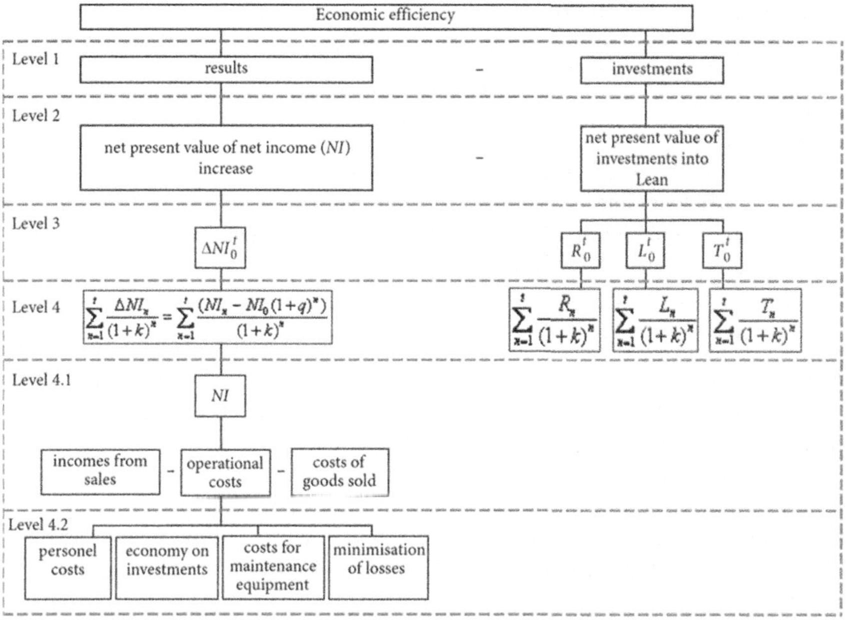

Fig. 5.6 The cash flow framework for the application of sustainability

Also, authors Chiarini and Brunetti (2019) that analysed the successful implementation of sustainability mentioned such factors as staff training on sustainable behaviour methods and principles. Bamber and Dale (2010), Chiarini and Brunetti (2019) reported a transformation from traditional to sustainable transport management performed by the freight owner. These researchers noted that the main difficulty was the lack of employees' education and training on sustainability ideology. Worley and Doolen (2006) highlighted investments into communication during the transition period, including in the communication not only employees but also suppliers, customers, and other partners (Worley & Doolen, 2006). Sacks et al. emphasized investments in visualization tools and safety management resources. Kindström et al. (2012) classified changes as changes in process; changes in coordination and control; changes in values and workforce behaviour.

Investments were classified as investments made into training, investments into resources other than the workforce, and investments into the workforce.

In Fig. 5.6 authors show a new cash flow model that measures results before (NI_0) and after using (NI_n) sustainable transport management. Investments needed to implement sustainable management, including training costs, resources other than workforce, and workforce, have been listed on the right side of Fig. 5.6. This model includes the net present value for the increase in net income during the period.

The sets of variables are:

T—Investments into training;
R—Investments into resources other than workforce;
L—Investments into the workforce;
k—discount rate;
t—time-period;
n—number of periods (years);
ΔNI_0^t—net current value of the increase in net income;
NI_0—net income before the implementation of the sustainability;
NI_n—net current value of net income per year;
E_t—economic efficiency after the period.

Usually, companies that invest in sustainable transport management are required to cover installation costs. However, daily costs have dropped significantly, leading to a decrease in average costs. With dropping costs which financial return, which means that costs of economic activity are bearable (Fig. 5.6).

The model implements various economic valuation levels (Fig. 5.6), where investments and results are achieved by applying sustainable transport management. The model can be used to evaluate the efficiency with one-time (single-period or initial investments) and follow-up investments (which can be called multi-period investments).

The concept of sustainability was used over the last decades to reach higher social sustainability. Such means extra efforts for businesses (DuBose et al., 1995). The term of sustainability born in 1960 sought to press businesses with the focus on the environment and cut CO_2 emissions.

Later on, the UN in 1972 emphasized the focus on CO_2 emission. Following such direction, researchers started to highlight that freight delivery hurts nature and focus in their studies on the environment from physical and social perspectives (Pirages, 1977). Researchers have expanded significantly from when studies started showing interest in it. Goodland and Ledec (1987) defined sustainable development (SD) aiming to reach benefits in coming years. Harwood promoted sustainable economics and called it the efficient usage of resources.

Initial studies were followed by conceptual studies. Robinson et al. presented a broader paradigm towards SD. Osorio et al. (2005) proposed a measurement method for ecological phenomena. Marchant (2010) introduced a three-stage sustainability model addressing economic, social, and environmental aspects.

In recent years, the number of studies is growing (Table 5.8). Herein transport is described as affecting standards of living (Furtado et al., 2013). Transport incorporates reuse, collection, disposal, and other activities that help to minimize adverse effects on the environment. Many individual studies focused on the development of green transport. Among many sustainability aspects, environmental sustainability was the main focus of studies in scientific transport literature, with ecological sustainability getting even more attention. In transport, environmental sustainability was used as an opportunity to reduce pollution. Economic sustainability relates to physical infrastructure of green corridor development and the length of the green corridor. Social responsibility direction deals with the resolution of conflicting

Table 5.8 Overview of the studies focusing on various sustainability directions in the freight delivery research area

Sustainability	Early phase	1990	2000	2010
Environmental	Sharp and Jennings (1976)	Ogden (1992) Murphy et al. (1995)	Verron and Friedrich (2004)	Marchant (2010); Caniato et al. (2012); Fichtinger et al. (2015)
Economical				Green et al. (2013)
Social				Sarkis et al. (2010)
Ecological				Joumard and Nicolas (2010); Jacyna et al. (2014)

situations focusing on achieving the optimization of the process, and direction oriented to ecology involves time perspective constraints. The authors focus on several sustainability directions but very few of them focused on a single aspect (Jeon et al., 2010). The authors list the studies that raise this concern in Table 5.8.

After 1980, still, the authors were focusing on the first sustainability direction highlighted by the UN. In 1990, new focus direction emerged, the authors were still discussing the definitions of various sustainability directions. Till now, many questions are not touched and are still actual.

For presenting sustainable transport, many simulation models have been developed, including system dynamics for revising resilience, discrete event simulations for revising uncertainties, and simulation games for applying in education (Rabelo et al., 2005).

Yeh and Li (2001) focus on sustainability in cities and routing. The study was provided for sustainable courier services delivered in a local city (Zachariadis, 2005). The authors Knaak et al. (2006) study the courier delivery service and clarified that orders for which drop-off location is near to pick-up location, must be shipped directly to save CO_2. Pawlewski and Greenwood (2014) conducted several studies revising the traffic issues. Checinski created a model analysing traffic spikes. Jacyna and Semenov (2020) presented possibilities to trip time, distance, and costs. Carteni (2014) simulated the planning effect on trip execution. Petranovic et al. (2015) were searching how to correspond to sustainability. Many more articles were written to introduce sustainable transport. The papers reached popularity as transport generates 20–40% of overall CO_2.

Apart from the transport system, several papers highlight freight transportation issues. García-Arca et al. (2017) presented research describing the packing importance in freight transportation. Tan et al. (2010) created studies about freight delivery companies. More of such studies will be presented in a further section.

Ocalir-Akunal (2016) identified that wastes are generated constantly by comparing various scenarios and revising the usage of resources. Most of such cases are presented in Fig. 5.7.

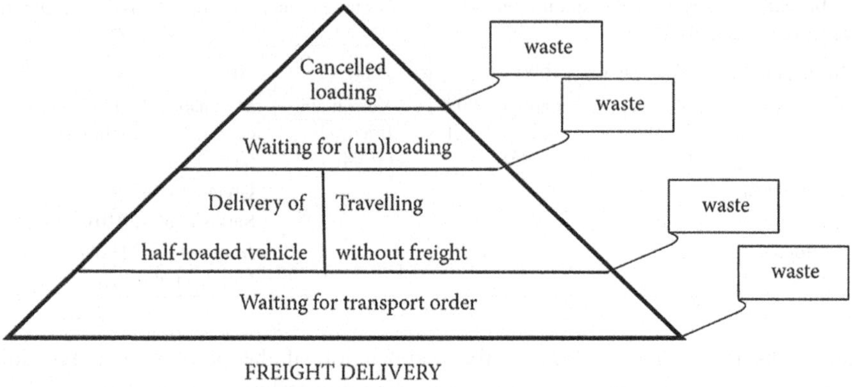

Fig. 5.7 The generation of the waste pyramid

5.8 Conclusions

The chapter is describing the freight delivery sector and recent trends in implementing electronic documents and reaching maturity in their usage. The application of ICT in the transport sector fosters new opportunities to reach synchro-modality, efficiency, and sustainability which are getting the highest attention by the authors recently. The problems that are named by the author help to focus on how to reach greener transportation and establish the links with main freight delivery processes.

The author investigates the application of advanced technologies opportunities by countries and by companies allowing to estimate costs and benefits in their use. These frameworks have no limitations on economic evaluations and waste eliminations and could be applied in a similar way for context-aware applications for freight delivery.

References

Academic Conferences and Publishing Limited. (2017). *ECDG 2017 17th European Conference on Digital Government* (pp. 18–134). Academic Conferences and Publishing Limited.

Achrol, R. S., & Stern, L. W. (1988). Environmental determinants of decision-making uncertainty in marketing channels. *Journal of Marketing Research, 25*(1), 36–50.

Age, L. J. (2011). Business manoeuvring: A model of B2B selling processes. *Management Decision, 49*(9), 1574–1591. https://doi.org/10.1108/00251741111173998

Ahuja, I. P. S., & Khamba, J. S. (2008). Total productive maintenance: Literature review and directions. *International Journal of Quality and Reliability Management, 25*(7), 709–756. https://doi.org/10.1108/02656710810890890

Anand, N., van Duin, J. H. R., & Tavasszy, L. (2016). Framework for modelling multi-stakeholder city logistics domain using the agent-based modelling approach. *Transportation Research Procedia, 16*, 4–15.

Ando, N., & Taniguchi, E. (2006). Travel time reliability in vehicle routing and scheduling with time windows. *Networks and Spatial Economics, 6*, 293–311.

Anholcer, M., Hinc, T., & Kawa, A. (2019). Losses in transportation—importance and methods of handling. In *SMART supply network* (pp. 111–128). Springer.

Asare, A. K., Brashear, T. G., Granot, A. E., & Kashyap, V. (2011). The role of channel orientation in B2B technology adoption. *Journal of Business and Industrial Marketing, 26*(3), 193–201. https://doi.org/10.1108/08858621111115912

Awasthi, A. (2015). Evaluating new business operation models for small and medium size logistics operators within low emission zones. In *Proceedings of the 9th International Conference on City Logistics, Tenerife, Spain, 17–19 June.*

Awasthi, A., & Proth, J.-M. (2006). A systems-based approach for city logistics decision making. *Journal of Advances in Management Research, 3*(2), 7–17.

Awasthi, A., Chauhan, S. S., Parent, M., & Proth, J. M. (2011). Centralized fleet management system for cybernetic transportation. *Expert Systems with Applications, 38*(4), 3710–3717.

Awasthi, A., Adetiloye, T., & Crainic, T. G. (2016). Collaboration partner selection for city logistics planning under municipal freight regulations. *Applied Mathematical Modelling, 40*(1), 510–525.

Awasthi, A., Sayyadi, R., & Khabbazian, A. (2018). A combined approach integrating gap analysis, QFD and AHP for improving logistics service quality. *International Journal of Logistics Systems and Management, 29*(2), 190–214.

Ballantyne, E. E. F., Lindholm, M., & Whiteing, A. E. (2013). A comparative study of urban freight transport planning: Addressing stakeholder needs. *Journal of Transport Geography, 32*, 93–101.

Bamber, L., & Dale, B. G. (2010). Lean production: A study of application in a traditional manufacturing environment. *Production Planning and Control, 11*(3), 291–298.

Batarliene, N., & Sakalys, R. (2020). Criteria impacting synchronisation of transport flows along international transport corridor. *Promet-Traffic and Transportation, 32*, 17–28.

Batty, M., Crooks, A. T., See, L. M., & Heppenstall, A. J. (2012). Perspectives on agent-based models and geographical systems. In *Agent-based models of geographical systems* (pp. 1–15). Springer.

Beckstrom, R. (2009). *A new model for network valuation* (pp. 1–10). National Cyber Security Center, Research paper.

Bergendahl, G. (2005). Models for investment in e-commerce—financial perspectives with empirical evidence. *Omega, 33*, 363–376. https://doi.org/10.1016/j.omega.2004.07.021

Bhuiyan, M. F. H., Awasthi, A., & Wang, C. (2015). Investigating the impact of access-timing-sizing regulations on urban logistics. *International Journal of Logistics Systems and Management, 20*(2), 216–238.

Buiel, E., Visschedijk, G., Lebesque, L., Lucassen, I., Van Riessen, B., Rijn, A. V., & te Brake, G. (2015). Synchro Mania-Design and evaluation of a serious game creating a mind shift in transport planning. In *Paper presented at the 46th International Simulation and Gaming Association Conference, ISAGA 2015, Kyoto, Japan (18–25 July, 2015)* (pp. 1–12).

Burinskas, A., & Burinskienė, A. (2020). Application of Gordon's growth model for the transport sector of Lithuania. *Journal of logistics, informatics and service science (JLISS)., 7*(1), 31–41.

Burrough, P. A. (1992). Development of intelligent geographical information systems. *International Journal of Geographical Information Systems, 6*(1), 1–11.

Button, K., Doyle, E., & Stough, R. (2001). Intelligent transport systems in commercial fleet management: A study of short term economic benefits. *Transportation Planning and Technology, 24*, 155–170.

Cagliano, A., Marco, A., & Rafele, C. (2017). E-grocery SCM enabled by mobile tools. *Business Process Management Journal, 23*(1), 47–70.

Caniato, F., Caridi, M., Crippa, L., & Moretto, A. (2012). Environmental sustainability in fashion supply chains: An exploratory case based research. *International Journal of Production Economics, 135*(2), 659–670.

Capgemini. (2012). Electronic freight documentation Maturity Model. Capgemini Oracle Blog. Retrieved August 21, 2017, from http://www.capgemini.com/blog/capgemini-oracle-blog/2012/11/electronic-freight-documentation-maturity-model

Carteni, A. (2014). Urban sustainable mobility. Part 1: Rationality in transport planning. *Transport Problems, 9*(4), 39–48.

Chang, E., Floros, E., & Ziliaskopoulos, A. (2007). *An intermodal time-dependent minimum cost path algorithm dynamic fleet management* (pp. 113–132). Springer.

Chiarini, A., & Brunetti, F. (2019). What really matters for a successful implementation of Lean production? A multiple linear regression model based on European manufacturing companies. *Production Planning and Control, 30*(13), 1091–1101.

Christopher, M. (1992). *Logistics and supply chain management—strategies for reducing costs and improving services*. Pitman Publishing.

Civelek, M. E., & Seckin, N. (2017). Paperless trade: evaluation of the current situation & towards the integrated single foreign trade document. *Journal of Management Research, 9*(2), 1–10.

Civelek, M. E., Çemberci, M., Uca, N., Çelebi, Ü., & Özalp, A. (2017). Challenges of paperless trade redesign of the foreign trade processes and bundling functions of traditional documents. *International Business Research, 10*(2), 12–35.

Clausen, I. U., Geiger, C., & Behmer, C. (2012). Green corridors by means of ICT applications. *Procedia-Social and Behavioral Sciences, 48*, 1877–1886.

Cohen, J. P. (2007). *Economic benefits of investments in transport infrastructure*. OECD.

Crainic, T. G., & Montreuil, B. (2016). Physical internet enabled hyperconnected city logistics. *Transportation Research Procedia, 12*, 383–398.

Crainic, T. G., Ricciardi, N., & Storchi, G. (2009). Models for evaluating and planning city logistics systems. *Transportation Science, 43*, 432–454.

Crooks, A. T., & Heppenstall, A. J. (2012). Introduction to agent-based modelling. In *Agent-based models of geographical systems* (pp. 85–105). Springer.

Davies, I., Mason, R., & Lalwani, C. (2007). Assessing the impact of ICT on UK general haulage companies. *International Journal of Production Economics, 106*, 12–27.

De, P. (2011). Trade facilitation in India: An analysis of trade processes and procedures (No. 95). In *ARTNeT working paper series*.

Downing, C. E. (2002). Performance of traditional and web-based EDI. *Information Systems Management, 19*(1), 49–55. https://doi.org/10.1201/1078/43199.19.1.20020101/31476.7

DuBose, J., Frost, J. D., Chamaeau, J. A., & Vanegas, J. A. (1995). Sustainable development and technology. In *The Environmentally Educated Engineer* (pp. 73–86).

Ehrler, V., & Hebes, P. (2012). Electromobility for city logistics—the solution to urban transport collapse? An analysis beyond theory. *Procedia—Social and Behavioural Sciences, 48*, 786–795.

European Commission. (2018). Transport in the European Union: Current trends and issues. https://ec.europa.eu/transport/sites/transport/files/2018-transport-in-the-eu-current-trends-and-issues.pdf

European Commission. (2019). *A European green deal: Striving to be the first climate-neutral continent*. European Commission.

Eurostat. (2021). Database. Retrieved July 10, 2021, from https://ec.europa.eu/eurostat/data/database.

Fatnassi, E., Chaouachi, J., & Klibi, W. (2015). Planning and operating a shared goods and passengers on-demand rapid transit system for sustainable city-logistics. *Transportation Research Part B: Methodological, 81*(2), 440–460.

Fichtinger, J., Ries, J. M., Grosse, E. H., & Baker, P. (2015). Assessing the environmental impact of integrated inventory and warehouse management. *International Journal of Production Economics, 170*, 717–729.

Furtado, P., Frayret, J.-M., Fakhfakh, R., & Biard, P. (2013). Simulation of a Physical Internet-Based transportation network. In *Paper presented at the Proceedings of 2013 International Conference on Industrial Engineering and Systems Management (IESM)*.

Fusari, A. (1996). Paths of economic development: modelling factors of endogenous growth. *International Journal of Social Economics, 23*(10/11), 164–191.

García-Arca, J., González-Portela Garrido, A. T., & Prado-Prado, J. C. (2017). Sustainable packaging logistics: The link between sustainability and competitiveness in supply chains. *Sustainability, 9*, 10–98. https://doi.org/10.3390/su9071098

Goodland, R., & Ledec, G. (1987). Neoclassical economics and principles of sustainable development. *Ecological Modelling, 38*(1–2), 19–46.

Gordon, M. J. (1962). *The investment, financing, and valuation of the corporation*. RD Irwin.

Green, J. L., de Weck, O. L., & Suarez, P. (2013). Evaluating the economic sustainability of sanitation logistics in Senegal. *Journal of Humanitarian Logistics and Supply Chain Management, 3*(1), 7–21.

Hajduk, S. (2017). Bibliometric analysis of publications on city logistics in international scientific literature. *Procedia Engineering, 182*, 282–290.

Heilig, L., Negenborn, R. R., & Voß, S. (2015). *Cloud-based intelligent transportation systems using model predictive control computational logistics* (pp. 464–477). Springer.

Ho, W., Zheng, T., Yildiz, H., & Talluri, S. (2015). Supply chain risk management: a literature review. *International Journal of Production Research, 53*(16), 5031–5069.

Hofman, W. (2014). Control tower architecture for multi-and synchromodal logistics with real time data. In *ILS2014, Breda*.

Hofman, W. (2015). Federated platforms for seamless interoperability in the Physical Internet. In *2nd International Physical Internet Conference, Paris*.

Hofman, W., Punter, M., Bastiaansen, H., Cornelisse, E., & Dalmolen, S. (2016). Semantic technology for enabling logistics innovations–towards Intelligent Cargo in the Physical Internet. *International Journal of Advanced Logistics, 5*, 1–12.

Hwang, B., Chang, S., Yu, H., & Chang, C. (2008). Pioneering e-SC integration in semiconductor industry: A case study. *The International Journal of Advanced Manufacturing Technology, 36*(7–8), 825–832.

Izadi, A., Nabipour, M., & Titidezh, O. (2020). Cost models and cost factors of road freight transportation: A literature review and model structure. *Fuzzy Information and Engineering*, 1–21. https://doi.org/10.1080/16168658.2019.1688956

Jacyna, M., & Semenov, I. (2020). Models of vehicle service system supply under information uncertainty. *Eksploatacja i Niezawodnosc—Maintenance and Reliability, 22*, 694–704. https://doi.org/10.17531/ein.2020.4.13

Jacyna, M., Wasiak, M., Jacyna-Gołda, I., Pyza, D., Merkisz-Guranowska, A., Merkisz, J., Lewczuk, K., Żak, J., & Pielecha, J. (2014). A holistic approach to modelling of the ecological domestic transport system. In *Materiały XVIII Congreso Panamericano de Ingeniería de Tránsito, Transporte y Logística. Spain* (pp. 34–98).

Jakob, A., Craig, J. L., & Fisher, G. (2006). Transport cost analysis: a case study of the total costs of private and public transport in Auckland. *Environmental Science and Policy, 9*(1), 55–66.

Jeon, C. M., Amekudzi, A. A., & Guensler, R. L. (2010). Evaluating plan alternatives for transportation system sustainability: Atlanta metropolitan region. *International Journal of Sustainable Transportation, 4*, 227–247. https://doi.org/10.1080/15568310902940209

Jetter, W. J., Johnson, P. A., Sill, C. B., Zanzot, M. D., ... & Keesey, J. (2011). U.S. Patent No. 8,050,975. Washington, DC.

Joumard, R., & Nicolas, J. P. (2010). Transport project assessment methodology within the framework of sustainable development. *Ecological Indicators, 10*(2), 136–142.

Jun, M., Cai, S., & Peterson, R. T. (2000). EDI use and participation models: from the inter-organizational relationship perspective. *Industrial Management and Data Systems, 100*(9), 412–420. https://doi.org/10.1108/02635570010358366

Kalai, E., Kamien, M. I., & Rubinovitch, M. (1990). Optimal service speeds in a competitive environment. *Management Science, 38*, 1154–1163.

Khayyat, M., & Awasthi, A. (2015). An intelligent multi-agent-based model for collaborative logistics systems. In *Proceedings of the 9th International Conference on City Logistics, Tenerife, Spain, 17–19 June*.

Kim, Y. J., & Sanders, G. L. (2002). Strategic actions in information technology investment based on real option theory. *Decision Support Systems, 33*(1), 1–11.

Kindström, D., Kowalkowski, C., & Nordin, F. (2012). Visualizing the value of service-based offerings: Empirical findings from the manufacturing industry. *Journal of Business and Industrial Marketing, 27*(7), 538–546.

Knaak, N., Kruse, S., & Page, B. (2006). An agent-based simulation tool for modelling sustainable logistics systems. In *International Congress on Environmental Modelling and Software, Vermont, USA* (pp. 161–388).

Ko, H. J., Ko, C. S., & Kim, T. (2006). A hybrid optimisation/simulation approach for a distribution network design of 3PLS. *Computer Industrial Engineering, 50*(4), 440–449.

Lagorio, A., Pinto, R., & Golini, R. (2017). Urban logistics ecosystem: A system of system framework for stakeholders in urban freight transport projects. *IFAC-PapersOnLine, 50*(1), 7284–7289.

Landschützer, C., Ehrentraut, F., & Jodin, D. (2015). Containers for the Physical Internet: requirements and engineering design related to FMCG logistics. *Logistics Research, 8*(1), 1–22.

Lin, I.-C., & Cheng, C.-Y. (2016). Case study of physical internet for improving efficiency in solar cell industry. *Journal of Ambient Intelligence and Humanized Computing, 9*, 1–10.

Mackevičius, J., & Senkus, K. (2006). The system of formation and evaluation of the information of cash flows. *Journal of Business Economics and Management, 7*(4), 171–182.

Marchant, C. (2010). Reducing the environmental impact of warehousing. In A. McKinnon (Ed.), *Green logistics: Improving the environmental sustainability of logistics* (pp. 167–192). Replika Press Pvt Ltd.

Marchet, G., Perego, A., & Perotti, S. (2009). An exploratory study of ICT adoption in the Italian freight transportation industry. *International Journal of Physical Distribution and Logistics Management, 39*, 785–812.

Marchet, G., Perotti, S., & Mangiaracina, R. (2012). Modelling the impacts of ICT adoption for inter-modal transportation. *International Journal of Physical Distribution and Logistics Management, 42*, 110–127.

McKinnon, A. C. (2007). Decoupling of road freight transport and economic growth trends in the UK: An exploratory analysis. *Transport Reviews, 27*, 37–64.

McKinnon, A. C., & Piecyk, M. I. (2009). Measurement of CO2 emissions from road freight transport: A review of UK experience. *Energy Policy, 37*, 3733–3742.

Mes, M. R., & Iacob, M.-E. (2016). *Synchromodal transport planning at a logistics service provider logistics and supply chain innovation* (pp. 23–36). Springer.

Mishra, D., Gunasekaran, A., Papadopoulos, T., & Hazen, B. (2017). Green supply chain performance measures: A review and bibliometric analysis. *Sustainable Production and Consumption, 10*, 85–99.

Montreuil, B., Meller, R. D., & Ballot, E. (2010). Towards a Physical Internet: the impact on logistics facilities and material handling systems design and innovation. *Progress in Material Handling Research, 472*, 305–327.

Montreuil, B., Meller, R. D., & Ballot, E. (2013). *Physical internet foundations Service Orientation in Holonic and Multi Agent Manufacturing and Robotics* (pp. 151–166). Springer.

Moon, H. C., Shim, S. R., & Kim, D. Y. (2011). Issues in the international standards of e-documents for global e-trade. *International Journal on Social Science, 1*(2), 164–168.

Muñuzuri, J., van Duin, J. H. R., & Escudero, A. (2010). How efficient is city logistics? Estimating ecological footprints for urban freight deliveries. *Procedia—Social and Behavioral Sciences, 2*(3), 6165–6176.

Murphy, P. R., Poist, R. F., & Braunschweig, C. D. (1995). Role and relevance of logistics to corporate environmentalism: an empirical assessment. *International Journal of Physical Distribution and Logistics Management, 25*(2), 5–19.

Nabais, J. L., Negenborn, R. R., Carmona-Benítez, R., & Botto, M. A. (2015). Cooperative relations among intermodal hubs and transport providers at freight networks using an MPC approach. In *International conference on computational logistics* (pp. 478–494). Springer.

Nabais, N., Behdani, B., Fan, Y., Wiegmans, B., & Zuidwijk, R. (2016). Multimodal schedule design for synchromodal freight transport systems. *European Journal of Transport and Infrastructure Research, 16*(3), 424–444.

Ngeno, K., & Kinoti, J. (2017). Effect of e-procurement on effective supply chain management process in energy sector in Kenya. *International Journal of Supply Chain Management, 2*(3), 18–37.

Ocalir-Akunal, E. V. (2016). Decision support systems in transport planning. *Procedia Engineering, 161*, 1119–1126.

Ogden, K. W. (1992). *Urban goods movement: A guide to policy and planning*. Routledge.

Osorio, L. A. R., Lobato, M. O., & Castillo, X. Á. D. (2005). Debates on sustainable development: Towards a holistic view of reality. *Environment, Development and Sustainability, 7*(4), 501–518.

Palsson, H., & Kovacs, G. (2014). Reducing transport emissions: A reaction to stakeholder pressure or a strategy to increase competitive advantage. *International Journal of Physical Distribution and Logistics Management, 44*, 284–304.

Pawlewski, P., & Greenwood, A. (2014). *Process simulation and optimization in sustainable logistics and manufacturing* (pp. 1–184). Springer.

Perego, A., Perotti, S., & Mangiaracina, R. (2011). ICT for logistics and freight transportation: A literature review and research agenda. *International Journal of Physical Distribution and Logistics Management, 41*, 457–483.

Petranovic, Z., Vujanović, M., & Duić, N. (2015). Towards a more sustainable transport sector by numerically simulating fuel spray and pollutant formation in diesel engines. *Journal of Cleaner Production, 88*(1), 272–279. https://doi.org/10.1016/j.jclepro.2014.09.004

Pirages, D. C. (1977). *Sustainable society: implications for limited growth*. Praeger.

Punnakitikashem, P., Somsuk, N., Adebanjo, D., & Laosirihongthong, T. (2009, December). A review of theoretical perspectives in lean manufacturing implementation. In *2009 IEEE international conference on industrial engineering and engineering management* (pp. 1204–1208). IEEE.

Purdy, J. E., & Wiegmann, J. D. (1987). Vehicle maintenance: Cost relationship and estimating methodology. *Transportation Research Record: Journal of the Transportation Research Board, 1140*, 1–17.

Putz, L., Haider, C., Haller, A., & Schauer, O. (2015). Identifying key enablers for synchromodal transport chains in Central Europe. In *Proceedings of the WCTRS SIGA2 2015 conference "the port and maritime sector: Key developments and challenges", Antwerp, Belgium.*

Quak, H., Nesterova, N., & van Rooijen, T. (2016). Possibilities and barriers for using electric-powered vehicles in city logistics practice. *Transportation Research Procedia, 12*, 383–398.

Rabelo, L., Helal, M., Jones, A., & Min, H. S. (2005). Enterprise simulation: A hybrid system approach. *International Journal of Computer Integrated Manufacturing, 18*(6), 498–508.

Reis, V. (2015). Should we keep on renaming a +35-year-old baby? *Journal of Transport Geography, 46*, 173–179. https://doi.org/10.1016/j.jtrangeo.2015.06.019

Richardson, B. C. (2005). Sustainable transport: Analysis frameworks. *Journal of Transport Geography, 13*, 29–39.

Rivera, A., & Mes, M. (2016). Service and transfer selection for freights in a synchromodal network. *Lecture Notes in Computer Science, 9855*, 227–242.

Salah, D., Paige, R., & Cairns, P. (2014). An evaluation template for expert review of maturity models. In A. Jedlitschka, P. Kuvaja, M. Kuhmann, T. Männistö, J. Münch, & M. Raatikainen (Eds.), *Product-focused software process improvement, LNCS 8892* (pp. 318–321). Springer.

Sallez, Y., Montreuil, B., & Ballot, E. (2015a). On the activeness of physical internet containers. In *Service orientation in holonic and multi-agent manufacturing* (pp. 259–269). Springer.

Sallez, Y., Berger, T., Bonte, T., & Trentesaux, D. (2015b). Proposition of a hybrid control architecture for the routing in a Physical Internet cross-docking hub. *IFAC-PapersOnLine, 48*(3), 978–1983.

Sarkis, J., Helms, M. M., & Hervani, A. A. (2010). Reverse logistics and social sustainability. *Corporate Social Responsibility Environmental Management, 17*, 337–354.

Sarraj, R., Ballot, E., Pan, S., Hakimi, D., & Montreuil, B. (2014). Interconnected logistic networks and protocols: simulation-based efficiency assessment. *International Journal of Production Research, 52*(11), 3185–3208.

Sharp, C., & Jennings, A. (1976). *Transport and the environment*. RSC Publishing.

Stefaniuk, T., & Chomicka, M. (2019). IT systems in customs service of transport companies–Case study. *Logistics and Transport, 42*, 399–408.

Sternberg, H., Prockl, G., & Holmström, J. (2014). The efficiency potential of ICT in haulier operations. *Computers in Industry, 65*(8), 1161–1168.

Suresh, S., & Vasantha, S. (2018). Influence of ICT in road transportation. *International Journal of Supply Chain Management, 7*, 49–58.

Suzuki, Y., & Williams, L. R. (1998). Analysis of EDI resistance behavior. *Transportation Journal, 37*(4), 36–44.

Tadić, S., Krstić, M., & Brnjac, N. (2019). Selection of efficient types of inland intermodal terminals. *Journal of Transport Geography, 78*(2019), 170–180.

Tan, K. S. L., Ahmed, M. D., & Sundaram, D. (2010). Sustainable enterprise modelling and simulation in a warehousing context. *Management Journal, 16*(5), 871–886. https://doi.org/10.1108/14637151011076511

Tavasszy, L. A., Janssen, G. R., van der Lugt, L. M., & Hagdom, E. (2010). *Verkenning Synchromodaal Transportsysteem*. TNO in opdracht van Ministerievan Infrastructuur en Milieu.

Tavasszy, L. A., Behdani, B., & Konings, R. (2015). Intermodality and Synchromodality. Available at SSRN 2592888.

The Lithuania Department of Statistics to the Government of the Republic of Lithuania. Lithuania in Europe. 2019. Available from Internet: <www.stat.gov.lt>.

Tob-Ogu, A., Kumar, N., & Cullen, J. (2018). ICT adoption in road freight transport in Nigeria—A case study of the petroleum downstream sector. *Technological Forecasting and Social Change, 131*, 240–252.

Tretola, G., Verdino, V., & Biggi, D. (2015). A Common Data Model for the Physical Internet. In *2nd international Physical Internet Conference, Paris*.

Tsertou, A., Amditis, A., Latsa, E., Kanellopoulos, I., & Kotras, M. (2016). Dynamic and synchromodal container consolidation: the cloud computing enabler. In *Paper presented at the Proceeding of the 6th Transport Research Arena, April 18–21, 2016, Warsaw, Poland*.

UNECE reported TIR Carnets 2001–2019. Available online: http://www.unece.org/fileadmin/DAM/tir/figures/TIRCarnets2001-2019.pdf (accessed on 20 August 2021).

Van Riessen, V., Negenborn, R. R., & Dekker, R. (2015a). *Synchromodal container transportation: An overview of current topics and research opportunities computational logistics* (pp. 386–397). Springer.

Van Riessen, V., Negenborn, R. R., Dekker, R., & Lodewijks, G. (2015b). Service network design for an intermodal container network with flexible transit times and the possibility of using subcontracted transport. *International Journal of Shipping and Transport Logistics, 7*(4), 457–478.

Van Riessen, V., Negenborn, R. R., & Dekker, R. (2016b). The cargo fare class mix problem-revenue management in synchromodal container transportation (working paper). Retrieved from Research Gate: DOI 10.13140.

Verron, H., & Friedrich, A. (2004). Environmentally sustainable transport in Germany. *European Journal of Transport and Infrastructure Research, 4*(1), 71–98.

Verweij, K. (2011). Synchronic modalities–Critical success factors. In P. J. van der Sterre (Ed.), *Logistics yearbook edition 2011* (pp. 75–88).

Villarreal, B., Garza-Reyes, J. A., Kumar, V., & Lim, M. K. (2017). Improving road transport operations through lean thinking: A case study. *International Journal of Logistics Research and Applications, 20*(2), 163–180.

Walha, F., Bekrar, A., Chaabane, S., & Loukil, T. M. (2016). A rail-road PI-hub allocation problem: Active and reactive approaches. *Computers in Industry, 81*, 138–151.

Walker, G. H., & Manson, A. (2014). Telematics, urban freight logistics and low carbon road networks. *Human Factors in Transportation, 37*, 74–81.

Wang, Y., Sanchez-Rodrigues, V., & Leighton, E. (2015). The use of ICT in road freight transport for CO2 reduction—An exploratory study of the UK's grocery retail industry. *The International Journal of Logistics Management, 26*, 2–29.

Winebrake, J. J., Green, E. H., Comer, B., Corbett, J. J., & Froman, S. (2012). Estimating the direct rebound effect for on-road freight transportation. *Energy Policy, 48*, 252–259.

Worley, J. M., & Doolen, T. L. (2006). The role of communication and management support in a lean manufacturing implementation. *Management Decision, 44*(2), 228–245.

Xu, Y., Cao, C., Jia, B., & Zang, G. (2015a). Model and algorithm for container allocation problem with random freight demands in synchromodal transportation. *Mathematical Problems in Engineering, 2015*, 13.

Xu, S. X., Cheng, M., & Huang, G. Q. (2015b). Efficient intermodal transportation auctions for B2B e-commerce logistics with transaction costs. *Transportation Research Part B: Methodological, 80*, 322–337. https://doi.org/10.1016/j.trb.2015.07.022

Yeh, A. G. O., & Li, X. (2001). A constrained CA model for the simulation and planning of sustainable urban forms by using GIS. *Environment and Planning B: Planning and Design, 28*(5), 733–753.

Zachariadis, T. (2005). Assessing policies towards sustainable transport in Europe: An integrated model. *Energy Policy, 33*(12), 1509–1525.

Zhang, M., & Pel, A. (2016). Synchromodal hinterland freight transport: Model study for the port of Rotterdam. *Journal of Transport Geography, 52*, 1–10.

Zhong, R. Y., Xu, C., Chen, C., & Huang, G. Q. (2015a). Big data analytics for Physical Internet based intelligent manufacturing shop floors. *International Journal of Production Research, 55*, 1.

Zhong, R. Y., Peng, Y., Fang, J., Xu, G., Xue, F., Zou, W., & Huang, G. Q. (2015b). Towards Physical Internet-enabled prefabricated housing construction in Hong Kong. *IFACPapersOnLine, 48*(3), 1079–1086.

Zilliaskopoulos, A., & Wardell, W. (2000). An intermodal optimum path algorithm for multimodal networks with dynamic arc travel times and switching delays. *European Journal of Operational Research, 125*(3), 486–502.

Chapter 6
Variables Important for Freight Delivery and Context Data Usage

Aurelija Burinskiene

Abstract For having actionable context, several primary variables, such as time and location, are important. Other variables, such as a trip distance, which is an indication between initial and final locations and the cargo position among them, are secondary ones. The delivery distance is a key variable, which is used for analysing alternatives, selecting the mode for freight delivery and discussing the service composition, which always responds to the customer's choose between the lower price and faster delivery. When the mode of transport is selected, the economic cargo delivery volume can be identified. Finally the contextual information could be classified according to different objects (infrastructure, environment, delivery) in static or dynamic presence, which later on is captured by various devices and is used in different users' applications.

Keywords Time variable · Location variable · Distance variable · Mode variable · Economic delivery volume · Service variable · Covariance analysis · Context capturing devices · Users' applications

6.1 Freight Delivery Data: Time Variable

Time constraint, also known as a planning constraint, is considered a soft constraint, because having violated it, vehicle routes are not separated and traffic is not impeded, but freight may arrive late to the place of destination. Violation of such features could be permitted, when this enlarges the probability to solve the problem and discourages the imposition of a fine (Gambardella et al., 2002). Goods successfully integrated for a joint trip should correspond to the maximum additional time that the parties involved find it acceptable.

Documents analysing transactions which are focusing on balance in customer choices when they look for services with the lowest price instead of worrying about

A. Burinskiene (✉)
Vilnius Gediminas Technical University, Vilnius, Lithuania
e-mail: aurelija.burinskiene@vilniustech.lt

D. Dzemydienė et al., *Development of Smart Context-Aware Services for Cargo Transportation*, International Series in Operations Research & Management Science 330, https://doi.org/10.1007/978-3-031-07199-7_6

delays. For example, literature on network traffic flows such as presented by Harker (1986) models customer choices based on prices and traffic-related congestion. Several papers on transport efficiency analyse the competitive supply of goods or services to time-sensitive customers. Lee and Cohen (1985) investigate the competitive assignment of customers to servers by agents seeking to reduce their waiting times.

Time-based identification services presuppose the existence of well-formed context data representing the performance of tasks. Contextual information can be used to describe the situation of the subject and is used for an interaction between the user and the object. To take advantage of contextual information in services, cargo delivery conditions in a specific time frame must be checked. Then decisions are made by the user of contextual data, who assesses the data and several other variables, and decides: (a) to accept delays in task performance (in observance of a predefined deadline) or (b) to reject the task fulfilment.

The planning activity helps to prioritize tasks over time. Ancillary methods are usually followed in priority work lists (Cook, 1998). Priority lists have been organized and presented to the manager, crossing out tasks upon their completion. Priority in order is given to a carefully thought-out, predetermined list of tasks that can quickly be updated in unforeseen situations. Having implemented a context-based business planning programme, two aspects are highlighted. First, the programme must be able to connect to the activity list automatically and efficiently, considering the relevant context. Second, the programme must be able to determine when to redesign the priority row after the situation with resources changes. There are ways to determine when the list of activities must be restructured based on a periodic review of its order. This approach increases the opportunity to take action when the order does not temporarily meet user needs.

Truck trips to deliver cargo with random breaks in long-distance journeys should also be taken into account. When travelling, vehicles need to stop at gas stations to refuel. Driver must take periodic rest breaks in rest areas. Vehicles may also need to stop at medical or police stations having suffered a medical or a legal accident. There may be a need to stop at a workshop in case of a vehicle failure. Moreover, vehicle speed may change depending on road and weather conditions. A freight ends its trip having reached the destination. During the trip, deviations from the schedule are reported to a shipment manager.

The time required for freight transfer lasts from the moment the freight leaves the warehouse (the stock business unit) till the moment it arrives at the customer's facility doors for reception. Transfer time is calculated between places of the warehouse and the customer. The time for the delivery is defined as a particular combination of these factors, programming the system to search for combinations to obtain a delivery time based on a particular order schedule.

The transfer time is calculated by following the scheduled date of dispatch and arrival when one of the required dates (date of dispatch or arrival) has been entered.

The transfer time is calculated based on arbitrary plans and rules. The system searches for all matching transportation time rules and sorts them according to the plan using the longest, shortest, or sequence of the plan. Having selected the planned sequence, the system sorts the rules alphabetically or by the longest or the shortest time. The system then uses the first delivery time rule in a list of valid dispatch and arrival dates.

Different types of customers have different service requirements and sensitivity acceptance to the delay time from order to delivery. Companies have a different distribution of costs and processing time and planning decisions (Karageorgos et al., 2003). Sequence theory is used in response modelling. Numerous articles in the literature on queueing address the issue of customer behaviour when faced with a single queuing system. Articles by Naor (1969), Yechiali (1970, 1972), Lippman and Stidham (1977), Bell and Stidham (1983), Mendelson (1985), and Mendelson and Whang (1990) are important. Customers cannot track the actual queue length. Thus, their choice focus on the average behaviour of the long-run queuing system. On the other hand, the model offered by Li and Lee (1989) analyses the case where customers monitor the length of a queue. Some results confirm the claims of the "time contest" literature, but others do not. By separating processing speed and delivery time and recognizing the external effects of congestion (Chęciński, 2014), a faster-operating company can take advantage of its ability to accept more orders and thus have a longer delivery time or accept relatively few orders, thus offering short delivery times.

The literation on "competition based on the time" does not recognize this difference and how this opportunity is used. Specifically, it is assumed that a faster manufacturer always has a faster delivery time. Indeed, a company that has a good processing speed but no cost advantage tends to compete by fast delivery and charges a fast response premium, while a company whose advantage is production costs rather than the processing speed tends to compete at a lower price, but has a slower delivery.

Scheduling and planning problems are complex problems for IT to deal with (Berghman et al., 2012), using time planning and combinatorial optimization to solve this issue. Collaborative search and the application of heuristics often lead to satisfactory but not optimal results. The following are the reasons why these tasks are so complex:

1. Effective heuristics is difficult to create.
2. The space for possible solutions to most real-world problems is too large.
3. Scheduling and/or planning problems are often encumbered by certain details of the scheduling/planning task (Bongaerts, 1998). The general algorithm-based approach to a situation may not be applied due to special constraints required for certain cases.
4. Real-time planning and/or scheduling cases often have constraints that are not accurately demonstrated or even accurately stated.

Value of time (VOT) is important. Transport time usually is not that important for goods (with an exception of vegetables, fruits, flowers, and fresh dairies, for example). The VOT for fresh salmon is higher than for frozen ones. Another benefit offered by shorter transportation time is access to the reached efficiency in the delivery chain. Production changes can also help to reduce delays and damages.

Studies confirm that higher capital interest expenses have been determined according to the value of delivered freight.

On-time and in-full (OTIF) is the main indicator in the industry and service sectors for cargo delivery. On-time and in-full is usually concerned with customer satisfaction and the order-to-cash cycle. The revision of on-time and in-full measures is quite often in the private sector and presents timeliness and completeness of delivery.

Delivered on Time Organizations set the target time (this KPI may show if the requested delivery was completed on the requested date).

Delivery Too Early We introduced the measure for late deliveries, but it is just as important that goods do not arrive too early. Too early deliveries are foreseen only in exceptional cases.

Delivered in-Full Delivery in-full is when a customer receives exactly the amount it has ordered. However, for example, the batch size produced may vary in the medical industry, thus all the criteria should be defined so that the customer receives the amount ordered at the least.

$$\text{OTIF} = (\text{eligible transport units delivered on date})/$$
$$(\text{the overall amount of transport units requested for delivery on date})$$
$$\text{On} - \text{time} = \text{difference between planned arrival day and actual arrival day}$$
$$\text{In} - \text{full} = \text{difference between the volume planned and volume transported.}$$

OTIF evaluates transportation of various size delivery service orders, assessing if services have been provided as agreed. It will be the data reported with the highest accuracy about freight shipment. OTIF approach could be applied to improve vehicle advance decision-making and could be a part of context-aware services.

The main goal of an organization is to drive the improvement of processes to gain business value. The organization has to agree on the delivery time with its customer without setting a lower limit, which means that early delivery of freight will cause no problems.

There are also other variations of KPIs, like the revision of order lines delivered versus order lines ordered for shipment, etc. Many of these KPIs help to improve the process (Becker et al., 2006).

The time-based context revision can help rising economic efficiency for the transportation systems and reduce the time for delivering single freight. The guidance system helps vehicles to make optimal decisions relating to time, congestion, and rest.

6.2 Freight Delivery Data: Location Variable

Location-based data is by nature contextual and presuppose the existence of well-formed data about the user environment. To take the advantage of contextual service information, data from the environment should be collected through appropriate mechanisms, then using it to develop end-user consumer applications.

The basic idea of the usage of the application is to project arrival to the next location according to the data of the current location (latitude and longitude):

- The projection takes the area into account, i.e., two points between the current space and the projected space, and vice versa.
- All points in the new space have the same "context", i.e., population density which is important for specific deliveries of freight (dangerous), or the environment, which is important for glass product deliveries, or temperature forecasts which are important for products with defined temperature conditions.
- The projection must not interfere with the location reporting service on a topological map and can be effectively calculated.

Systems deal with location information. Different locations used different solutions. For example, after a warehouse is equipped with a card reader, the system will know that a cargo is somewhere inside the warehouse if GPS resolution is up to 10 m. Badge resolution maybe 1 m, which is indicating that the object is in the sensory zone. This probability occurs in sensors that register an object, and if there are multiple sensors, the information is fused obtaining one value. Freshness of data is fixed by the last sensor readings and each sensor type that follows a temporary degradation function that, based on time of record retrieval undermines the last information.

6.3 Freight Delivery: Types of Distances

The mode is selected by following freight delivery data. According to Eurostat (Eurostat, 2021a, 2021b), cargo shipment by road heavy trucks is separated into international (long distance) and inside the country (short distance). In the EU territory, a long-distance delivery is around 600 km (Statista Research Department, 2020). Most freights are carried at the distance between 300 km and 999 km but there are exceptional cases. The trips with distances higher than 1000 km are declining recently (Eurostat, 2021a, 2021b). For such trips, a combination of several modes of transport has been used. The length of short distance highly depends on the perimeter of the country and possibilities to travel over it. The average distance in this category is around 149 km. The length of such trips is also affected by the size of the islands. Such trips are becoming more popular.

Table 6.1 Classification of distances (Eurostat, 2021a, 2021b)

Delivery, km	Distance type
0–149 km	Short distance
150–299 km	Intermediate distance
300–1000 km	Long distance

The author notes that the literature analysed does not cover the range between 150 and 299 km. Such distance should be called intermediate-distance delivery. Such trips are becoming less popular (Eurostat, 2021a, 2021b).

Following these investigations, the author offers the classification of distances and presents it in Table 6.1.

The distance between two locations measured as straight line is called Euclidean. And the distance of between two locations is constructed following the road network and is known as path-based one.

6.4 Freight Delivery: Mode Selection

Identification of a mode of transport from travelling trajectories has become increasingly interesting in recent years in various fields such as contextual computer, location services, city understanding dynamics, travel demand research, traffic monitoring, and travel behaviour analysis. Context-aware services must reflect user selection preferences. Users could consider the mode and green transport-related aspects (Elektrodegalines, 2021).

The selection of electric trucks has been growing recently. A seven-time increase in the reported number was observed (Eurostat, 2021a, 2021b).

Charging infrastructure is critical and has to reflect the growth. Researchers have analysed the demand for charging points and electric vehicles and provided different models based thereon. Charging infrastructure scenarios in cities connected to a grid assume infrastructure costs. The model investigates for proper locations to install chargers. These stations for charging are linked to the hubs in urban areas. If infrastructure for charging is not sufficient, drivers have to wait to be able to charge their trucks. In general, the recommendation is to have charging points within a distance of 60 km (European Parliament, 2021).

There are many different business models for charging electric trucks presented in the literature, including swapping of batteries, sharing of the truck, grids for vehicles, charging option (public, non-public), choice of intermodality, etc., as stated by Laurischkat et al. (2016). Context-aware services have to react to the location of electric charging stations and the time required for charging, assuming battery shipping times in planning freight delivery terms.

The author observed that there is no relation between the adoption of context-aware services in green transport, as presented in Table 6.2. Thus, quantitative modelling is needed. The energy-optimal route for electric vehicles in the context of freight delivery has higher complexity than for fuelled trucks, where the total

Table 6.2 Models and studies with a focus on new transport modes

Type	Orientation to	Model	Study
Infrastructure	Infrastructure for charging scenarios in gridded city	Simulation	Chen et al. (2016); Von Meier (2018)
	Model of metropolitan governance	Transaction model	Lowery (2000)
	Electric mobility models	Holistic instrument	Kley et al. (2011); Knapen et al. (2012); Madina et al. (2016)
Vehicle	Vehicle quantity estimation	Probability	Brownstone et al. (1996)
	Route selection behaviour models	Trip model	Bohnsack et al. (2014)
	Delivery mode selection models	Preference models	Brownstone et al. (1996)
Context-aware services	Quantitative modelling	Activity model	Masuch et al. (2011)

travel distance is the most important indicator. Total energy consumption for electric trucks depends on the weight of a consignment and its changes over time, which directly affects battery performance. Therefore, the optimal route solution cannot be defined using traditional routing algorithms, which are based on the distance as a metric in calculations (Pach et al., 2014). Truch charging activity should be planned on time, location and schedule, providing for charging intervals are also recommended.

In addition to green transport, considering modality selection topics and links among different mode pairs when it comes to the cargo path from destination to origin (Fahimnia et al., 2015) is also important. The geospatial model uses the long-distance delivery network to combine different modes (rail, road, water) through terminals (Lee & Cohen, 1985). Trips consider time and length of route, costs, emission characteristics, and energy. The model is applied to research trade-offs between different route choices based on modal combinations and the identification of optimal routes.

A multimodal freight transport system includes (1) consignors who are responsible for the transport of freights and who choose modes of transport and logistics; (2) consignees; (3) shippers (for example, transferring a container from the main port to an off-site freight company for loading the container on a long-distance lorry; (4) custom clearance company; (5) companies managing terminals; (6) operators of the road network; and (7) policy makers and transport analysts interested in revision of the impact of policy on freight transport.

Geographically, such a system includes nodes indicating the origin, destinations, ports, railway stations, multimodal terminals, and the arcs connecting nodes (Patel & Hudson-Smith, 2012). Technically, the system consists of trucks, trains, ships, and equipment enabling intermodal terminals to operate effectively. In summary, a multimodal freight transport system is about understanding the network of locations

Table 6.3 Covariance analysis for transport modes in pairs

Road-rail		
Probability	Road	Rail
Road	1	0.72
Rail		1
Maritime-rail		
Probability	*Maritime*	*Rail*
Maritime	1	
Rail	0.75	1

that is intertwined with technologies and is designed for supplying freights nationwide involving an efficient, timely, and environment-friendly manner.

The transfer of freight from one mode of transport to another impacts the volumes delivered by different modes of transport. The author researched the relationship between different modes and presented the results. An analysis of the relationships of volume transfer by rail or road networks (the results are provided in Table 6.3).

Empirical research can be used to estimate the physical and informational flows created by delivering cargo by network type. Data on the delivery of goods by sea, road, rail infrastructure are revised in 53 states (EU28, Asia, USA, CIS, and others) for 18 years (2000–2018). Data is taken from the official statistical database of the Organisation for Economic Co-operation and Development (OECD, 2020).

Further on, the author used covariance and cointegration procedure. This method facilitates the dynamism of the selected variables representing the networks. The system follows the procedures of time-series revision, incorporates the approach of equilibrium, and follows long-run equilibrium among nonstationary variables. The review of cointegration follows the logic that economic variables could reflect the directions of the trends. In particular, the linear setup of analysed patterns is converted during the time to reach equilibrium.

In addition, the model of vector error correction (VEC) is constructed by incorporating the revision of lags as the variables interact together. The case of multiple cointegration relationships is also considered. Thus, implemented VEC model has a long-term β-vector, α-vector responsible for adjustment process and cross-vectors necessary for interactions.

In addition, a method that takes into account data management in the area of the VEC model may examine the appropriate channels responsible for the procedure of adjustment of deviations to reach equilibrium in a longer perspective.

The procedure of cointegration facilitates the stationary pattern of time series. The procedures demonstrate that products are delivered by rail for a long distance, first are carried by heavy road vehicles and not the opposite way (Fig. 6.1).

The author of the study also analysed other modes of transport (Fig. 6.2). At first, the covariance analysis (Fig. 6.3) shows the relationship between freight deliveries by maritime and rail networks (the results are provided in Table 6.3).

The link between maritime and rail transport could be defined by the volume of containers transferred by maritime transport and rail transport.

This link could be presented as a linear relationship in the Figure below (Fig. 6.4).

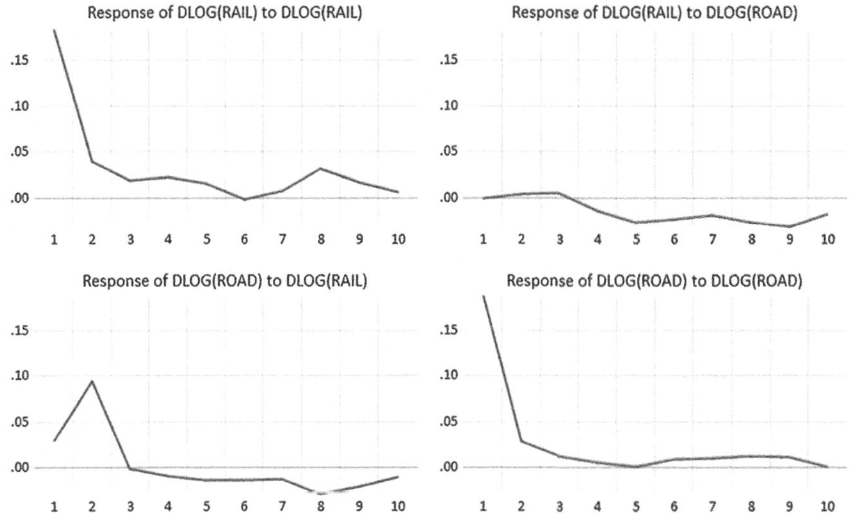

Fig. 6.1 Relationship between freight transportation by road network per tonne-kilometres and freight deliveries by rail network (left side) and vice versa (right side)

The results are also presented by countries below (Fig. 6.5).

The detailization of these results are provided in Fig. 6.6.

Finally, the author researched the impact of the three modes of transport on the volume of freight which they carry and found that freight carried by maritime transport is first brought by road and then carried by sea. It is carried by rail after that, finally reloading cargo from rail onto a road vehicle. The table above (Table 6.4) illustrates this process as statistical analysis.

Time series are integrated if they show the same degree of integration, and the linear combination of these variables is stationary (Crooks & Castle, 2012). In addition, the application of time series at their levels guarantees that the loss of information because of the normal use of the first differences will be avoided. The process of dynamic adjustment by adjusting variables towards long-run equilibrium could be represented by the vector error correction (VEC) model. In such cases, relationships of long-term equilibrium are linked with the dynamics evident in the short term.

The results show that the physical and informational flow is between road and rail transport modes is more important and is not other way around. Assessment of the flows reveals that in the case of multimodal delivery the information about the first part of delivery by road has to be carried to the rail to continue with delivery.

Synchro-modality can finally be considered as a further development of these transport paradigms. Synchronous logistics should cover a diverse and highly integrated mode of transport. The goal is to ensure sustainability and time-efficiency, reducing environmental impact and optimizing resource use. Synchro-modality is inherited from intermodality for flexible corridor management to meet partner needs.

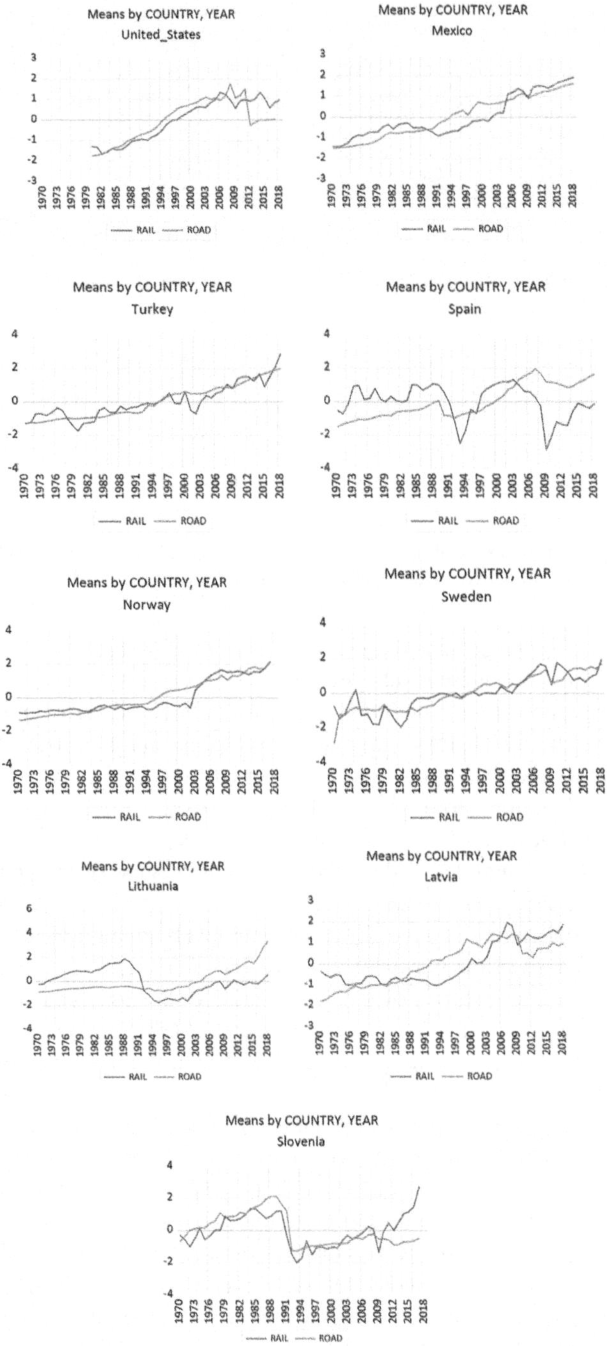

Fig. 6.2 Sample states: the relationship among volumes delivered by rail and road transport modes, in ton-kilometres

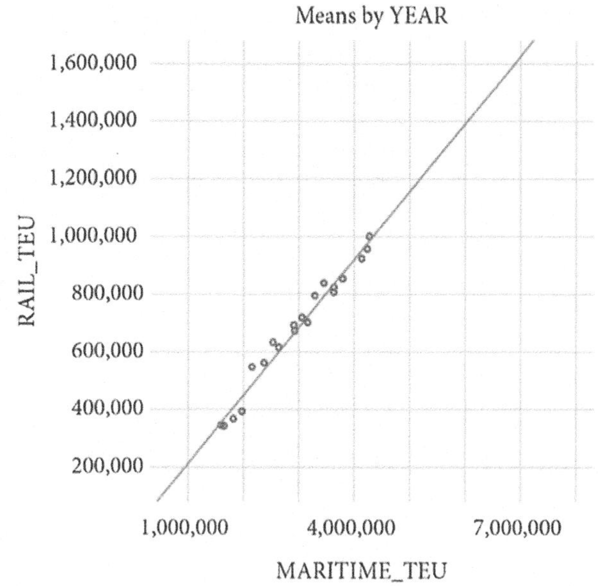

Fig. 6.3 Relationship between freight carried by maritime and rail transport

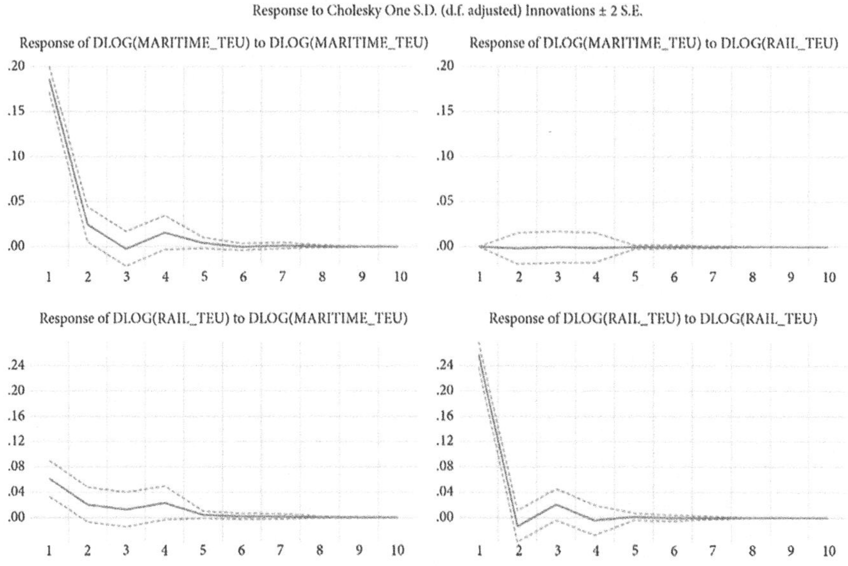

Fig. 6.4 Response between freight carried by maritime and rail transport

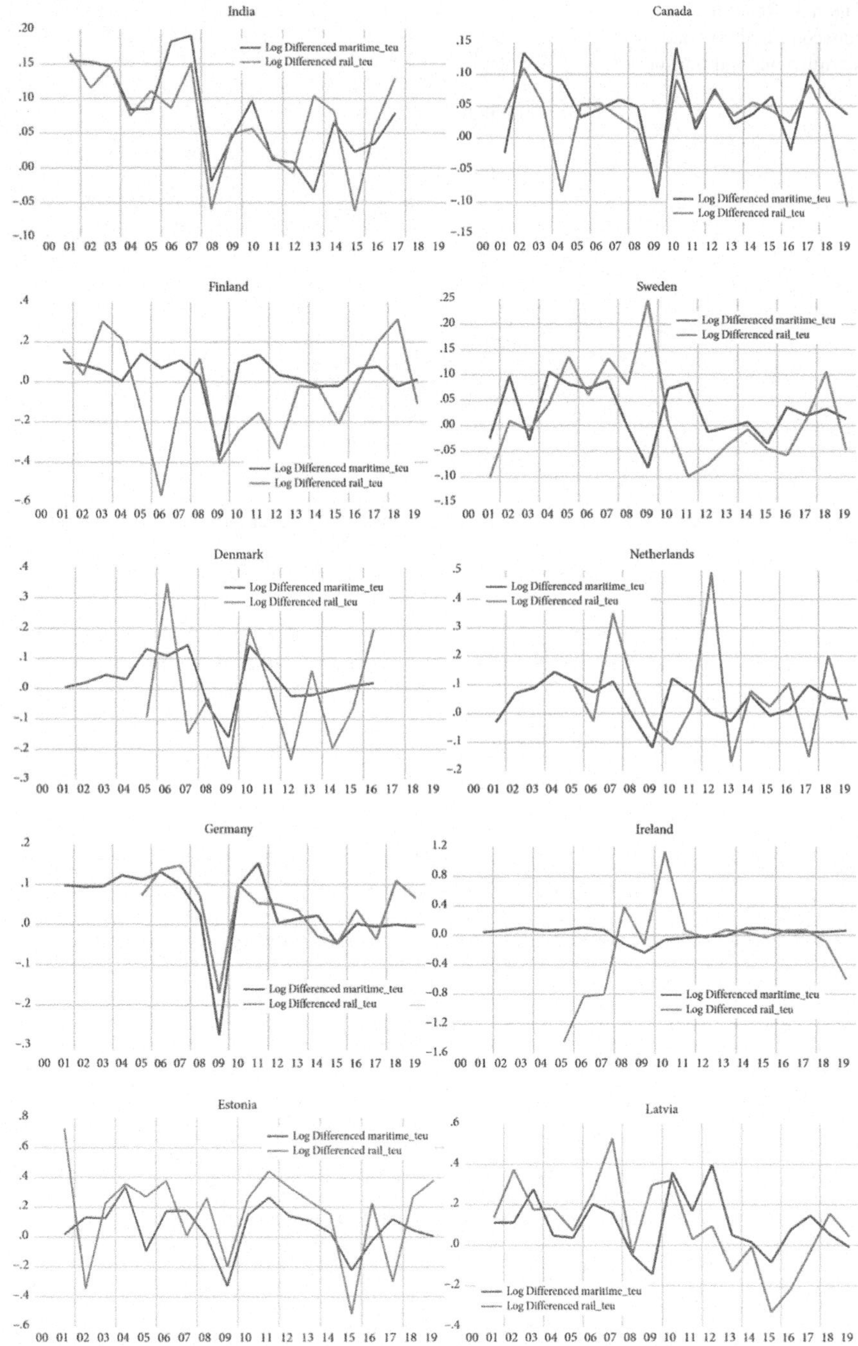

Fig. 6.5 Sample countries: the long-term relationship between freight deliveries by maritime and rail transport (Red colour—freight carried by rail (teu), blue colour—freight carried by maritime transport (teu))

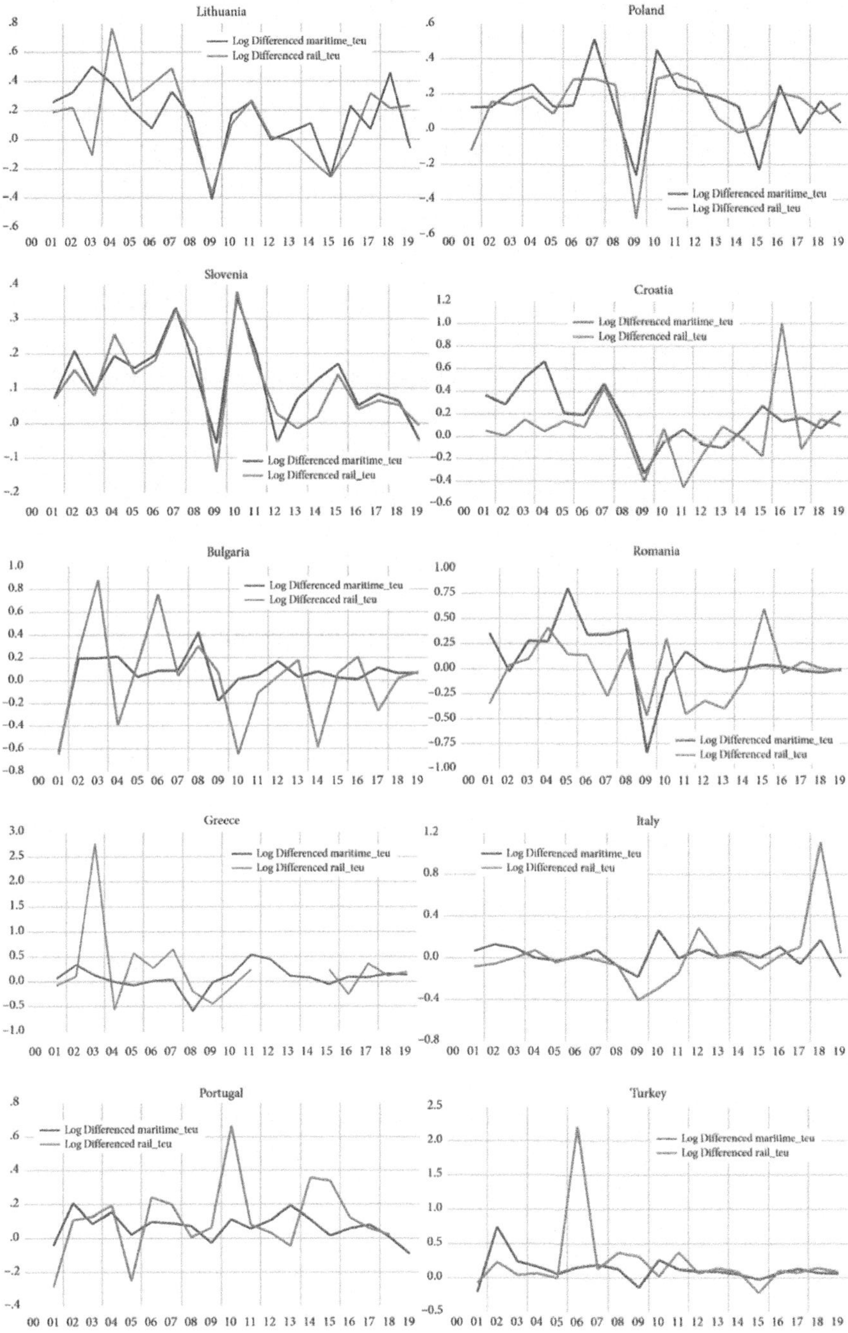

Fig. 6.5 (continued)

Fig. 6.6 VAR Granger causality tests (for the 2000–2018 period) performed by EViews software

VAR Granger Causality/Block Exogeneity Wald Tests
Sample: 2000 2018
Included observations: 364

Dependent variable: DLOG(RAIL_TN_KM_MLN)

Excluded	Chi-sq	df	Prob.
DLOG(ROA	7,158994	6	0,3064
DLOG(MAI	28,87193	6	0,0001
All	36,97035	12	0,0002

Dependent variable: DLOG(ROAD_TN_KM_MLN)

Excluded	Chi-sq	df	Prob.
DLOG(RAII	14,50786	6	0,0244
DLOG(MAI	9,694021	6	0,1381
All	21,57779	12	0,0425

Dependent variable: DLOG(MARITIME_TEU)

Excluded	Chi-sq	df	Prob.
DLOG(RAII	3,433495	6	0,7528
DLOG(ROA	17,30535	6	0,0082
All	21,56168	12	0,0427

Table 6.4 Covariance analysis for maritime and rail transport modes

Probability	Rail	Road	Maritime
Rail	1		
Road	0.77	1	
Maritime	0.46	0.65	1

The particularities of synchro-modality allow synchronization, smart use of resources, dynamism, coordination, and cooperation. Real-time information is a key for sychromodal freight delivery operations. Based on real-time information, other features are initiated.

6.5 Freight Delivery Data: Economic Delivery Volume

The model of the economic order quantity (called EOQ) defines the size of freight and reduces delivery-related costs (Fig. 6.7). For assumptions, it includes loading and unloading costs, delivery costs, backhauling costs, and administration costs for transport order handling and routing. The minimum of total costs shows the number of transport units that has to be allocated for delivery seeking to optimize costs.

In the case of an economic delivery volume, each pallet is assumed to be homogeneous; however, delivery costs for heterogeneous pallets will be a bit higher.

A truckload is considered to be fully optimized when it is 100% loaded for all routes with homogeneous pallets. There are several types of models:

1. Models that assume full truckloads. Studies have searched for the highest possible level of truck use. Some studies highlight a single-echelon case as the best way to utilize vehicle capacity. In some of these models, back-orders are permitted. If back-orders are not allowed, the vehicle capacity is insufficient for delivery to the customer.
2. Models that assume mixed loads. In these models, the number of orders carried in a route depends on the capacity of trucks, which work more than one shift a day.

All these models affect costs. Truck occupancy is highest in round-trip transport order delivery and lowest in the case of one-side transport order delivery (Fig. 6.8). Considering that the height of a truck is 2 m, even having filled a truck 100% in one direction, the overall truck occupancy will be 50% in a two-way trip, if it comes back empty (Fig. 6.8).

Costs for each delivery type are highly dependent on pallet height assembled during shipping when multiple products are placed on a pallet. When the delivery of heterogeneous products is organized, the height of a pallet is 1.2 m, and, when the delivery of homogeneous products is placed, it is—1.5 m (or 1.25 times higher).

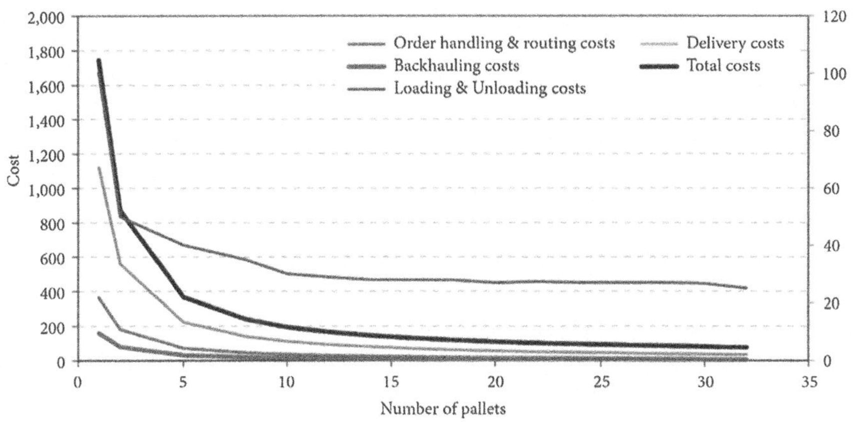

Fig. 6.7 Minimizing total costs

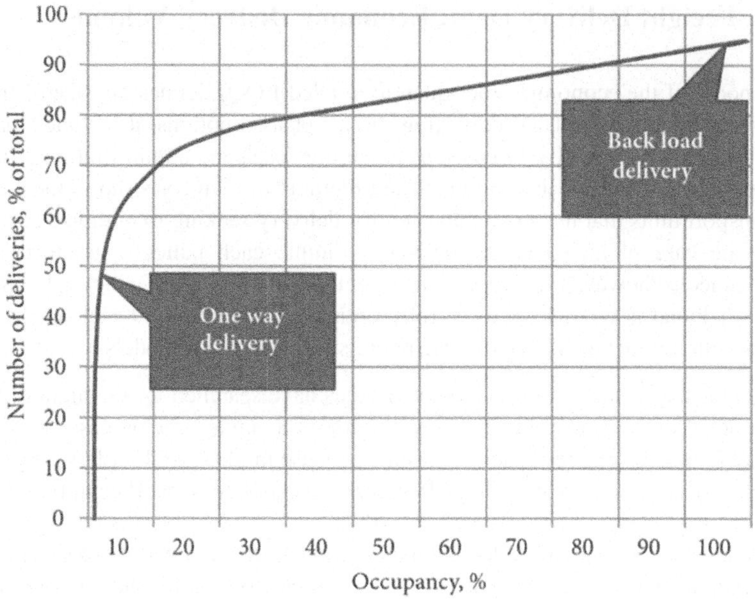

Fig. 6.8 Illustration of delivery strategy and truck occupancy

All the above-mentioned models mean different delivery costs. The costs depend on the delivery type—whether it is a round-trip or a one-side delivery. If a truck carries pallets round trip, its occupancy is higher (as presented in Fig. 6.8).

In these cases, the volume of carried freight depends on the capacity of trucks. In logistics, capacity constraints ensure that the number of transport units carried matches transport capacity (Savelsbergh & Van Woensel, 2016). These restrictions are applied when goods are carried by several or single vehicles (Li et al., 2014). Integrated models, when goods are transported together, may need to consider capacities for different freight types that could not be combined in the same compartment.

In addition to planning, some papers suggest keeping vehicle capacity 30% higher than production capacity, which would help to guarantee that products fit in a truck and customers receive their goods according to the schedule.

6.6 Freight Delivery Data: Service Variable

Competition between companies producing products that are sensitive to delays has been growing (European Commission, 2011). Companies compete by product prices and production limits. Companies may group customers by the cost of delay. More insights can be obtained from the analysis of several cases. For example, one company offers different costs, average turnaround times and variability in

Table 6.5 The focus of the service

Service focus on	Response time	Service coverage by a single-mode
Efficiency	No	Yes
Availability	Yes	Yes
Accessibility	No	Yes

turnaround time. This company may give quite high prices and faster trips, or lower prices but slower trips. The other company has a different cost and average turnaround time, but its customers also differ in demand and latency sensitivity. Therefore, customers with higher wait for freight delivery time pay lower prices, and the companies charge a higher price but make faster deliveries to customers who cannot wait (Table 6.5).

Integrated delivery of goods can reduce freight delivery costs (Crainic & Montreuil, 2016). Most researches of complex systems leave to travel and transportation costs unresearched or see them as the constraints of the system instead of reviewing how to reduce them.

Seeking to benchmark several efficiency increases approaches the identification of costs and their analysis are suggested. The author selected labour and transport mean expenses for the analysis (Fig. 6.9).

Freight delivery includes several operations: routing, loading, delivery, unloading, and backhauling.

Each trip of a truck could be expressed in costs by computing trip distance with labour and equipment costs (6.1):

$$U_{jr} = d_{Pe} \cdot (c_L + c_F), \tag{6.1}$$

herein: c_L—labour costs per kilometre; c_F—equipment costs per kilometre, e—vehicle.

Context-aware service could come in handy in both scenarios (i.e. faster delivery and higher prices or longer delivery and lower prices). Leading companies strive to deliver products quickly and fulfil customer orders in the shortest possible time (Shah & Khanzode, 2015). Stalk (1988) called enterprises competing with lean production, fast reaction, expanded diversity and focusing on "time competitors". The study by Stalk and Hout (1990) proves the time strategy that reflects the impact of response on customers.

Performance measures, which indicate the effectiveness of the delivery service, include customer complaints, on-time delivery, order fill rate, and outbound freight costs. These measures were used by 77–87% of the respondents to the Boon's (2002) survey. The study revealed that companies did not measure the following indicators: invoice accuracy (48%), cycle time (38%), back-orders (36%), and on-time delivery (21%) (Boon, 2002).

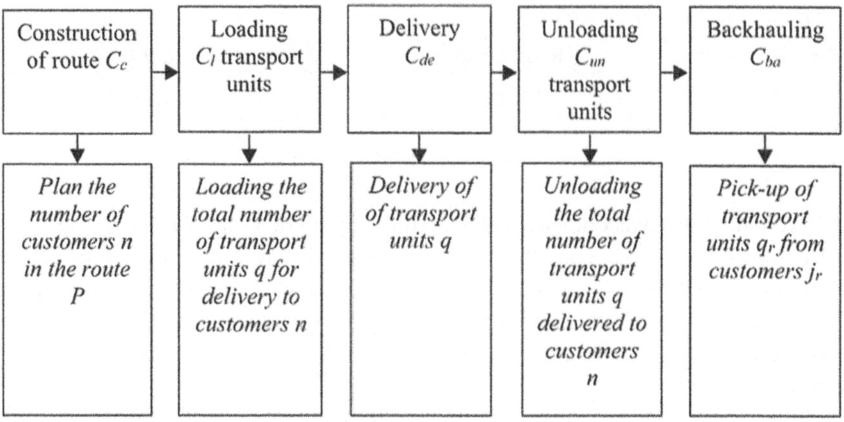

Fig. 6.9 Freight delivery distance components. Herein: *P*—route, *r*—return cases, C_c—routing, C_l—loading, C_{de}—delivery, C_{un}—unloading, C_{ba}—backhauling. The route starts with a customer (j) and ends with a customer (n)

6.7 Scope of Context Data Usage

All the variables mentioned before could be used as context data. First of all the data could be static and dynamic, operating and showing contextual relationship. Static context data in transport could be describing transport mode, identifying truck and driver, regulations on work time, insurance data, capacity of truck, maximum speed limit, and regularly visited locations (depots, clients, repair service centre, fuel refilling, or battery charging stations). Dynamic context data means reporting of real-time information, such as physical location, actual time, speedometer, engine and cooler data, status of truck, accident, seekness of driver, etc. Operating data could calculate based on real-time information how long operations will take and forecast arrival time. Contextual relationship data could help to create main relationships, for example, dependence between weather conditions and speed, or traffic jumps and arrival time and so on. In some cases, complex relations could be described aiming to support user, i.e. driver, transport planner or manager.

In Table 6.6, the author links the context data via data structures with devices capturing the context data, context-aware services, and user applications. Devices are constantly capturing the data from multiple sources. After this the data is structured and transformed into actionable by using more intelligence. The services ensure that data has been exchanged between heterogeneous systems and later on information that is being exchanged ensure that the meaning of exchanged data is understood precisely by communicating entities, with the purpose of creating value out of it. This valuable information is visible in different user applications.

As defined above in the table infrastructure context include such examples of road characteristics, custom borders, terminals, parking slots.

Table 6.6 Context data and the scope of data usage

Context categories	Context data	Devices for context capture	Context-aware services	User applications
Infrastructure context	Road characteristics	Road condition reporting system	Plan arrival date	Infrastructure monitoring tool
	Custom borders, terminals	GIS	Plan occupation at places	Electronic terminal system
	Parking slots	Parking availability sensors	Book parking slot	Parking booking system
Environment context	Traffic jumps places	Traffic control device	Plan route avoiding jumps	Traffic analytic software
	Current time	Truck clock	Display time	Time management setup
	Weather conditions	Weather reporting sensors	Present weather forecast	Weather reporting system
Delivery context	Transport order, route	Electronic seal	Inform delivery details	Smart transport application
	Delivery conditions	Temperature control device	Post e-documents with data	System posting e-documents
	Volume	Weight scales device	Plan loading	Fleet planning application
Object static context	Transport mode	Transport mode detector	Plan charging or refiling	Application reporting CO2
	Truck, trailer	Gate tracking device	Book gate	Loading and unloading system
	Driver	Security card device	Plan drive and rest time	Personnel management system
Object dynamic context	Real-time location	GPS	Show current location	Integrated navigation
	Accident events	eCall device	Allocate emergency vehicle	Emergency vehicle approaching
	Forecast of arrival	Mobile device	Forecast arrival time	Scheduling application
User static context	Role	Authentication device	Present user preferences	Transport planning application
	Resources in-use	Transport handler	Update status	Resource planning application
	Available resources	Position sensors	Tracking resources	Asset management application
User dynamic context	Task fulfilment overview	Portable electronic device	Display task	Intelligent decision support system
	Issues requiring decisions	Web-services	Inform about issues	ITS
	Future transport orders	e-services	Plan new orders	AI application

When environmental context includes traffic jumps, cover time, weather conditions variables; and delivery context focus on transport order, route, delivery conditions and volumes variables that are linked with service price, pickup and destination places, total route distance. The object static variables hold fixed information for certain period of time, such as transport mode, truck, trailer, and driver combination. And finally, object dynamic variables represent real-time information like real-time location, accidents, and forecasting task execution. The same is with the user context, which is separated into static and dynamic one and is used for transport planning, decision-making, and support cases and includes context data about resources and their statuses and new transport orders which have no resources assigned.

6.8 Conclusions

The author of the chapter focused on the freight delivery components and context-aware service engineering. The freight delivery process is quite complex and involves many variables (time, location, distance, mode, order, and freight delivery service) which are combined into one system.

In the last section of this chapter, the author proposes classification of the above-mentioned context according to the nature of its capture and the contextual information it presents. The categories are presenting different scope of data usage across the various user applications supporting transport operations. The author tried to present the way how from context data valuable information could be gathered in users' applications.

References

Becker, M., Wenning, B.-L., Görg, C., Gehrke, J. D., Lorenz, M., & Herzog, O. (2006). Agent-based and discrete event simulation of autonomous logistic processes. In *Proceedings of: Borutzky, W.; Orsoni, A.; Zobel, R. (eds.)* (pp. 566–571).

Bell, C. E., & Stidham, S., Jr. (1983). Individual versus social optimization in the allocation of customers to alternative servers. *Management Science, 29*, 831–839.

Berghman, L., Briand, C., Leus, R., & Lopez, P. (2012). The truck scheduling problem at cross-docking terminals. In *Proceedings of international conference on project management and scheduling. Louvain* (pp. 67–74).

Bohnsack, R., Pinkse, J., & Kolk, A. (2014). Business models for sustainable technologies: Exploring business model evolution in the case of electric vehicles. *Research Policy, 43*(2), 284–300. 10.1016/j.respol.2013.10.014.

Bongaerts, L. (1998). Integration of scheduling and control in holonic manufacturing systems. Ph. D. Thesis, KU Leuven.

Boon, J. (2002, May). *Effective benchmarking strategies*. Catalyst International.

Brownstone, D., Bunch, D. S., Golob, T. F., & Ren, W. (1996). A transactions choice model for forecasting demand for alternative-fuel vehicles. *Research in Transportation Economics, 4*, 87–129. https://doi.org/10.1016/S0739-8859(96)80007-2

Chęciński, S. (2014). Simulation analysis of traffic congestion in mineral mining transport. In P. Pawlewski & A. Greenwood (Eds.), *Process simulation and optimization in sustainable logistics and manufacturing* (pp. 99–109). Springer.

Chen, T. D., Kockelman, K. M., & Hanna, J. P. (2016). Operations of a shared, autonomous, electric vehicle fleet: Implications of vehicle and charging infrastructure decisions. *Transportation Research Part A: Policy and Practice, 94*, 243–254. 10.1016/j.tra.2016.08.020.

Cook, M. J. (1998). *Time management: Proven techniques for making the most of your valuable time*. Adams Media Corporation.

Crainic, T. G., & Montreuil, B. (2016). Physical internet enabled hyperconnected city logistics. *Transportation Research Procedia, 12*, 383–398.

Crooks, A. T., & Castle, C. J. (2012). The integration of agent-based modelling and geographical information for geospatial simulation. In *Agent-based models of geographical systems* (pp. 219–251). Springer.

Elektrodegalines. (2021). The list of electric charging stations. http://elektrodegalines.lt

European Commission. (2011). *White paper on transport: Roadmap to a single european transport area: Towards a competitive and resource-efficient transport system*. Publications Office of the European Union.

European Parliament. (2021). Briefing: Electric road vehicles in the European Union Trends, impacts and policies. Retrieved October 6, 2021, from http://www.europarl.europa.eu/RegData/etudes/BRIE/2019/637895/EPRS_BRI(2019)637895_EN.pdf

Eurostat. (2021a). Average distance on which goods are carried. Retrieved April 11, 2021, from https://ec.europa.eu/eurostat/statistics-explained/index.php/Road_freight_transport_by_journey_characteristics#Average_distance_travelled

Eurostat. (2021b). Database. Retrieved July 10, 2021, from https://ec.europa.eu/eurostat/data/database

Fahimnia, B., Sarkis, J., & Davarzani, H. (2015). Green supply chain management: A review and bibliometric analysis. *International Journal of Production Economics, 162*, 101–114.

Gambardella, L. M., Rizzoli, A. E., & Funk, P. (2002). Agent-based planning and simulation of combined rail/road transport. *Simulation, 78*(5), 293–303.

Harker, P. T. (1986). Alternative models for spatial competition. *Operations Research, 34*, 410–425.

Karageorgos, A., Mehandjiev, N., Weichhart, G., & Hämmerle, A. (2003). Agent-based optimisation of logistics and production planning. *Engineering Applications of Artificial Intelligence, 16*(4), 335–348.

Kley, F., Lerch, C., & Dallinger, D. (2011). New business models for electric cars—A holistic approach. *Energy Policy, 39*(6), 3392–3403. 10.1016/j.enpol.2011.03.036.

Knapen, L., Kochan, B., Bellemans, T., Janssens, D., & Wets, G. (2012). Activity-based modeling to predict spatial and temporal power demand of electric vehicles in Flanders, Belgium. *Transportation Research Record, 2287*(1), 146–154. 10.3141/2287-18.

Laurischkat, K., Viertelhausen, A., & Jandt, D. (2016). Business models for electric mobility. *Procedia CIRP, 47*, 483–488. 10.1016/j.procir.2016.03.042.

Lee, H., & Cohen, M. (1985). Multi-agent customer allocation in a stochastic service system. *Management Science, 31*, 752–763.

Li, R. J., & Lee, I. S. (1989). Analysis of fuzzy queues. *Computers and Mathematics with Applications, 27*, 1143–1147.

Li, B., Krushinsky, D., Reijers, H. A., & Van Woensel, T. (2014). The share-a-ride problem: People and parcels sharing taxis. *European Journal of Operational Research, 238*(1), 31–40. https://doi.org/10.1016/j.ejor.2014.03.003

Lippman, S. A., & Stidham, S. (1977). Individual vs. social optimization in exponential congestion system. *Operations Research, 25*, 233–247.

Lowery, D. (2000). A transactions costs model of metropolitan governance: Allocation versus redistribution in urban America. *Journal of Public Administration Research and Theory, 10*(1), 49–78. 10.1093/oxfordjournals.jpart.a024266.

Madina, C., Zamora, I., & Zabala, E. (2016). Methodology for assessing electric vehicle charging infrastructure business models. *Energy Policy, 89*, 284–293. 10.1016/j.enpol.2015.12.007.

Masuch, N., Lützenberger, M., Ahrndt, S., Heßler, A., & Albayrak, S. (2011). A context-aware mobile accessible electric vehicle management system. In *2011 federated conference on computer science and information systems (FedCSIS)* (pp. 305–312). IEEE.

Mendelson, H. (1985). Pricing computer services: Queueing effects. *Communications of the ACM, 28*(3), 312–321.

Mendelson, H., & Whang, S. (1990). Optimal incentive-compatible priority pricing for the M/M/1 queue. *Operations Research, 38*, 870–883.

Naor, P. (1969). The regulation of queue size by levying tolls. *Econometrica, 37*, 15–24.

OECD. (2020). Goods transport database. Retrieved August 20, 2020, from https://stats.oecd.org/Index.aspx?DataSetCode=ITF_GOODS_TRANSPORT

Pach, C., Sallez, Y., Berger, T., Bonte, T., Trentesaux, D., & Montreuil, B. (2014). Routing management in physical internet crossdocking hubs: Study of grouping strategies for truck loading advances in production management systems. In *Innovative and knowledge-based production management in a global-local world* (pp. 483–490). Springer.

Patel, A., & Hudson-Smith, A. (2012). Agent tools, techniques and methods for macro and microscopic simulation. In *Agent-based models of geographical systems* (pp. 379–407). Springer.

Savelsbergh, M., & Van Woensel, T. (2016). 50th anniversary invited Article'city logistics: Challenges and opportunities. *Transportation Science, 50*(2), 579–590. https://doi.org/10.1287/trsc.2016.0675

Shah, B., & Khanzode, V. (2015). A comprehensive review and proposed framework to design lean storage and handling systems. *International Journal of Advanced Operations Management, 7*(4), 274–299.

Stalk, G., Jr. (1988). Time—The next source of competitive advantage. *Harvard Business Review, July–August*, 41–51.

Stalk, G., Jr., & Hout, T. M. (1990). *Competing against time*. The Free Press.

Statista Research Department. (2020). Total length of the road network in Lithuania in 2019, by road type. Retreived November 30, 2020. https://www.statista.com/statistics/ 449922/lithuania-length-of-roadnetwork-by-road-type

Von Meier, A. (2018). EV infrastructure planning and grid impact assessment: A case for Mexico. Retrieved October 6, 2019, from http://www2.eecs.berkeley.edu/Pubs/TechRpts/2018/EECS-2018-70.html

Yechiali, U. (1970). On optimal balking rules and toll charges in the GI/M/1 queue process. *Operations Research, 19*, 348–370.

Yechiali, U. (1972). Customers' optimal joining rules for the GI/M/1 queue. *Management Science, 18*, 434–443.

Chapter 7
Automated Freight Brokerage Based on Multi-Agent Semantic Web Service Composition

Arūnas Miliauskas

Abstract Semantic web technologies and multi-agent system are expected to affect brokers' role in freight transportation. Services provided by freight brokers can be seen as automated service composition problems. This problem is popular, and many research results published. Major elements of this problem are described and presented. The service composition in the freight domain resulted in multiple papers with a strong focus on optimization. By focusing on interoperability the ontology-based approach is taken and based on this baseline extensible solution architecture is presented.

Keywords Web service composition · Ontology · Freight forwarding · Freight brokerage · Multi-agent system · Software architecture

Semantic web technologies and the multi-agent system are expected to affect the role of a broker in freight transport. Services provided by a freight broker can be seen as a problem of automated service composition. This problem is popular and was distinguished in some published study results. Major elements of such issue have been described and presented. Many articles have been written on service composition in the freight domain with a strong focus on optimization. By focusing on interoperability, the ontology-based approach is followed, offering an extendable solution architecture on this basis.

A freight broker coordinates transportation on behalf of consignees, consignors, and freight forwarders. Usually, all transport-related activities are outsourced. His revenue is based on commissions. The key skill of freight brokers is the ability to efficiently match freight to the available means of transport based on customer requirements and other constraints. To some extent, such a match can be found on full-time equivalents (FTE) platforms, such as TimoCom or Trans EU.

FTEs market is widespread in Europe, with multiple platforms and strong competition. However, Main Europe freight exchange platforms serve as information

A. Miliauskas (✉)
Vilnius Gediminas Technical University, Vilnius, Lithuania

D. Dzemydienė et al., *Development of Smart Context-Aware Services for Cargo Transportation*, International Series in Operations Research & Management Science 330, https://doi.org/10.1007/978-3-031-07199-7_7

boards on available freight for road transport. The assessment of capacity maturity shows that FTE is less mature than the energy market (Jain et al., 2019), and multimodal transportation is even less mature. Currently, communication in FTE markets is managed manually for the most part.

New technologies and trends can disrupt the traditional freight brokerage model. Artificial agents, semantic web service-based integration can increase automation to a level where decisions can be made without (or with a minimal) human intervention. Humans will be needed to supervise transactions and to create models for artificial agent solutions only. Additionally, communication technologies, together with new user engagement technologies, such as chatbots/voice assistants, can keep users connected and reduce the time needed for human response.

The following technologies and architectures that are at the core of these transitions will be reviewed:

- Web services and web service composition.
- Semantic web.
- Agents and multi-agent systems.

Finally, a conceptual solution architecture that lists the main components of automated freight brokerage will be offered.

7.1 Web Services

Business-to-business integration allows companies to optimize costs (Perego et al., 2011). Information technology offers many different ways for the implementation of this integration. Currently, web services have become the principal approach to integrating systems. Web services are defined as: a service that is software accessed via a published API and which is a part of a service contract (Erl, 2005).

Web service is a service that provides functionality using such web protocols as HTTP. Common web service methods include:

- SOAP (simple object access protocol)—XML based message exchange format. It is provided in the description language for web service (WSDL) (Chinnici et al., 2007), where information received and provided is described in the contract.
- RESTfull services, where services are designed with the orientation to the representational state transfer (REST) software architecture style. The common REST service operations method is suitable for HTTP methods, such as GET, POST, PUT, and DELETE. There are several machine readable RESTFul service descriptions, such as Web Application Description Language (WADL) or OpenAPI specification.
- Microservices. This architectural style solution has been broken down into multiple services, each of which covers a relatively small set of functionalities. The Micro-service approach does not constrain communications protocols and

formats. Usually, the focus is on simple methods, focusing on intra-solution reusability rather than the provision of services to external parties.

7.2 Semantic Web Services

The integration approach based on traditional web services (SOAP, RESTful) mostly focuses on data syntax (rather than semantics). For example, WSDL data are exchanged following the XML scheme. This supports automation developments tasks, such as the generation of stub classes, but people are the ones who make interpretations (of what each element and attribute means). A contract is usually the scope of information, and elements with the same names can have different meanings in different WSDL documents.

Semantics web (Shadbolt et al., 2006) brings the vision of moving from web of documents to web of data. This requires an ability to associate the same type of information in various sources and comes with a stack of technologies and protocols. Semantic web services are based on the same set of technologies for extending the traditional web service-based approach.

A resource description framework (RDF) (Resource Description Framework (RDF): Concepts and Abstract Syntax, 2014) is a basic framework for presenting information in the semantic web, where the core is the data model based on RDF-graph structure, which has a triple <subject, predicate, object> structure. An RDF has many serialization formats for storing and exchanging data, such as Turtle, N-Triples, N-Quads, JSON-LD, N3, and RDF/XML.

Ontology web language (OWL) adds semantic interpretations to RDF graphs. Using ontology applications, additional information (RDF triples) can be obtained from existing explicitly stated triples. Moreover, ontology can serve as the conceptualization of the selected domain.

OWL-S (Martin et al., 2004) provides ontology to describe semantic web services. It defines the upper service ontology. WSMO (Holger et al., 2005) was created to describe web services semantically (Fig. 7.1).

7.3 Web Service Composition

When a business provides a web service interface which could be applied to arrange the provision of business services, the entire organization may be seen as a problem of web service composition.

Web service composition means creating (manually or automatically) a composite service that would be possible to deliver value added by using the existing web services. Several dimensions could be used to categorize all composition approaches (Lemos et al., 2015), with centralization vs distribution perspective being one of the main categories:

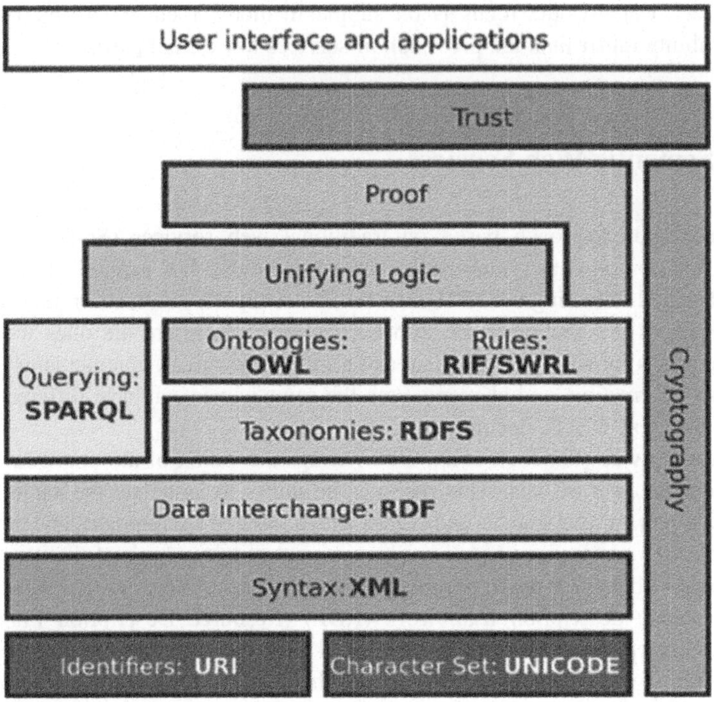

Fig. 7.1 The displaying of layers of semantic web stack

- Orchestration—a software application that coordinates the interaction of web services to achieve the required composite functionality (Peltz, 2003). Web services in the Business process execution language (WS-BPEL or just BPEL) (Holger et al., 2005) is a well-known language for orchestration implementation. This is a centralized approach.
- Choreography—is used when there is no single coordinator (Peltz, 2003). This is a distributed approach, where participants adhere to a common communication protocol. Instead of a single executable process, participants implement fragments individually. A multi-agent system, when there is no single coordinating agent, and composite behaviour is achieved solely by the interaction of autonomous agents, is an example of choreography.

There might be different ways to build an orchestration:

- Manual, where a software developer created orchestration based on the provided requirements;
- Automated, where a person having software development skills can create, modify, or at least configure orchestration.
  - Workflow-based—where that person takes responsibility to create a composition workflow. In workflow-based methods, the composition is considered a

workflow. Therefore, web services are composed by creating control and data flow among them. Workflow-based approaches can be divided into two main groups: static and dynamic workflow generation. The difference between them is the level of user involvement. Using a static method, the user creates an abstract process model, and the selection and linking of the performance of web services are automatized. By the way, with such an approach, the process design model and the selection of service are automatically performed.
  - Template-based—when there are predefined templates that simplify the composition. User empowerment is defined by template extensibility.
  - Automated—when the composition is synthesized based on high-level goals.

The problem of service composition was broken down into two subproblems:

- Service binding. Choosing services of best candidates from several functionally equivalent services based on the selected criteria (such as price). All functionally equivalent services are considered instances of the same abstract service. 31 illustrates them as elements within alternatives. These services can be abstract and comprise other services.
- Arranging process. This means identifying the order in which services must call. A simple scenario involves steps to call out instance services, and a scenario—abstract services.

7.4 Development of Multi-Agent Systems

A multi-agent system (MAS) is an AI system consisting of two and more agents performing under the same conditions. The agent is seldom in an environment alone, and MAS is the most common situation of management of agents. Usually, agents in MAS interact with the environment and with each other. This inter-agent communication gives new levels of complexity that form by combining several or more autonomous entities.

7.4.1 Communication Between Agents

Communication is the central aspect in MAS. This interaction is based on a certain agent communication language. There are many standardization proposals for agent communication language. Most of them share the idea that communication involves agents performing speech acts. Boella et al. (2006) has distinguished two traditions in agent communication, where semantics is based on:

- *The mental attitudes* which are *beliefs* and *intentions*. These so-called mentalist approach example languages are the FIPA standard (Foundation for Intelligent Physical Agents, 2002) for agent communication and the Knowledge Query and

Manipulation Language (KQML) (Finin et al., 1994). These *mental attitudes* are a part of preconditions and effects of speech acts such as *informing* or *requesting*.
- *The social commitments*. Communication acts are considered to be social commitments. These approaches were drawn from the limitation of mental attitude in the non-cooperative situation, where individual mental states cannot be viewed and verified. So instead of focusing on the mental state of agents, a shift is made to the public commitments of an agent.

Here a greater focus is placed on mental approaches.

KQML was developed by AI researchers to improve interoperability between virtual knowledge bases (KB). A message language has syntax similar to LISP. Each message has two main components:

- The *performative*, which indicates speaker intention. There are many performatives in KQML and each of them has a specific meaning. For example, *ask-one* shows that *sender wants one recipient to answer the question on the content parameter.*
- *Set of parameters,* which are key/value pairs. Content is one of the parameters, which is a place for a message content (beliefs and desired states). Other parameters are used to provide additional information about a message, such as a recipient, the sender, the language, or the ontology.

The KQML has no restrictions on content provision format, and it may have many languages. A knowledge interchange format (KIF) (Genesereth et al., 1992) can be one from them (Fig. 7.2).

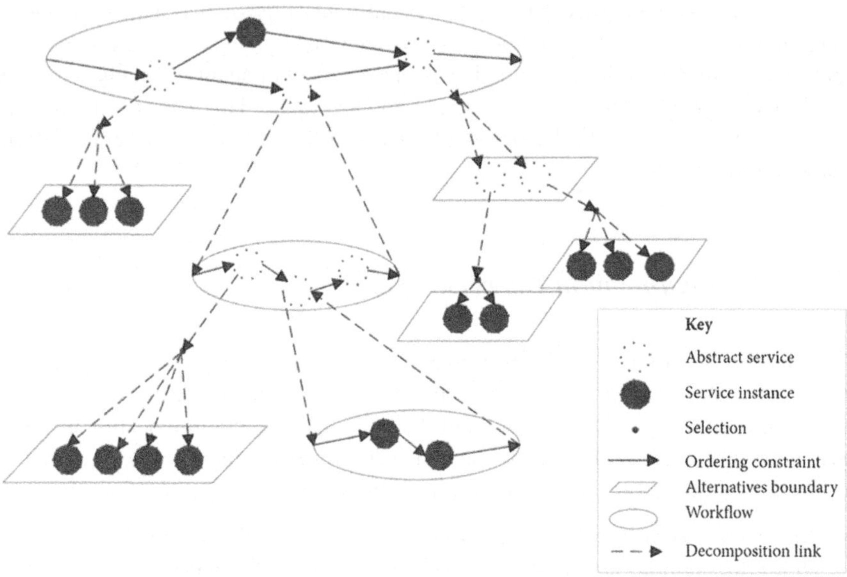

Fig. 7.2 Description of work flow possibilities for solving of the service composition problem

FIPA Agent Communication Language (ACL) illustrates standardization efforts by FIPA, which is a non-profit association for promoting agent-based applications and services. FIPA ACL has many similarities with KQML. It uses the most common structure of the message. Nevertheless, a set of performatives differs. FIPA ACL defines 20 performatives, such as to inform, confirm, query-if. But parameters are very similar to those of KQML, including the receiver, the sender, the ontology, the language, or the content. Like KQML, it allows using of different content languages. Communication can be extended so support semantic web services (Canito et al., 2019; Gan et al., 2018).

7.4.2 Coordination of Services between Artificial Agents

Being able to communicate with each other modelled artificial agents and these activities lead to question: how to ensure coherent communication between agents, i.e. how to coordinate actions and communication between them. FIPA ACL comes with predefined interaction protocols, for particular cases, such as FIPA-request protocol or FIPA-contract-net protocol. It is the last version of the contract net protocol. The basic idea is to choose, using the contracting mechanism, agents which will be able to complete the requested task.

Another approach is to view the coordination problem as multi-agent planning. To have consistent actions and interactions, a plan for an agent can be created. The plan will define when and what actions need to be taken for agents to achieve their goals. Multi-agent planning can be either:

- *centralized*, where the central agent coordinates planning, manages inconsistencies and conflicts;
- *distributed*, when agents have models of goals of each other and resolve inconsistencies and conflicts by communicating with each other.

7.4.3 Implementation of Types of Ontology

In literature there are provided multiple definitions for description of multi-layered ontology. In this section, it has been approached from the computational perspective, using the definition by Gruber (1993) referring to it as "explicit specification of a conceptualization".

The definition can further be narrowed to OWL 2 (OWL 2 Web Ontology Language, 2012), where ontology is formalized vocabularies of terms that cover a specific area and are shared by users/agents. The main focus is on the support interoperability when the exchange of information is based on it. The system based on ontologies can ensure that participant agents have the same meaning of data.

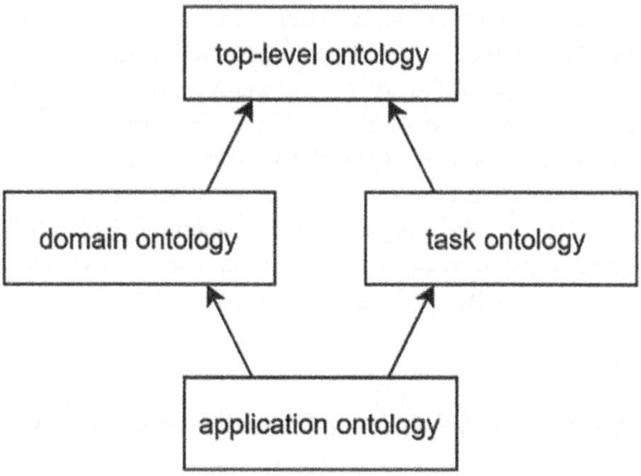

Fig. 7.3 Types of ontologies based on generality and domain

Different (Guarino, 1998) ontologies can be distinguished based on their similarities, purpose, and application field:

- Top-level ontologies, which define the top level, most general concepts (problems or independent domains) like matter, object, time, space, etc.
- Ontologies of domains—ontologies of some general areas.
- Ontologies of tasks describe vocabulary of some general task or activity.
- Ontologies of applications define concepts that belong to a particular domain or task. Often, these concepts depict roles played by domain entities while performing a specific task.

These types of ontologies are presented in Fig. 7.3, where arrows represent the specialization relationship.

This approach, when multiple specialized ontologies are used, should minimize ontology integration problems.

7.5 Service Composition in Freight Transport Management

The problem faced by freight brokers in matching freight to transport, which is the automation of corresponding processes, can be seen as a service composition problem, when a broker selects the most suitable set of services from the available transport service providers and provides a composite service to a customer, additionally providing for additional information services providers. When planning transportation, multiple service providers can be optimized. Then the services provider can assess the optimized transportation plan and choose the best candidate. Figure 7.4 illustrates this process. The diagram was created using the ArchiMate language (ArchiMate 3.1 Specification, 2019). To support interoperability, all interfaces presented in the diagram can be implemented as semantic web services.

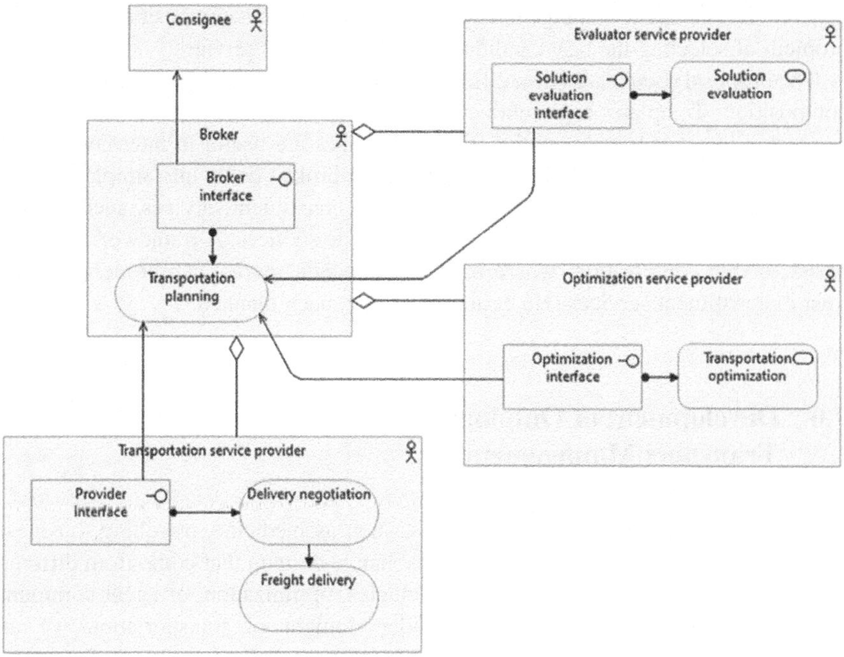

Fig. 7.4 Business service model in ArchiMate

This approach is similar to the blackboard architecture, where a request is placed on the blackboard and multiple knowledge sources are updated, enriching information. In this scenario, all actors can be designed as agents, leading to multi-agent system architecture. Then, the agent-based approach enables qualities of autonomy, social skills, reactivity, and proactiveness.

This conceptual model comes in hand with other studies. There were attempts to develop a system for cargo brokering services that are web based (Ji et al., 2004). Other authors see agent-based systems and multi-agent systems for building composite freight brokerage services (Luncean et al., 2014).

Leon and Bădică (2017) provide an example where matching customers and carriers have been centralized into a single service. Matching is treated as an optimization problem that is attempted in two ways in parallel: using an exact solver (for smaller problems) and the heuristic technique with an evolutionary algorithm to find an approximation. Our vision is similar, however, we do not confine ourselves to a particular algorithm. We see a platform as the environment where multiple knowledge services might compete. Then software providers could develop additional services, where the best solution will be used and they will earn commissions. Another example of the application of evolutionary algorithms to optimize composition is Artificial Bee Colony (ABC) algorithm (Lartigau et al., 2015) for the cloud generation. The use of composition optimization means that composition is seen as a problem of selecting the best candidates among identical services.

The proposal (Leukel & Kirn, 2011) uses the template-based approach for service composition. Templates represent actually the transportation network model: hubs connected like a hub and a spoke. These templates are useful in intermodal transportation since the transportation network can be broken down into simple patterns.

Another important moment is dimensions of constituent services, such as time, price, quality, and trust aggregation into composite services. A framework for trust-based service composition, where trust in composite service can be derived from trust in constituent services (Hu et al., 2020), has been planned.

7.6 Development of Ontology for System of Freight Transport Management

Ontologies are created for many domains, such as medicine, travelling, business. There are several concepts in the freight exchange domain that come from different domains (such as sales, road networks, vehicles, optimization, or agent communication) and for the description of wider domain of transportation we are recommending to integrate some of them. Luncean and Badica (2014) define four ontologies: transport request, transport resource, freight, and message ontology. Hendi et al. (2014) present 3 types of ontologies: logistics ontology, optimization ontology, and software suite ontology.

The freight transport planning domain contains several participants. Therefore, it is very important (as in real life) that communication is based on a common knowledge model. This model should be created having analysed the domain and capturing the most important concepts and their relations.

However, current ontologies in freight transport differ in their scope, level of detail, and generality (Katsumi & Fox, 2018). Ontologies have also been created for different applications, which makes the combination of these ontologies problematic, as the same concepts have different semantics. This is an example of an ontology integration problem (mentioned in the Chapter on Ontologies).

Ontologies are applied for the systems that are multi-agent such as the cooperative platforms for communicating and the exchange of messages (Anand et al., 2012, 2014). However, when ontologies are not based on any accepted top-level ontology, reuse or integration may be impossible, especially if they focused on specific concepts.

There are ways to develop the ontology based on the existing top-level ontology, such as the (Ameri & Kulvatunyou, 2019) supply chain reference ontology.

To avoid the ontology integration problem, the ontology for automated freight brokerage should be based on a set of specialized ontologies, when all the specialized ontologies should be based on a single top-level ontology. These ontologies are presented in Fig. 7.5.

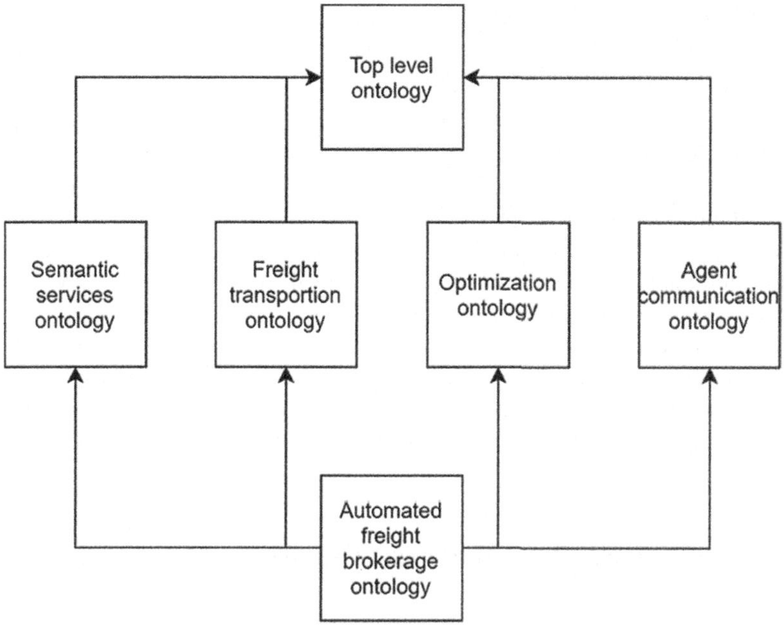

Fig. 7.5 Ontology integration for automated freight brokerage

Basic formal ontology (BFO) (Barry, 2015) is a potential candidate, as it is small, has 35 classes, and is easy to learn and use. Other alternatives to be considered include the Unified Foundational Ontology (UFO) (Guizzardi, 2005) or the gist (Gist, 2021).

The optimization ontology is based on Hendi et al. (2014) and contains the following basic concepts:

- Optimization component. These are the Objective, the Parameter, the Constraint, and the Data. They are used to describe the optimization problem;
- Optimization problem which covers the following subclasses: stochastic optimization problem, deterministic optimization problem, multi-objective optimization problem, and discrete optimization problem;
- Optimization method defines a set of optimization methods. Some examples of the methods are linear programming method, non-liner programming method, network programming method, an approximate method.

The already-mentioned OWL-S and WSMO are semantic service ontologies. However, these ontologies do not use concepts from the upper ontology, requiring an additional integration.

The freight transportation ontology should cover:

- transport networks,
- transportation methods,
- modes of transport,
- transportation resources, for example, vehicles,

- logistics objects, such as freight,
- transportation services,
- carriage contracts and parties thereto.

Luncean and Badica (2014) can be the basis for the freight transportation ontology. GenCLOn (Anand et al., 2012) also defines an ontology for freight transport, but it is confined to cities. GenCLOn has been developed from the public administrator's perspective to balance private and public benefits. Osmonto's (Codescu et al., 2011) OpenStreetMap tag ontology can be useful. However, when writing the paper, no freight ontology was found that would be developed as a specialization of some top-level ontology.

Since composition can be seen as the interaction of agents, the agent communication ontology supports interoperability. FIPA-ALC-based ontology is presented in Gan et al. (2018). This is an attempt to apply the semantic web for MAS to increase interoperability. SW is used as the encoding for communication.

7.7 Proposed Solution of Service Provision System Architecture

In Fig. 7.6, the proposed architecture shows the main components and their interfaces, which are required for installing the multi-agent system (MAS) for automated brokerages.

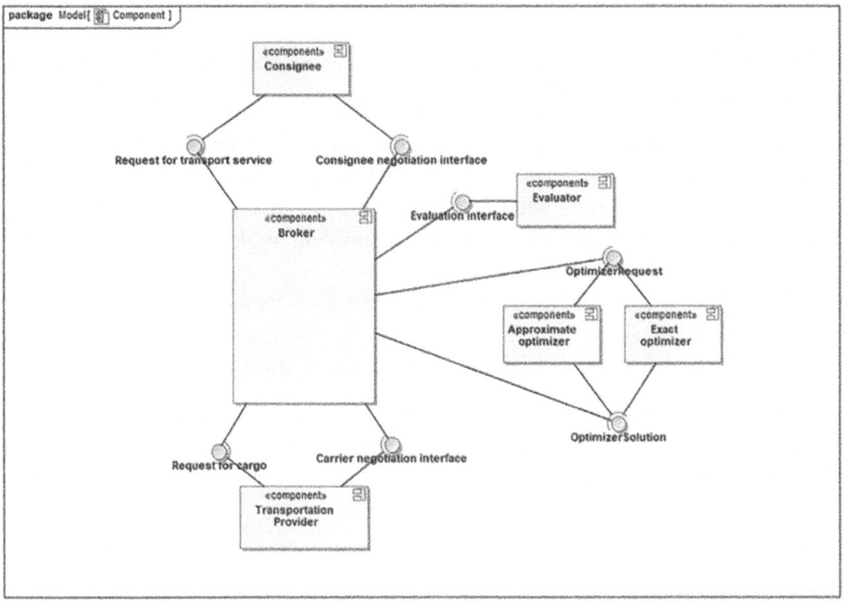

Fig. 7.6 Illustration of solution of subsystem with interfaces and main components of modules for MAS

The proposed architecture is based on semantic web services which simplifies the reuse process, because the accessing services that have already been deployed are usually easier than using those functionalities via libraries or components. This also enables specialization, when certain capabilities can be shared as services. All components in the diagram have been implemented as agents.

A broker-agent has exposed interfaces that allow other agents to make requests: a Consignee—for transportation services, a provider of transportation services—for freight, and an Optimizer—for optimization. Other interfaces have been used for interaction in transportation planning and optimization. The Evaluator agent used the Evaluation interface to participate in planning by evaluating solutions delivered by optimizers.

This approach allows adding of multiple agents, except the Broker and the Evaluator. This structure allows extending the distributed freight planning and optimization.

7.8 Conclusions

The implementation of an automated freight broker can be viewed as an automated service composition problem. Since the problem is well known in academia, the result can be applied in the freight brokerage context. The main approaches to web service composition are summarized and main requirements are identified. Semantic web services with multi-agent systems are core elements for web service compositions. The communication of agents in a semantic web service environment should be based on common ontology. The approach is taken based on ontology integration, rather than designing specific stand-alone ontology. This allows reuse ontologies from multiple domains when all ontologies are based on simple top-level ontology.

The conceptual architecture is proposed for this type of system. It is based on multi-agent systems, where multiple agents collaborate in the search for a solution. Agent communication is based on presented shared ontology. The structure is based on having multiple Consignee, Transportation Provider, and Optimizer agents. Only Broker and Evaluator agents are singleton. The approach with multiple Optimizer agents allows extensions for distributed freight planning and optimizations.

References

Ameri, F., & Kulvatunyou, B. (2019, August). Modeling a supply chain reference ontology based on a top-level ontology. In *ASME 2019 international design engineering technical conferences and computers and information in engineering conference*. https://doi.org/10.1115/DETC2019-98278

Anand, N., Yang, M., van Duin, J. H. R., & Tavasszy, L. (2012). GenCLOn: An ontology for city logistics. *Expert Systems with Applications, 39*(15), 11944–11960. https://doi.org/10.1016/j.eswa.2012.03.068

Anand, N., van Duin, R., & Tavasszy, L. (2014). Ontology-based multi-agent system for urban freight transportation. *International Journal of Urban Sciences, 18*(2), 133–153. https://doi.org/10.1080/12265934.2014.920696

ArchiMate 3.1 Specification. (2019). The Open Group. https://publications.opengroup.org/archimate-library/archimate-standards/c197

Barry, S. (2015). Basic formal ontology 2.0.

Boella, G., Damianoa, R., Hulstijn, J., & van der Torre, L. (2006). A common ontology of agent communication languages: Modeling mental attitudes and social commitments. *Applied Ontology, 3*, 1–3.

Canito, A., Santos, G., Corchado, J. M., Marreiros, G., & Vale, Z. (2019). Semantic web services for multi-agent systems interoperability. In P. Moura Oliveira, P. Novais, & L. P. Reis (Eds.), *Progress in artificial intelligence* (pp. 606–616). Springer.

Chinnici, R., Moreau, J. J., Ryman, A., & Weerawarana, S. (Eds.). (2007). Web Services Description Language (WSDL) Version 2.0 Part 1: Core language. W3C. https://www.w3.org/TR/2007/REC-wsdl20-20070626/

Codescu, M., Horsinka, G., Kutz, O., Mossakowski, T., & Rau, R. (2011). Osmonto-an ontology of OpenStreetMap tags. *State of the Map Europe (SOTM-EU), 2011*, 23–24.

Erl, T. (2005). Service-oriented architecture: Concepts, technology, and design. Prentice Hall PTR.

Finin, T., Fritzson, R., McKay, D., & McEntire, R. (1994). KQML as an agent communication language. In *Proceedings of the third international conference on information and knowledge management* (pp. 456–463).

Foundation for Intelligent Physical Agents. (2002). FIPA communicative act library specification. http://www.fipa.org/specs/fipa00037/SC00037J.html

Gan, K. S., Chin, K. O., Anthony, P., & Hamdan, A. R. (2018). A FIPA-ACL ontology in enhancing interoperability multi-agent communication. In R. Alfred, H. Iida, A. A. A. Ibrahim, & Y. Lim (Eds.), Computational science and technology (pp. 151–160). Springer.

Genesereth, M. R., Fikes, R. E., et al. (1992). Knowledge interchange format-version 3.0: Reference manual.

Gist. (2021). Semantic arts. https://www.semanticarts.com/gist/

Gruber, T. R. (1993). A translation approach to portable ontology specifications. *Knowledge Acquisition, 5*(2), 199–220. https://doi.org/10.1006/knac.1993.1008

Guarino, N. (1998). Formal ontology in information systems. In *Proceedings of the first international conference (FOIS'98), June 6–8, Trento, Italy* (Vol. 46). IOS Press.

Guizzardi, G. (2005). Ontological foundations for structural conceptual models. PhD Thesis–Research UT, graduation UT. Telematica Instituut/CTIT.

Hendi, H., Ahmad, A., Bouneffa, M., & Fonlupt, C. (2014). Logistics optimization using ontologies. In *ICCSA 2014-international conference on complex systems and applications.*

Holger, L., Axel, P., & Dumitru, R. (Eds.). (2005). *Web service modeling ontology (WSMO).* W3C. https://www.w3.org/Submission/WSMO/

Hu, C., Wu, X., & Li, B. (2020). A framework for trustworthy web service composition and optimization. *IEEE Access, 8*, 73508–73522. https://doi.org/10.1109/ACCESS.2020.2984648

Jain, A., Bruckmann, D., van der Heijden, R. E. C. M., & Marchau, V. A. W. J. (2019). Towards rail-related multimodal freight exchange platforms: Exploring regulatory topics at EU level. *Competition and Regulation in Network Industries, 20*(2), 138–163. https://doi.org/10.1177/1783591719847543

Ji, C., Li, M., & Li, L. (2004). Freight transportation system based on web service. In *IEEE international conference onservices computing, 2004. (SCC 2004). Proceedings. 2004* (pp. 567–570). https://doi.org/10.1109/SCC.2004.1358065

Katsumi, M., & Fox, M. (2018). Ontologies for transportation research: A survey. *Transportation Research Part C: Emerging Technologies, 89*, 53–82. https://doi.org/10.1016/j.trc.2018.01.023

Lartigau, J., Xu, X., Nie, L., & Zhan, D. (2015). Cloud manufacturing service composition based on QoS with geo-perspective transportation using an improved Artificial Bee Colony optimisation algorithm. *International Journal of Production Research, 53*(14), 4380–4404. https://doi.org/10.1080/00207543.2015.1005765

Lemos, A. L., Daniel, F., & Benatallah, B. (2015). Web service composition: A survey of techniques and tools. *ACM Computing Surveys, 48*(3). https://doi.org/10.1145/2831270

Leon, F., & Bădică, C. (2017). An optimization web service for a freight brokering system. *Service Science, 9*(4), 324–337. https://doi.org/10.1287/serv.2017.0191

Leukel, J., & Kirn, S. (2011). A service-oriented approach to freight routing in intermodal transport systems. In *10th international conference on Wirtschaftsinformatik A. Bernstein and G. Schwabe* (pp. 292–301) https://www.zora.uzh.ch/id/eprint/60505/1/20120227114120_merlin-id_2353.pdf#page=314

Luncean, L., & Badica, C. (2014). Semantic modeling of information for freight transportation broker. In *2014 16th international symposium on symbolic and numeric algorithms for scientific computing* (pp. 527–534). https://doi.org/10.1109/SYNASC.2014.76

Luncean, L., Bădică, C., & Bădică, A. (2014). Agent-based system for brokering of logistics services – Initial report. In N. T. Nguyen, B. Attachoo, B. Trawiński, & K. Somboonviwat (Eds.), *Intelligent information and database systems* (pp. 485–494). Springer.

Martin, D., Burstein, M., Hobbs, J., Lassila, O., McDermott, D., McIlraith, S., Narayanan, S., Paolucci, M., Parsia, B., Payne, T., Sirin, E., Srinivasan, N., & Sycara, K. (2004). *OWL-S: Semantic markup for web services*. W3C. https://www.w3.org/Submission/OWL-S/

OWL 2 Web Ontology Language. (2012). W3C. https://www.w3.org/TR/owl2-overview/

Peltz, C. (2003). Web services orchestration and choreography. *Computer, 36*(10), 46–52. https://doi.org/10.1109/MC.2003.1236471

Perego, A., Perotti, S., & Mangiaracina, R. (2011). ICT for logistics and freight transportation: A literature review and research agenda. *International Journal of Physical Distribution and Logistics Management, 41*(5), 457–483. https://doi.org/10.1108/09600031111138826

Resource Description Framework (RDF): Concepts and Abstract Syntax. (2014). https://www.w3.org/TR/2004/REC-rdf-concepts-20040210

Shadbolt, N., Berners-Lee, T., & Hall, W. (2006). The semantic web revisited. *IEEE Intelligent Systems, 21*(3), 96–101. https://doi.org/10.1109/MIS.2006.62

Chapter 8
Transport Network Design Methods and Context-Aware Service Specifics

Aurelija Burinskiene

Abstract This chapter presents network design models. Then it describes different network types, finally reviewing variables important for the functioning of the network. Context-aware service systems replicate the functioning of supply networks in four aspects: network structure, traffic in the network, management of connections, and service architecture. Below the description of supply network is presented. The supply network "is an organizational form". The network structure is examined from the point of view of horizontal and vertical integrations, centralization, and geographic coverage. Supply network flows are described by the level of integration. The management of supply flows involves coordination and information transfer related to the delivery of freights. Traffic is managed successfully when actions linked with moving products, and data about order processing for delivery or return have been coordinated and integrated. The nature of companies' relationships and the degree to which companies affect each other are important. Moreover, exchange management has been included in the network structure that encompasses a broader set of relationships with other supply network members. Voss and Hsuan highlight the architecture of exchange service as "the system broken down into individual elements". Every service element/module refers to a particular function.

Keywords Network design · Network types · Service-oriented transportation · Shared transportation services · The hierarchy of qualitative methods

A. Burinskiene (✉)
Vilnius Gediminas Technical University, Vilnius, Lithuania
e-mail: aurelija.burinskiene@vilniustech.lt

D. Dzemydienė et al., *Development of Smart Context-Aware Services for Cargo Transportation*, International Series in Operations Research & Management Science 330, https://doi.org/10.1007/978-3-031-07199-7_8

8.1 Transport Network Design Studies

Transportation networks have access to raw materials and products, and the possibility to deliver them by different transportation modes, namely, by land, rail, or water. Entities use networks to cross physical distances to be able to engage in freight transport activities. The authors have modelled physical transportation networks, the infrastructure of public capital, and the market of transportation services.

The first transportation network was introduced by Euler (1736). Later on, Garrison and Marble (1962) described the Ireland railway network. Since 1960, transportation networks have received more attention. Transportation networks could be described as two types of systems: first, a transportation network is a physical system; and, second, a transportation network is a supply system.

A physical system consists of infrastructure (roads) and system components (vehicles) that carry products and create flows. A supply system requires understanding network users and service providers. As the network expands, network users gain extra benefits from accessibility to new places.

The term "network" is applied to identify the structure, which is a physical one and has the conceptual design. The network includes the elements—sets of places and segments which connect all locations. Graph theory is used to describe the networks. Graphs are used for modelling and presenting the systems of transport networks. Although the content for these schemes differs, they are also common properties and reflect certain elements that are interrelated. So, a transport network graph can involve production centres and road and rail links directly connected to those centres. In the road network, each arc can be assigned two parameters, such as the price and time. The fastest way can be too expensive. Nevertheless, the decision maker must select one solution, which may not be the best one in terms of both criteria.

In most studies, a network with multi-tiers could be represented mathematically. Programmes have been created to analyse equilibriums in networks. Frank–Wolfe method (Takayama & Judge, 1964) is the method used to determine equilibrium flows for transportation networks. Programmes become more important for large size networks. In addition, the authors are searching for ideas allowing to review the limitations of network capacity (a solution was first proposed by Wigan, 1971) and for the shortest and longest routes in the network.

The studies focusing on networks fall within several directions: structure analysis, geographical features, and optimization studies. The studies oriented to network design are searching for an optimal design for a period. The review of network typology was initiated after 1960 and is not able yet to provide mechanisms for network development and growth. Optimization research has been evident since the 1970s. The models focus on minimum setup costs, network maintenance, and transportation changes. Models used to solve the problem of minimum costs are designed to find the cheapest way (minimum cost) to deliver a certain volume of freight through a certain road network. Optimization solutions use actual and projected delivery amounts. The results of the design of networks have various use

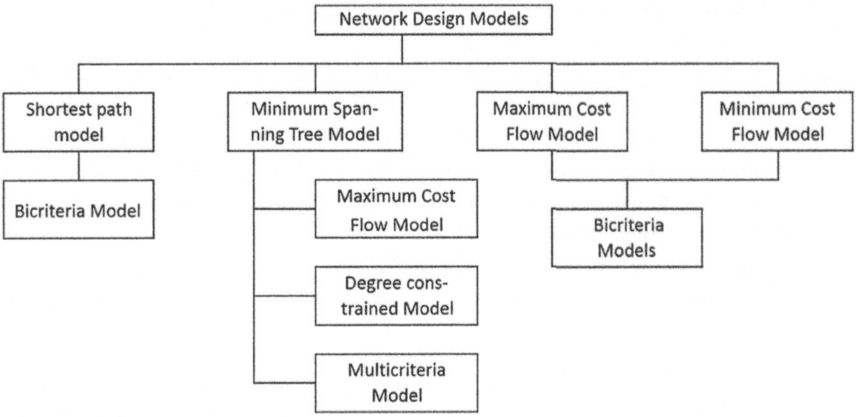

Fig. 8.1 Application of models for transport network design

cases in practice, such as a search for the shortest route, a travelling salesman and a Chinese postman, the Konigsberg bridge, one-side streets, and resource distribution, etc.

The shortest route model is associated with the arc length. The article by Cooke and Halsey (1966) is the first document that talks about the fastest road problem. Bellman's principle of optimality (Bellman, 1958) has been extended through the fastest routes from each network hub to a single destination hub, i.e. combining all the fastest roads into one. Dijkstra's Algorithm has been used to find the shortest route having the lowest price between two pinnacles. Dreyfus (1969) proposed to change Dijkstra's static shortest route algorithm by an algorithm of the same complexity as the Dijkstra algorithm to calculate the fastest routes between two hubs for a certain departure time. Shortest route problems are widespread and cover a variety of practical problem identification methods, both as individual models and as additional problems in identifying more complex problems. The shortest route problem is identified as the most common problem in studies of combinatorial operations focusing on the problem of a road of maximum capacity, the fastest route problem, the most reliable route problem, the minimum cost-time ratio problem, and the reliability ratio problems. Accurate algorithms are effective for solving the shortest single-purpose route.

There are various documents, which are related to time dependence but have one target function (Fig. 8.1). Here, the time-dependent shortest route can be broken down to the fastest route and the lowest-cost route. The most likely road problem is the cost of an arc and the travel time in that arc. The aim is then to find the minimum route length, taking into account that the journey is time dependent. To solve the minimal costs issue, the authors have searched for roads of minimal length, taking into account certain costs and average speed to travel from one hub to another.

The two-criteria shortest route model differs from common shortest route problems in two aspects: first, two contradictory and incompatible objective functions change the one function of a target, and secondly, data may change over time, when there is a time-dependent application.

The model of Minimum Spanning Tree can apply the Kruskal's algorithm to search for all the connecting vertices with the lowest total costs.

Capacitated model is used when the volume is indicated and is stable, and the incapacitated approach is used in case of random demand. The model without capacity restrictions is more convenient for meeting future business development needs and responding to fluctuations in demand.

The degree-constrained spanning tree model sets the limited minimum coverage.

Multi-criteria model is used to compare networks according to various criteria (economic, technical, social, and environmental), which later helps to choose the network.

The maximum flow model (MXF) focuses on finding the network that reaches the maximum possible traffic constraints. The algorithms of Goldberg (1985, 1986) were new deviations from these methods in developing multi-layer networks.

The minimum cost flow model (MCF) is used to find the cheapest possible way to transmit some traffic through the network when there is a link between costs and capacity constraints in the network. The capacity of a network is treated as the maximum possible flow when the demand for transportation services is calculated by estimating the number of network users. If the network has capacity constraints, there are more congestions and higher transportation costs in it. There is an obvious non-cooperative behaviour case, when a user selects the route from loading to destination, which minimizes its route costs, but not the total costs in the network, and intensifies traffic. Specific models have been created for shared networks that guarantee the sharing of resources. In short distances, freight delivery companies rarely share vehicles, but in large distances, they often cooperate to achieve sharing efficiency.

Companies directly communicate with interconnected network users. They benefit from these connections. The value of a connection depends on the distance, so the companies have to weigh the benefits of a particular connection against connection costs. There are two types of specific models. Strategic interaction models analyse strong connections among companies.

The three-player (consignor, freight carrier, and infrastructure owner) model proves that they cooperate. All infrastructure hubs in the model have considered efficient in case these hubs are not empty compared to others. A game theory model, which looks for the best relationships and the overall approach, could be used for network design. According to Beckstrom, the gain of each entity is proportionate to the number of cooperating users (from two to several thousand users). The use of the network can be assessed by evaluating enterprise costs and benefits of all enabled transactions (Beckstrom, 2009).

The two-criteria MXF/MCF model is used when researchers search for two-criteria optimization solutions for the network, including maximum traffic (MXF) and minimum cost flow (MCF) problems. The compromise points ensure maximum traffic on the network at a minimal cost. The use of the network has two effects. If the use of network capacity intensifies due to more intense traffic, then the transportation costs decrease.

8.2 Classification of Transport Networks

The design studies are aimed at network structure, which could be classified by distance and mode. The management of data flow using large volumes of data, service architectures is important for supply networks (Harris et al., 2015). Moreover, all delivery models such as long-distance, last mile, mode related, and shared transportation use network data. Up-to-date information about dynamic events (for example, traffic jams, traffic violations, incoming applications, and route blocks) and delivery status is available here. The transport network is mapped to the central station that continuously tracks the implementation of transportation plans inside the network (Bock, 2010a). In case of congestions scheduled deliveries travelling back and front the same road will be late in all directions. The situation requests for the careful control of cargo flows.

Long-Distance Shipment

Long-distance shipments are usually loaded as one truckload and part truckload. The second case can also include transshipment during the trip. After 1980, fleet sharing was introduced as the new delivery model for long distances (Royo et al., 2016). Due to that, enterprises often collaborate to reach fleet sharing and get added value from this activity. This cooperation requires the integration of information and joint decision-making between various stakeholders, improvement of transport activities, and the reduction of fuel consumption. This concept takes into account the route selection of the carriers organizing shipments in the particular road network. In the case of less than a truckload, a review of the physical network characteristics, line operations, and load planning is required. As a part of the routing strategy, freight reloads at terminals should be identified.

Last-Mile Shipment

Under the last-mile shipment case, carriers use a single transport mode.

Crainic and Montreuil (2016) presented the concept of "last mile" shipping which is a "concept used to design highly efficient and sustainable transport network" with the introduction of information flows, which in future could emerge into the shared-fleet model. Some cases could be evident:

1. Transport operator that performs last-mile shipment allows other carriers to use the free and available vehicle capacity, making slight modifications to the routes to meet customers' needs;
2. Transport operator delivers the freight using the shared-fleet carriers' service.

Provided examples show additional ways to connect routes and increase capacity in the last-mile delivery model (Fan et al., 2019). This could be organized through the exchange of information and where enough individual carriers support more efficient management (Sternberg et al., 2014).

Mode-Related Shipment

Mode-related shipment involves various modes and intermodal facilities (Cepolina & Ghiara, 2013; Sörensen et al., 2012). All of these components are important for creating an efficient mode-of-transport delivery model (Tadić et al., 2019). This type of shipment starts from the mode(s) selection (Janic, 2008). Mode selection for freight delivery also includes integration, if we are talking about the connectivity where rail deliveries account for 20% of inland deliveries.

The shipment needs to be organized by investigating different circumstances. The mode-related shipment includes causal links with transit time and passed distance (Kengpol et al., 2012). The highest issue to perform such shipments is information transfer, which is the most complex. However, it is not just about one time transfer but it is about the tracking of objects and about the real-time information when freight is transferred from one mode into another one (Bock, 2010b). To this end, investigations must be repeated over and over given the permeability (freight delivery capacity), delays concerning the delivery and delivery time interval.

Shared Transportation Shipment

Shared-fleet is the new way for delivery more or less linked with reduction of CO_2 emission and avoidance of truck travelling with no freight. Innovations in sharing platforms, location services, and Internet technologies have allowed for the rapid development and expansion of shared mobility services, with improved efficiency, cost savings, and revenue from underutilized resources (Furtado et al., 2013). Smartphone apps and information about vehicle positions increased the popularity of shared-fleet usage.

Closer cooperation between companies is an important way of combating inefficiency of the freight delivery in cities (Anand et al., 2012). Here we talk about horizontal logistics cooperation, i.e. integration of transport systems and subcontracting.

Such logistics cooperation can be defined as cooperation between two or more undertakings operating at the same level of the transport system and performing a similar freight delivery-related function (Hosni et al., 2014). These companies serve the same segment of transport services, provide the same services, and are treated as competitors in the highly competitive segment (Herbawi & Weber, 2012).

However, there are many other reasons why companies work together, such as:

- To increase productivity and reduce operating costs;
- To offer a better quality of service;
- To expand services;
- To sign contracts for higher volumes of carriage, etc.

Moreover, traffic and its effects can be reduced as a side effect of increased productivity. Coming back to urban freight transport, there are three types of sharing that comes from the cooperation of companies:

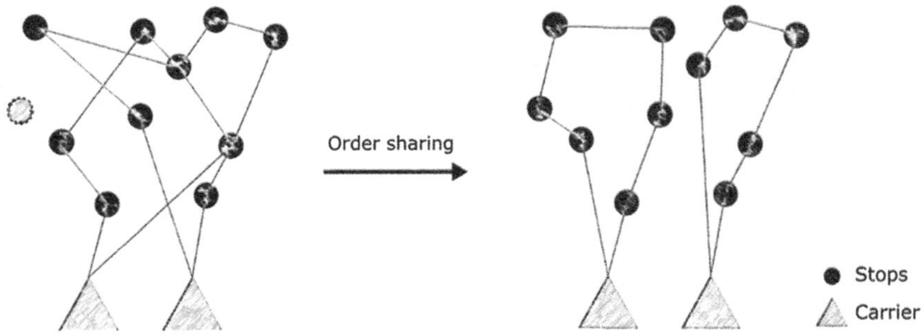

Fig. 8.2 Distribution of customer orders among competitors

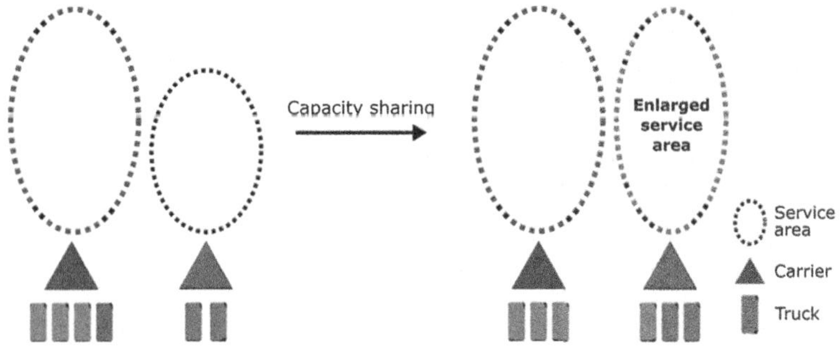

Fig. 8.3 Distribution of transport capacity among beneficiaries

- Order sharing;
- Capacity sharing;
- Information sharing.

Cooperating carriers could combine, share, and exchange customer orders or requests having optimized their transportation capabilities and still offering the same services. An acceptable model of fair distribution of costs and revenues is important for such a service approach.

Figure 8.2 shows vehicle traffic without and with order sharing. The route can be improved by distributing the nearest support points among couriers, resulting in shorter travelling distance and better courier delivery service.

For capacity sharing, transport units, such as vehicles or drivers, and temporary transport freight units (i.e. containers) could be used. Figure 8.3 shows how the capacity of transport units can be shared. In addition, to increase the use of assets of the parties involved, the service area of the beneficiary has increased and the income associated with high investment costs has risen. A vehicle-sharing party can indicate the maximum travel costs for other parties.

In sharing-fleet case delivery costs have to be attractive for parties participating in freight transportation, compared with alternatives.

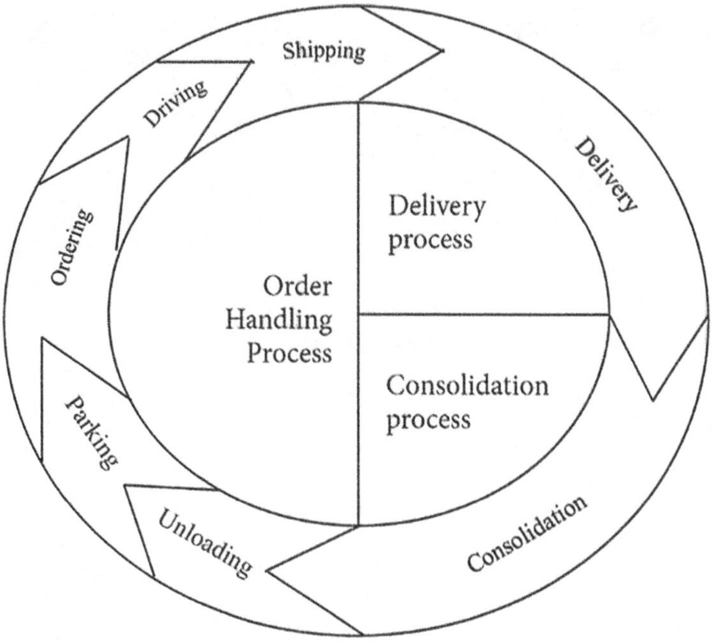

Fig. 8.4 Standard information-sharing model

Information sharing helps to improve the overall efficiency of operations. Information sharing between carriers follows legal protection and business interests. Information could be collected, stored, and processed through an IT partner. Strategically, partnership with service providers, freight carriers, and partners using direct, collaborative, electronic data exchanges could be defined as external. Transport service agreements define specific roles and responsibilities in the planning process. Partners' deliveries to customers should be performed on agreed service levels. External communication frequency and the overall number of elements communicated at the same time are documented and understandable. Demand and supply decisions and information are shared among partners. Improvements in the process are typically made by comparison against historical performance.

Figure 8.4 illustrates a delivery by a partner at the agreed service level and a consolidation that could be done by customers having less than truckloads, who seek to minimize the delivery price, or by business partners, which work on behalf of a consignee and a consignor. Information could be shared in all stages of the process, including freight delivery, traceability, and empty truck driving. Information sharing can also include other types of sharing, such as order or capacity sharing.

This would also allow sharing more data. 79% of consignees share their forecasts with their service providers and 38% of them electronically link their forecasts with their service providers' forecasts. In other words, by using the electronic system, changes in customer's demand are automatically reflected in forecasts of service providers (ECR, 2004).

8.3 Service-Oriented Transportation: Revision of Research Methods

The ability to deliver context-aware service for customers can require the interoperability of their networks characterized using various aspects: accessibility (taxes for using the network, network connection to communication infrastructure), timeliness, and security. The network interoperability is classified (Fig. 8.5) into security layer, real-time communication layer, interaction possibilities (integrations and the strength of connections).

The frequency of interactions depends on the context of information. Interoperability often affects many organizations, for example, in terms of organization of activities, transportation processes requiring several modes of transport, such as long-distance journeys. All carriers need access to the information which is constantly reported. Interactions should follow security requirements to connect to information flows within the network, and physical and electronic protection of the information exchanged. Authorities involved in transport-related services, for example, the police or the ambulance, need such interactions. Real-time interaction is important for roles that rely on the information.

Today, the environment can be characterized by a large heterogeneous number of services and contextual applications that dynamically connect to and leave the network. Such an environment provides a large-scale online work infrastructure, offering a new level of openness and dynamism. A computerized environment requires contextual data provided by seamlessly connected devices or a mobile phone, embedded in the surrounding physical environment and unobtrusive to the

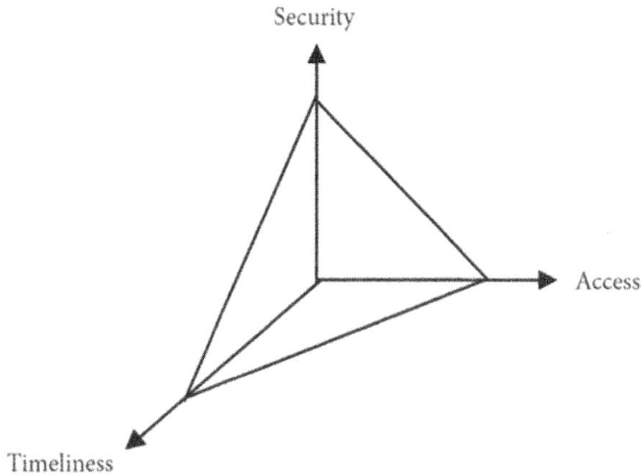

Fig. 8.5 Dimensions of context-aware service interactions

user. By getting the information sets, the enterprise has to react and select preferences or manage uncertainty cases and take timely decisions.

Service-oriented freight transport is related to (i) the management of time answering to the question of "when freight has to be delivered"; (ii) the management of content related to the question "which freight to transport"; (iii) the management of communication, telling "how to reach the destination"; and (iv) the management of context, which concerns "why to transfer freight and to which place". Finding the right information-based solutions requires a common approach to both systems and services. However, usually, it focuses on a part of the system only, with minor attention to the service. This is a pity because services are the reason why such systems are developed altogether.

A service can be analysed from the life cycle perspective and can also be classified as important or as enabling because it is a set of features offered to the user in the organization. As regards transport as a service of interest, the capacity of transport, the status of the environment during freight delivery, and timing constraints are important. Requirements of increasing traffic flow, services (e.g. data transfer and validation) to be met seeking to improve the quality of freight delivery carriage.

Strong growth in international trade and fierce competition have been increasing pressure on logistics companies. The selection of various transport modes and better decision-making can help to tackle issues. Such means closer cooperation between logistics service providers and system-oriented service delivery which facilitates closer cooperation between logistics companies and lets partners operate in synchromodality (Prandtstetter et al., 2016).

Service-oriented transport systems could be analysed from various directions. To follow recent trends in mathematical formulations, you can apply a hierarchical representation that defines the model type, method, and specific methods in the category.

There are five main categories of methods: mathematical programming, modelling, heuristical, analytical, and hybrid. Modelling methods are different and deal with single and multiple targets.

Much research is focused on the application of optimization techniques (Table 8.1). Also, various methods have been applied in service-oriented transportation, including multi-objective mixed-integer linear programming, variational inequality, system dynamics, discrete events, fuzzy logic, Markov chains, life cycle analysis, and input–output analysis, but also others which are not yet explored.

In literature, for the analysis of service-oriented transportation, methods of mathematical programming are used. Artificial intelligence methods such as Markov chains, Petri nets, case-based reasoning, Grey system, and coarse Neighbourhood sets have been introduced in studies analysing service-oriented transport (Russell et al., 2003).

The authors use rarely metaheuristic methods in process studies; however, the popularity of metaheuristic methods is rising in transportation systems research. It should be noted that the authors have already recognized evolutionary, genetic, and

Table 8.1 Hierarchy of qualitative methods and models for researching service-oriented transportation

Model type	Model technique	Solution method	Authors researching service-oriented transportation
Mathematical programming method	Single objective	Bi-level linear programming, multi-objective mixed-integer linear programming, fuzzy-goal programming, polynomial dynamic programming, queuing model, non-linear programming, variation inequality	Kuo and Han (2011) Wang et al. (2020) Selim and Ozkarahan (2008) Rasti-Barzoki and Hejazi (2015) Zhou and Min (2011) Hifi and M'Hallah (2009) Wang et al. (2012)
	Multiple objectives		
Simulation methods	System dynamic (SD)		Samuel et al. (2010) Sidola et al. (2011)
	Discrete event (DES)		Schmitt and Singh (2009)
	Business game	Simulation game	Anderson and Morrice (2000)
Heuristic methods	Simple heuristic	Simulated annealing heuristics	Daniel and Rajendran (2005)
	Artificial intelligence techniques	Markov chains, object-oriented petri nets, case-based reasoning, Bayesian network modelling, fuzzy logic, artificial neural network, Grey system and rough sets, neighbourhood rough sets	Raghavan and Viswanadham (1999) Dong and Chen (2001) Kwon et al. (2007) Ojha et al. (2018) Krikke et al. (2005) Kuo et al. (2010) Bai and Sarkis (2010) Bai and Sarkis (2012)
	Metaheuristic	Genetic algorithm, multi-objective evolutionary algorithm, multi-objective differential evolution algorithm, hybrid Taguchi particle swarm optimization ant colony optimization, greedy randomized adaptive search procedure	Altiparmak et al. (2006) Bandyopadhyay and Bhattacharya (2014) Babu and Gujarathi (2007) Bachlaus et al. (2008) Silva et al. (2009) Lee et al. (2010)
Hybrid models	Hybrid simulation	SD-DES	Rabelo et al. (2007)
Analytical models	Multi-attribute decision-making	Analytical hierarchy process	Rabelo et al. (2007)
	Systematic models	Life cycle analysis, input/output analysis, performance metrics	Albino et al. (2002) Cho et al. (2012)
	Game theory		Cachon and Netessine (2006)

Table 8.2 Hierarchy of qualitative methods used for service-oriented transportation efficiency evaluation

Classes of methods	Method	Authors researching service-oriented transportation
Methods for cost-efficiency analysis	Framework for assessment	Alsyouf et al. (2011)
Methods for cost–benefit analysis	Primitive method	
	Cash-flow discounting method	
Methods for making multi-attribute solutions	Process of analytical hierarchy (AHP)	Thanki et al. (2016)
	Simple multi-attribute rating technique	Barla (2003)
	Data envelopment analysis (DEA)	Meza and Jeong (2013)
	Integrated approach	Anvari et al. (2014)
Metaheuristic methods	Genetic algorithms (GA)	Cai et al. (2015) Karakatic and Podgorelec (2015) Ahmadizar et al. (2015)
	Evolutionary algorithms (EA)	Erbao and Mingyong (2009)
	Differential evolution algorithms (DE)	
	Particle swarm optimization (PSO)	Marinakis and Marinaki (2010)
Methods for measuring effects	Simulation-based approaches	Baradaran and Akhavan (2019)
	Graphical approaches	
Methods for cost-utility analysis	Performance evaluation	Kainuma and Tawara (2006)
Combined approach methods	Balanced scorecard (BSC) and data envelopment analysis (DEA)	Bazrkar and Iranzadeh (2017)

other algorithms as suitable for addressing service-oriented transportation efficiency issues (Vergara et al., 2002).

All of the above methods include forward and reverse flow analysis and environmental considerations. Service-oriented transportation could be studied from a variety of perspectives. The research focuses on several applications of the methods, but new research could be conducted considering the methods used for research efficiency evaluation of service-oriented transportation (Table 8.2).

The efficiency increase approach could be used for sharing transportation services.

8.4 Methods Proposed for Shared Transportation Service Design

The application of the concept of sharing in real transport systems is expected to bring many benefits, i.e. reduced congestion and pollution levels and lower cost of freight carriage.

The idea to create common mobility schemes for freight transport is established seeking to reduce congestions and CO_2 emission. These schemes could be named delivery sharing, vanpooling, and others. These deliveries allow companies with a similar route and times to share vehicles in a journey aiming to reduce travel expenses.

Joint travel planning can be divided into "pre-prepared", "static", or "dynamic". Shared trips must be pre-planned knowing the needs in advance (i.e. the origin, the destinations, the departure and arrival times). These pre-arranged frameworks are usually applied in arranging suburban routes and routes between cities (Diziain et al., 2012). But there are usually more long-distance trips with a more flexible schedule (Delaitre et al., 2009). Dynamic sharing focuses on the arrangement during a trip.

Vanpooling operates on a daily or long-term basis and provides regular and economic vehicles, disregarding unexpected schedule changes. Vanpooling is related to optimization models and decision algorithms. A van driver goes to an intermediate place known as a park, picks up a van, and drives all the freight together to the destination place.

In multi-level systems, goods are carried on scheduled transport and are picked on a trailer fleet. Such systems must also provide inquiries, available vehicles, and regular departures on different directions. Freight synchronization helps to check if all loaded transport units are delivered. The freight could be delivered using several trucks with reloading in satellite places before it reaches the clients (Grangier et al., 2016). The resource-oriented synchronization means that shared vehicles and drivers for freight transportation are used depending on availability only (Drexl, 2012). Companies share information about drivers, capacities, and possible parking places, but before that, data must be synchronized.

Route construction. When it comes to general mobility, vehicles are routed accordingly between places of loading and delivery, while meeting requirements for service itself. Carrier gets requirements and manages the available vehicle resources (vehicles, fleets, drivers, etc.). This shared travel planning is a key challenge for shared mobility. Multiple service sharing requests may be serviced using the same source (e.g. trucks) in the same time slot. To create such generic services, various features and limitations must be revised.

In demand-side transport systems (including vehicle sharing) and many other systems where vehicle fleet is arranged in certain places (warehouses) and is ready for operation, there are extra routing constraints that requires a special vehicle and driver's knowledge. Having completed a trip, each vehicle should return to the warehouse location.

8.5 Technologies and Data Management to Support Shared Transportation

It is not easy to build an environment for shared transport, including planning and communication. Companies need a more adaptive and context-aware approach to network security, but many have not analysed the legal framework around identity access management and asset management allowing to introduce such an approach.

Context-aware security embodies functions, and while it was just a popular word some time ago, now context-based security is a reality. Most network security vendors send products using some form of these functions. However, with the growing need for a larger context and the technologies that allow it to move forward, efforts to implement a context-based approach often come to a halt due to operational challenges.

The shared service may be organized in several ways. The first way is the traditional shared service centre, which supports activities and processes. Herein, digital activities can help to organize continuous resource monitoring activities using sensors and advanced technologies (Büyüközkan & Göçer, 2018).

The second way, i.e. outsourcing, is when activities and processes of a company are supported by another company (called an outsourcer). Thus, artificial intelligence algorithms are used to perform complex processes. Artificial intelligence allows sharing companies to pre-plan their resources.

The information should have such features as interactivity, convergence, and hypertextuality.

Interactivity includes the ability of users to communicate with each other, which some authors call communicative interactivity (Boullier, 2000). According to Lister et al. (2008), interactivity is defined by several elements:

- Hypertext navigation provides performance (selective content usage) and content control capabilities. Hypertext navigation is exhaustive—the user selects information from the database and extracts it in special forms.
- Immersive navigation. The user uses the database to analyse spatial information.

Other researchers (cit. from McQuail, 1997) distinguish the following five dimensions of interactivity:

- bidirectional and multidirectional communication;
- flexible timeliness of information use and changing roles of participants (a consignor turns into a consignee, and vice versa);
- a sense of place in the communication environment;
- a higher level of communication application (the content of communication could be individual for each company);
- perceived purpose of communication (orientation towards its change). The information available to the user is stored in a database.

Interactivity is essential to enable communication with transport management and to quickly and conveniently receive information flows from different sources.

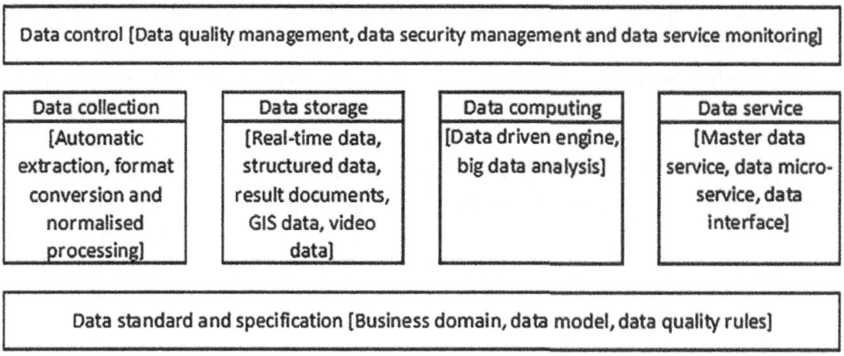

Fig. 8.6 Intellectual IT data management for shared transportation services

Second, convergence links statistics, forms, and functions of electronic calculations into a single whole. Convergence manifests from a technological point of view, i.e. analysing the same content using different technologies.

Thirdly, hypertextuality is not aimed at a more awake communication but expresses the connectivity of information. Thinking about hypertextuality is driven by the problem of information overload—a huge (excessive) amount of information that systems have to deal with. The recommendation is to store data in the system allowing to find associations rather than alphabetical and numbering methods of data processing. Associative relationships between data are preferable and provide a model for the processing of information using conventional alphabetical or similar methods.

Figure 8.6 presents Intellectual IT data management for shared transportation services. First, the size of the network where shared resource services are available is expected to increase.

Four data collection modes have been created for the handling and archiving of information: (I) automatic up-to-time collection; (II) manual data collection; (III) information, which is received from different sources, unification; and (IV) e-documents archive.

The application of ITS in freight transport improves the reporting of data. The transport information is grouped to be most commonly applied for supported ITS linked to general resource services that are provided for road transport only:

- Infrastructure related traffic data: Up-to-time data which reports vehicles flows and the availability of traffic lines (Fukui et al., 2009). Data about the place of the road, the status in the streets, restrictions and congestion, collisions, and additional events must be relevant to this mode of transport.
- Information concerning vehicle and cargo locations: the information allows to track and trace goods over the transport network (Mirzabeiki & Sjöholm, 2012). In some cases, reporting could include notifications about freight arrival to the consignee, pick-up and drop-off data for the carriage of goods, the position and the status of monitoring systems.

- Information on the condition of the load: such information relates to physical characteristics of goods during transport operations, including up-to-date data about freight delivery conditions, the time of storage of the product during its carriage by vehicles of different transport network operators (Sundarakani et al., 2012).
- Cargo location data: which relates to location and sequence of freights important for shipping and storing. Automatic identification systems help to maintain such information and reduce the number of errors associated with it (Poon et al., 2009).
- Information about vehicle identification: such information as vehicle type and class, vehicle registration number and other information are used in transport resource management operations. This information is used by shared resource service companies for planning vehicle maintenance or security checks. The recognition of licence plates is an important area of transport research and is also extensively studied (Thome et al., 2011).
- Delivery reservation systems: these systems allow to order a load of goods for a specific vehicle over some time. They contribute to efficiency aspects seeking to maximize the occupancy of the trailer (contributing to the efficient use of the fleet).
- Route planning tool: Such tool is used for planning transport routes and to respond to situations on the road. The tool helps to reduce freight delays and reach higher freight delivery standards. With such systems, operators can plan their transport resources better. A dynamic vehicle route planning is beneficial for carriers by reducing their costs, guaranteeing better service to customers, and being more environmentally friendly by reducing traffic congestion.
- Weight movement review: The tool is designed to monitor the weight of the trailer aiming to maintain safety on roads and minimize the damage done by overweighted trailers. The system helps to eliminate mistakes. Some programmes include laws, regulations, and transportation administration. In addition to road maintenance costs, infrastructure failures cause environmental damages. The authors provided the benefits of the system in improving freight delivery (Jacob et al., 2011).
- Driving behaviour revision: speed and acceleration data is reported, giving feedback to the driver thereon. The feedback concerns vehicle fuel consumption and greener transportation, therefore, supporting transport resource management. Also, the feedback could improve the concentration during driving seeking to reduce accidents and improve operational safety (Marell & Westin, 1999).
- Accident prevention systems: such systems use sensors capable to minimize the number of accidents. Established in the truck accident prevention systems send a warning to a driver as he approaches the object. Another accident prevention system is possible to measure the distance between the truck and the detected object and to take action. Such a system improves transportation security and reduces the likelihood of accidents.
- Fuel management systems: track fuel consumption in vehicles and identify cases that require fuel. They use a variety of techniques to follow and check the level in the fuel tank, report refills, and emissions. Systems for fuel revision ensure that

fuel supply information allows organizing operational controls and simplifies tax accounting for purchased fuel.

8.6 Conclusions

The chapter focus on various transportation networks classified according to the distance per trip and the usage of main resources. The study highlights service-oriented transportation and leads the reader to the application of intelligent data management systems and their provision in sharing transportation cases. This is investigated in several directions: as overall freight delivery activity and as the use in each stage of freight delivery integrating the main logic in the software that is applied for this purpose.

The author also presents the dimensions of context-aware service interactions helping to find real-time information-based solutions, which are important for shared transportation supporting efficiency approach. Finally the classification of intellectual services is presented which is demonstrating complexity of freight delivery system.

References

Ahmadizar, F., Zeynivand, M., & Arkat, J. (2015). Two-level vehicle routing with cross-docking in a three-echelon supply chain: A genetic algorithm approach. *Applied Mathematical Modelling, 39*(22), 7065–7081.

Albino, V., Izzo, C., & Kühtz, S. (2002). Input–output models for the analysis of a local/global supply chain. *International Journal of Production Economics, 78*(2), 119–131.

Alsyouf, I., Al-Aomar, R., Al-Hamed, H., & Qiu, X. (2011). A framework for assessing the cost effectiveness of lean tools. *European Journal of Industrial Engineering, 5*(2), 170–197. https://doi.org/10.1504/EJIE.2011.039871

Altiparmak, F., Gen, M., Lin, L., & Paksoy, T. (2006). A genetic algorithm approach for multi-objective optimisation of supply chain networks. *Computers and Industrial Engineering, 51*(1), 196–215.

Anand, N., Quak, H., van Duin, R., & Tavasszy, L. (2012). City logistics modelling efforts: Trends and gaps – A review. *Procedia-Social and Behavioural Sciences, 39*, 101–115.

Anderson, E. G., Jr., & Morrice, D. J. (2000). A simulation game for teaching service-oriented supply chain management: Does information sharing help managers with service capacity decisions? *Production and Operations Management, 9*(1), 40–55.

Anvari, A., Zulkifli, N., Sorooshian, S., & Boyerhassani, O. (2014). An integrated design methodology based on the use of group AHP-DEA approach for measuring lean tools efficiency with undesirable output. *The International Journal of Advanced Manufacturing Technology, 70*(9), 2169–2186. https://doi.org/10.1007/s00170-013-5369-z

Babu, B. V., & Gujarathi, A. M. (2007, September). Multi-objective differential evolution (MODE) for optimisation of supply chain planning and management. In *2007 IEEE congress on evolutionary computation* (pp. 2732–2739). IEEE.

Bachlaus, M., Pandey, M. K., Mahajan, C., Shankar, R., & Tiwari, M. K. (2008). Designing an integrated multi-echelon agile supply chain network: A hybrid Taguchi-particle swarm optimisation approach. *Journal of Intelligent Manufacturing, 19*(6), 747.

Bai, C., & Sarkis, J. (2010). Integrating sustainability into supplier selection with grey system and rough set methodologies. *International Journal of Production Economics, 124*(1), 252–264.

Bai, C., & Sarkis, J. (2012). Supply-chain performance-measurement system management using neighbourhood rough sets. *International Journal of Production Research, 50*(9), 2484–2500.

Bandyopadhyay, S., & Bhattacharya, R. (2014). Solving a tri-objective supply chain problem with modified NSGA-II algorithm. *Journal of Manufacturing Systems, 33*(1), 41–50.

Baradaran, V., & Akhavan, B. (2019). Determining number of withdrawal Kanban using bi-level optimization and simulation approaches. *International Journal of Industrial Engineering, 26*(2), 70–83.

Barla, S. B. (2003). A case study of supplier selection for lean supply by using a mathematical model. *Logistics Information Management, 10*, 25–34. https://doi.org/10.1108/09576050310503420

Bazrkar, A., & Iranzadeh, S. (2017). Choosing a strategic process in order to apply in Lean Six Sigma methodology for improving its performance using integrative approaches of BSC and DEA. *Journal of Business and Retail Management Research, 11*(4), 45–53. https://doi.org/10.24052/jbrmr/v11is04/caspiotailssmfiipuiaobad

Beckstrom, R. (2009). *A new model for network valuation* (pp. 1–10). National Cyber security Center, Research paper.

Bellman, R. E. (1958). On a routing problem. *Quarterly of Applied Mathematics, 16*, 87–90.

Bock, S. (2010a). Real-time control of freight forwarder transportation networks by integrating multimodal transport chains. *European Journal of Operational Research, 200*(3), 733–746.

Bock, S. (2010b). Real-time control of freight forwarder transportation networks by integrating multimodal transport chains. *European Journal of Operational Research, 200*(3), 733–746.

Boullier, D. (2000). 649 60206 COMPIEGNE CEDEX.

Büyüközkan, G., & Göçer, F. (2018). Digital supply chain: Literature review and a proposed framework for future research. *Computers in Industry, 97*, 157–177.

Cachon, G. P., & Netessine, S. (2006). Game theory in supply chain analysis. In *Models, methods, and applications for innovative decision making* (pp. 200–233).

Cai, Y. G., Tang, Y. L., & Yang, Q. J. (2015). An improved genetic algorithm for multi-depot heterogeneous vehicle routing problem with simultaneous pickup & delivery time windows. *Applied Mechanical Material, 738*, 361–365.

Cepolina, S., & Ghiara, H. (2013). New trends in port strategies. Emerging role for ICT infrastructures. *Research in Transportation Business and Management, 8*, 195–205.

Cho, D. W., Lee, Y. H., Ahn, S. H., & Hwang, M. K. (2012). A framework for measuring the performance of service supply chain management. *Computers and Industrial Engineering, 62*(3), 801–818.

Cooke, K., & Halsey, E. (1966). The shortest route through a network with time-dependent internodal transit times. *Journal of Mathematical Analysis and Applications, 14*, 493–498.

Crainic, T. G., & Montreuil, B. (2016). Physical internet enabled hyperconnected city logistics. *Transportation Research Procedia, 12*, 383–398.

Daniel, J. S. R., & Rajendran, C. (2005). Determination of base-stock levels in a serial supply chain: A simulation-based simulated annealing heuristic. *International Journal of Logistics Systems and Management, 1*(2–3), 149–186.

Delaitre, L., Molet, H., Awasthi, A., & Breuil, D. (2009). Characterising urban freight solutions for medium sized cities. *International Journal of Services Sciences, Special Issue on Methods, Tools and Applications in Transport and Logistic Services, 2*(3/4), 281–302.

Diziain, D., Ripert, C., & Dablanc, L. (2012). How can we bring logistics back into cities? The case of Paris metropolitan area. *Procedia – Social and Behavioural Sciences, 39*, 267–281.

Dong, M., & Chen, F. F. (2001). Process modeling and analysis of manufacturing supply chain networks using object-oriented petri nets. *Robotics and Computer-Integrated Manufacturing, 17*(1–2), 121–129.

Drexl, M. (2012). Synchronization in vehicle routing—A survey of VRPs with multiple synchronization constraints. *Transportation Science, 46*(3), 297–316.

Dreyfus, S. (1969). An appraisal of some shortest-path algorithms. *Operations Research, 17*, 395–412.

ECR. (2004). Efficient consumer response. http://www.ecr-baltic.org/en/about-ecr/ecr-efficient-consumer-response

Erbao, C., & Mingyong, L. (2009). A hybrid differential evolution algorithm to vehicle routing problem with fuzzy demands. *Journal of computational & applied mathematics, 231*(1), 302–310.

Euler, L. (1736). Solutio problematis ad geometriam situs pertinentis. *Commentarii Academiae Scientiarum Petropolitanae, 1741*, 128–140.

Fan, Y., Behdani, B., Bloemhof-Ruwaard, J., & Zuidwijk, R. (2019). Flow consolidation in hinterland container transport: An analysis for perishable and dry cargo. *Transportation Research Part E: Logistics and Transportation Review, 130*, 128–160.

Fukui, M., Nishinari, K., Yokoya, Y., & Ishibashi, Y. (2009). Effect of real-time information upon traffic flows on crossing roads. *Physica A: Statistical Mechanics and its Applications, 388*(7), 1207–1212.

Furtado, P., Frayret, J.-M., Fakhfakh, R., & Biard, P. (2013). Simulation of a Physical Internet-Based transportation network. In *Paper presented at the proceedings of 2013 international conference on industrial engineering and systems management (IESM).*

Garrison, W. L., & Marble, D. F. (1962). The structure of transportation networks. Technical Report.

Goldberg, D. E. (1985, July). Genetic algorithms and rule learning in dynamic system control. In *Proceedings of the first international conference on genetic algorithms and their applications* (pp. 8–15). Lawrence Erlbaum.

Goldberg, D. E. (1986). The genetic algorithm approach: Why, how, and what next? In *Adaptive and learning systems* (pp. 247–253). Springer.

Grangier, P., Gendreau, M., Lehuédé, F., & Rousseau, L. M. (2016). An adaptive large neighborhood search for the two-echelon multiple-trip vehicle routing problem with satellite synchronization. *European Journal of Operational Research, 254*(1), 80–91.

Harris, I., Wang, Y., & Wang, H. (2015). ICT in multi-modal transport and technological trends: Unleashing potential for the future. *International Journal of Production Economics, 159*, 88–103.

Herbawi, W., & Weber, M. (2012). The ridematching problem with time windows in dynamic ridesharing: A model and a genetic algorithm. In *2012 IEEE congress on evolutionary computation* (pp. 1–8). IEEE. https://doi.org/10.1109/CEC.2012.6253001

Hifi, M., & M'Hallah, R. (2009). Beam search and non-linear programming tools for the circular packing problem. *International Journal of Mathematics in Operational Research, 1*(4), 476–503.

Hosni, H., Naoum-Sawaya, J., & Artail, H. (2014). The shared-taxi problem: Formulation and solution methods. *Transportation Research Part B: Methodological, 70*, 303–318. https://doi.org/10.1016/j.trb.2014.09.011

Jacob, R., Márton, P., Maue, J., & Nunkesser, M. (2011). Multistage methods for freight train classification. *Networks, 57*(1), 87–105.

Janic, M. (2008). An assessment of the performance of the European long intermodal freight trains (LIFTS). *Transportation Research Part A: Policy and Practice, 42*(10), 1326–1339.

Kainuma, Y., & Tawara, N. (2006). A multiple attribute utility theory approach to lean and green supply chain management. *International Journal of Production Economics, 101*(1), 99–108. https://doi.org/10.1016/j.ijpe.2005.05.010

Karakatic, S., & Podgorelec, V. (2015). A survey of genetic algorithms for solving multi depot vehicle routing problem. *Applied Soft Computer, 27*, 519–532.

Kengpol, A., Meethom, W., & Tuominen, M. (2012). The development of a decision support system in multi-modal transportation routing within Greater Mekong sub-region countries. *International Journal of Production Economics, 140*(2), 691–701.

Krikke, H., van Nunen, J. A., Zuidwijk, R., & Kuik, R. (2005). E-business and circular supply chains: Increased business opportunities by IT-based customer oriented return-flow management. In *Distribution logistics* (pp. 221–239). Springer.

Kuo, R. J., & Han, Y. S. (2011). A hybrid of genetic algorithm and particle swarm optimisation for solving bi-level linear programming problem–A case study on supply chain model. *Applied Mathematical Modelling, 35*(8), 3905–3917.

Kuo, R. J., Wang, Y. C., & Tien, F. C. (2010). Integration of artificial neural network and MADA methods for green supplier selection. *Journal of Cleaner Production, 18*(12), 1161–1170.

Kwon, O., Im, G. P., & Lee, K. C. (2007). MACE-SCM: A multi-agent and case-based reasoning collaboration mechanism for supply chain management under supply and demand uncertainties. *Expert Systems with Applications, 33*(3), 690–705.

Lee, D. H., Chen, J. H., & Cao, J. X. (2010). The continuous berth allocation problem: A greedy randomised adaptive search solution. *Transportation Research Part E: Logistics and Transportation Review, 46*(6), 1017–1029.

Lister, M., Dovey, J., Giddings, S., Grant, I., & Kelly, K. (2008). *New media: A critical introduction*. Routledge.

Marell, A., & Westin, K. (1999). Intelligent transportation system and traffic safety–drivers perception and acceptance of electronic speed checkers. *Transportation Research Part C: Emerging Technologies, 7*(2–3), 131–147.

Marinakis, Y., & Marinaki, M. (2010). A hybrid multi-swarm particle swarm optimization algorithm for the probabilistic traveling salesman problem. *Computers and Operations Research, 37*(3), 432–442.

McQuail, D. (1997). *Audience analysis*. Sage.

Meza, D., & Jeong, K. Y. (2013). Measuring efficiency of lean six sigma project implementation using data envelopment analysis at NASA. *Journal of Industrial Engineering and Management, 6*(2), 401–422. https://doi.org/10.3926/jiem.582

Mirzabeiki, V., & Sjöholm, P. (2012). Sharing tracking and tracing data to improve operations of supply chain actors. *International Journal of Logistics Systems and Management, 13*(1), 81–95.

Ojha, R., Ghadge, A., Tiwari, M. K., & Bititci, U. S. (2018). Bayesian network modelling for supply chain risk propagation. *International Journal of Production Research, 56*(17), 5795–5819.

Poon, T. C., Choy, K. L., & Lau, H. C. W. (2009). A RFID case-based logistics resource management system for managing order-picking operations in warehouses. *Expert Systems with Applications, 36*(4), 8277–8301.

Prandtstetter, M., Putz, L., Pfoser, S., Haller, A., Lenz, G., & Ponweiser, W. (2016). Introduction to synchromodal networks in Austria. In *Tagungsband FFH 2016, Vienna, Austria, 2016* (p. 6).

Rabelo, L., Eskandari, H., Shaalan, T., & Helal, M. (2007). Value chain analysis using hybrid simulation and AHP. *International Journal of Production Economics, 105*(2), 536–547.

Raghavan, N. S., & Viswanadham, N. (1999, December). Performance analysis of supply chain networks using Petri nets. In *Proceedings of the 38th IEEE conference on decision and control (Cat. No. 99CH36304)* (Vol. 1, pp. 57–62). IEEE.

Rasti-Barzoki, M., & Hejazi, S. R. (2015). Pseudo-polynomial dynamic programming for an integrated due date assignment, resource allocation, production, and distribution scheduling model in supply chain scheduling. *Applied Mathematical Modelling, 39*(12), 3280–3289.

Royo, B., Fraile, A., Larrodé, E., & Muerza, V. (2016). Route planning for a mixed delivery system in long distance transportation and comparison with pure delivery systems. *Journal of Computational and Applied Mathematics, 291*, 488–496.

Russell, S. J., Norvig, P., Canny, J. F., Malik, J. M., & Edwards, D. D. (2003). *Artificial intelligence: A modern approach* (Vol. 2). Prentice Hall.

Samuel, C., Gonapa, K., Chaudhary, P. K., & Mishra, A. (2010). Supply chain dynamics in healthcare services. *International Journal of Health Care Quality Assurance, 23*, 631–642.

Schmitt, A. J., & Singh, M. (2009). Quantifying supply chain disruption risk using Monte Carlo and discrete-event simulation. In *Proceedings of the 2009 winter simulation conference* (pp. 1237–1248). IEEE.

Selim, H., & Ozkarahan, I. (2008). A supply chain distribution network design model: An interactive fuzzy goal programming-based solution approach. *The International Journal of Advanced Manufacturing Technology, 36*(3–4), 401–418.

Sidola, A., Kumar, P., & Kumar, D. (2011, June). System dynamic methodological approach for design and analysis of risk in supply chain. In *First international technology management conference* (pp. 495–500). IEEE.

Silva, C. A., Sousa, J. M. C., Runkler, T. A., & Da Costa, J. S. (2009). Distributed supply chain management using ant colony optimisation. *European Journal of Operational Research, 199*(2), 349–358.

Sörensen, K., Vanovermeire, C., & Busschaert, S. (2012). Efficient metaheuristics to solve the intermodal terminal location problem. *Computers and Operations Research, 39*(9), 2079–2090.

Sternberg, H., Prockl, G., & Holmström, J. (2014). The efficiency potential of ICT in haulier operations. *Computers in Industry, 65*(8), 1161–1168.

Sundarakani, B., Tan, A. W. K., & van Over, D. (2012). Enhancing the supply chain management performance using information technology: Some evidence from UAE companies. *International Journal of Logistics Systems and Management, 11*(3), 306–324.

Tadić, S., Krstić, M., & Brnjac, N. (2019). Selection of efficient types of inland intermodal terminals. *Journal of Transport Geography, 78*(2019), 170–180.

Takayama, T., & Judge, G. G. (1964). Equilibrium among spatially separated markets: A reformulation. *Econometrica, 32*, 510–524.

Thanki, S., Govindan, K., & Thakkar, J. (2016). An investigation on lean-green implementation practices in Indian SMEs using analytical hierarchy process (AHP) approach. *Journal of Cleaner Production, 135*, 284–298. https://doi.org/10.1016/j.jclepro.2016.06.105

Thome, T., Vacavant, A., Robinault, L., & Miguet, S. (2011). A cognitive and video-based approach for multinational license plate recognition. *Machine Vision and Applications, 22*(2), 389–407.

Vergara, F. E., Khouja, M., & Michalewicz, Z. (2002). An evolutionary algorithm for optimising material flow in supply chains. *Computers and Industrial Engineering, 43*(3), 407–421.

Wang, P., Hunter, T., Bayen, A. M., Schechtner, K., & González, M. C. (2012). Understanding road usage patterns in urban areas. *Scientific Reports, 2*(1001), 1. https://doi.org/10.1038/srep01001

Wang, X., Xie, J., & Fan, Z. P. (2020). B2C cross-border E-commerce logistics mode selection considering product returns. *International Journal of Production Research, 59*, 1–20. https://doi.org/10.1080/00207543.2020.1752949

Wigan, M. R. (1971). *Benefit assessment for network traffic models and application to road pricing*. Transport and Road Research Laboratory.

Zhou, G., & Min, H. (2011). Designing a closed-loop supply chain with stochastic product returns: A genetic algorithm approach. *International Journal Logistics Systems Management, 9*(4), 397–418.

Chapter 9
Context-Aware Service Support Efficiency Improvement in the Transport System

Aurelija Burinskiene

Abstract The advance of the Internet of Things (IoT) technology and the development of sensors have allowed transport services to apply the Intelligent Transport Systems (ITS). The use of object-to-object communications has led to large amounts of contextual data sets generated for ITS. This basic information must be handled to obtain context and to suggest services that can communicate with various types of trucks about traffic, accidents, forecasting, etc. Since 2014 researchers are paying more attention to the transport systems. It covers processes including routing, clients clustering, backhauling management, and scheduling. The author targets in this chapter to revise the use of context in ITS.

Keywords Transport system · Routing algorithms · Clustering · Costs metrics · Backhauling · Scheduling · Intelligent Transport System

9.1 Context-Aware Service in the Transport System

The intelligent system used in transport plays a key role in the transformation of traditional transportation, increasing the services for users, supporting the efficiency of the freight delivery process. All countries of the world adopted and begun to implement ITSs as a complimentary technique to modern management of traffic. Study in the field of intelligent transport is aimed at providing services related to various modes of road transport and assisting users in raising awareness and improving safety, coordinating, using transport comfortable and reasonable (Perego et al., 2011).

Land transport, which is now practised, encompasses more communication between people. Turning it into intelligent transport would lead to many technical problems in providing communication between people and machines operating in the transport sector or only machines. However, technologies, sensors, the IoT

A. Burinskiene (✉)
Vilnius Gediminas Technical University, Vilnius, Lithuania
e-mail: aurelija.burinskiene@vilniustech.lt

D. Dzemydienė et al., *Development of Smart Context-Aware Services for Cargo Transportation*, International Series in Operations Research & Management Science 330, https://doi.org/10.1007/978-3-031-07199-7_9

enforced new possibilities which are available for the freight delivery process (Svarnamugi & Chinnaiyang, 2018). The systems of intelligent transport have expanded the possibilities of studying the issues existing in land transport (Erl, 2005). For solving the issues, several objects interact in communication. The collected information provides a deep understanding of the application of intelligent transport systems.

The application of Intelligent transport and the Internet of Things support the development of existing transport systems. Perera et al. (2014) addressed the importance of contextual data collected from the Internet of Things. The researchers considered data revision approaches, starting from interaction in communication, retrieving data, machine learning and finalizing with users' applications (Erl, 2005).

The authors studied the importance of machine learning in ITS (Veres & Moussa, 2019). Their research revealed the application of machine learning in freight delivery. The researchers pointed out the issues and announced questions on the topic. Another author considered the BigData analysis in freight transport (Zhu et al., 2019). Their article proposes a method for delivering BigData analysis in intelligent transport, discussing various learning methods and ontology. New authors provided the application of context examples in various sectors (Vahdat-Nejad et al., 2016). They draw the architecture which focuses on the surroundings, the ITS, applications, and contextual investigations. By following this structure, the classification of used ITS was provided. Guarino proposes the way for using the Internet of Things technology in context-related cases. The researchers emphasized the necessity to improve the quality of context data employment for freight delivery processes.

Above-mentioned authors provide insights into the contemporary studies published in the field of transport.

Both terms connected to context and its awareness were present in studies from the 1990s, evolving from initial computer-based applications, internet portals, truck driver apps, to the cloud solutions integrating the Internet of Things in recent years. Both terms were first announced by Schilit et al. (1994a) and were initially promoted in the area of information technologies. A lot of authors identified and provided various descriptions of those terms. The authors defined "context" as data retrieval from the objects describing places, resources, and entities (Schilit et al., 1994b). The researchers referred to context as describing entities, such as position, time, and environment (Brown et al., 1997). Scientists defined context as vehicle position, personality, surroundings and time (Ryan et al., 1997). Other authors described context as a vehicle situation (Franklin & Flaschbart, 1998). And finally, the "context" was named as "the application state" by Ward et al. (1997).

All explanations are handled in various study directions, although the data is changing in a dynamic environment, nevertheless, the status has to be reported about places, resources, and entities due to varying conditions over time (Otebolaku & Lee, 2018).

The term "context-awareness" refers to knowledge as a whole that determines decisions made by the system taking into account the data. In the literature, "context-awareness" is defined as the possibility of a service focusing on detection and interpretation and response to the surroundings (Hull et al., 1997). Day (1998) emphasizes that "context is conscious as an implementation to the automatization of a system following the contextual data". One more definition is given to "context-awareness" as to "the capability of a system to offer computing services following contextual information" (Salber et al., 1998). Brown et al. (1997) defined the term "context–awareness" as "the capability of a system to respond automatically to the contextual data perceived by multiple sensors". In addition, other authors described "context–awareness" as "the use of context to provide task-relevant information to the planner" (Abowd et al., 1997).

Both terms rely on the data about the objects and the application for which the system has been designed. Out of all the definitions, we refer to context–awareness as the opportunity of service and application to react appropriately considering varying runtime surroundings. The systems of intelligent services for transport work under dynamic environments and focus on constant communication among the objects located in different places. For such a purpose huge volume of information is created. Sensor technologies are treated as multiple sources of information, forwarding it to intelligent transport applications, and they play a vital role in large-scale ITS deployment. Currently, the costs for sensor technologies are lower, but the quality of data generated is higher. This leads to extensive reactions to constant changes. Before adding new structured information from sensors, it is necessary to revise the benefits of this action. Data will have no value unless it is analyzed, interpreted and understood. The raw data retrieved from sensors is passed to services able to apply techniques helping to simplify interpretations (Woerndl et al., 2007).

Further on, the author would like to provide investigations on how the raw data could be processed in the application of ITS. In particular, the answers are important for understanding the capabilities of ITS and suggesting the further developments helping to improve the collection of data from sensors, suggesting contextual modelling, and providing management investigations following the results of such modelling (Meyer et al., 2006).

The system of intelligent transport must process a vast amount of records to present on-site solutions addressing situation-describing issues. Techniques dedicated for information processing revise data size and limits when the number of records ranges from a trillion bytes to a petabyte. This is the most challenging task for the above-mentioned system. Such allows solving the problem and ensures IT services that support efficient freight transportation. Revising, obtaining, modelling, and reasoning activity with the data, as well the use of data for decision-making is vital in the application of ITS.

The authors define three contextual functions which could support the proper work of all ITS applications. There is the number of functions such as a demonstration, performance and provision of investigations (Abowd et al., 1997). The critical aspect of the contextual data is knowledge, which meets the capabilities of the system. Main characteristics could be proposed for application on the system of intelligent transport to revise the behaviour of the objects providing information from the environment accordingly. In addition, any characteristic helps to define various scenarios of context-related ITS applications (Worndl et al., 2011).

Demonstration relies on services that should be provided for the decision makers. In particular, any time when a driver (who is a user) delivers freight, the system should propose a less time consuming trip to get to the destination, neglecting the crossing of overcrowded streets. According to the case, ITS applications must receive global positioning data about vehicle location, simulate, revise the records, and give insights for users. Here, contextual data, such as the position, timing and identity of the users, is used.

Implementation The system of intelligent transport incorporates both a connection between humans and machines or only machines connection. Sensor technologies used in a transport interact among each other. The automation of this interaction between sensors is the most important invisible background work before the data is moved to ITS for further processing. Considering this scenario where driver begins delivering freight to a retailer shop, the ITS except searching for the route proposal, should also initiate the reservation for park place accordingly and in advance following the estimated arrival time.

Marking The records generated by a sensor technology are not able to present enough context, which could be required for higher quality output, so all records have to be retrieved, analyzed, and combined with other records coming from various sources. The marking has synonym terms such as annotation of data records. For the implementation of the reporting about traffic density and measuring its level, the data records from street cameras and combined with data records retrieved from the global positioning system.

Functional Setup ITS must continuously adjust several functions linked to the system applications. In particular, in street light control, when a highly important service van, such as an ambulance travelling to an accident place, approaches a lane with a red traffic light signal on, the system must dynamically adjust the hardware functions switching the light signal to the one which allows the passing the crossroad.

Various authors classified contextual information into primary and secondary ones. Perera et al. (2014) analyse different types of context and categorization:

Initial Contextual Data The data retrieved about positions, resources, timing, etc., which is not combined with any other information is treated as raw data. For example the trailer, where the freight is loaded, licence plate, freight position according to GPS data or place of destination.

Processed Data The raw data, which is further revised using various techniques and transformed, is known as processed or secondary data. For example, trailer licence plate numbers are initially reported and later on are linked with the loaded freight data, particular, with freight dimensions and so on. Below the author placed Table 9.1, which gives the understanding of initial and transformed data specifying the one that is important for the system of intelligent transport.

Table 9.1 Classification of contextual data

Timeline	Studies	System using context	ITS application	Primary context	Secondary context
In 2007s	The study of Woerndl and co-authors (2007)	Adds contextual records to Vehicle Ad Hoc Networks seeking a higher quality. An example of application is presented—recommendations of gas stations taking into account the context system.	Advanced Traveller Information System	Time to determine the location of the vehicle identifier	Acceleration, fuel volume in the truck, trip length, destination
In 2008s	The study of Fuchs and co-authors (2008)	Gives the recommendations for vehicle drivers upfront the decision-making	Improved management of heavy cargo trucks	Resources (trucks, trailers, drivers)	The length of the route, the limits for speed, the particularities of roads that are crossed during the trip
In 2009s	The study of Sun and co-authors (2009)	Provides the context linked to the vehicle status	Improved control of freight delivery vehicles	Identification of vehicles Identification of driving drivers Status of surroundings	Work condition of mechanisms, accelerometer and fuel level Status (driving, resting, under emergency case) Street lights, speed limits on the roads and other information, meteorological conditions
In 2010s	The study of Chang and co-authors (2010)	Predicts access points using context	Operation of commercial vehicles	Date and time, location of commercial vehicles	The exact names of streets, meteorological conditions, distances among urban places, an overview of trip events
In 2011s	The study of Vieira and co-authors (2011)	Accesses to not private contextual data important for freight transport	Advanced Public data Transport System	Identification of object, its position, routing and stopping data, timing	Meteorological data, nature emergency issues
In 2012s	The study of Alhamdia and co-authors (2012)	Assists the heavy vehicle drivers in avoiding collisions	Improved systems used for vehicle management	Resources positions, road signs	Driving style of the driver (average speed), road signs

(continued)

Table 9.1 (continued)

Time-line	Studies	System using context	ITS application	Primary context	Secondary context
In 2013s	The study of Ramesh and co-authors (2013)	Detects accidents using context data	Advanced system for vehicle control	Resources, position, accidents	Severity (e.g. no people injured) presented about each occurrence
In 2014s	The study of Fan and co-authors (2014)	Detects driver behaviour using context	Advanced Public Transport System	Resources, position, timing	Status and behaviour (when, where, how)
In 2015s	The study of Guo and co-authors (2014)	Assists drivers in crossing cities	System assisting drivers	Vehicle, position, street/lane	Position
In 2016s	The study of Younes and co-authors (2016)	Includes an algorithm for changing traffic lights	Advanced System for emergency vehicles management	Salvage vans (priorities defined)	Density and flow of the traffic
In 2017s	The study of Bohmlander and co-authors (2017)	Provides information about detected potential collision	Advanced Systems used for preventing collisions	Resources, positions, surroundings	Status, location, speed and its change
In 2018s	The study of Iqbal and co-authors (2018)	Analyses contextual data for The Internet of Vehicles (IoV)	Improved Vehicle Management System	Resources, positions, status	Trip length
In 2019s	The study of Chavhan and co-authors (2019)	Integrates the Internet of Things	Advanced systems for transport	Resources, surroundings, identification of trip	Contextual data about resources, metrological data, data about the trip, the status of streets
In 2020s	The study of Doush and co-authors (2020)	Presents interaction models	Proposed Multi-Layer context-aware system framework	Surroundings	Geospatial mapping service
	The study of Dinh and co-authors (2020)	Manages big data	Multi-layer context models	Physical context	Situational context

	The study of Liu and co-authors (2020)	Revises location and accessibility	Transit-oriented development	Transit flow	Level of density
	The study of Jnr and co-authors (2020)	Manages data services	Development for electric trucks	Position	Smart city services (identification of charging stations, etc.)
	The study of Kembro and co-authors (2020)	Manages context for operations	Degree of process automation	Primary data	Timing of operations
In 2021s	The study of Love and co-authors (2021)	Conceptualizes the context of cost performance	Forecasting schedules	Series of performance	Pre-planned schedule
	The study of Parola and co-authors (2021)	Revisiting traffic forecasting	Context of port planning	Historical data	Forecasted traffic
	The study of Brum-Bastos and co-authors (2022)	Analyses Context-Aware Movements	Semantic trajectory model	Space, time, movement data	Potential Path of animal
	The study of Chinnaiyan and co-authors (2021)	Analyses big data and lifecycle of context	ITS assists different transport modes	Infrastructure	Adjusted services
	The study of Liu and co-authors (2021a, b)	Plans urban transit	Urban traffic planning	Locations	Maps urban dynamics
In 2022s	The study of Kalatian and co-authors (2022)	Supports automated vehicles	Predicting the future pedestrian trajectory	Environment, route	Pedestrian and vehicle interaction
	The study of Zampou and co-authors (2022)	Focus on green management	Environmentally aware decision-making	Environment, energy, fuel consumption	Carbon measures

(continued)

Table 9.1 (continued)

Time-line	Studies	System using context	ITS application	Primary context	Secondary context
	The study of Antoine and co-authors (2022)	Focus on road damage detection	Integrates smart environment	Infrastructure	Driver advisory warning
	The study of Deveaux and co-authors (2022)	Supports automated vehicles	An automated risks detection system	Trip, trajectory	Collision risk in entry, drive and exit roundabout
	The study of Kachousangi and co-authors (2022)	Integrates nested logit model	Preference in selecting the driver-less mode	Trips, time slots	Transit costs and mode availability

Deriving context data requires the collection of records, management, evaluation, including dissemination. The authors define the lifecycle phases of a common context for an IoT application (Perera et al., 2014).

ITS uses a variety of algorithms to obtain records with raw data reported from multiple sources.

9.1.1 Onboard Applications

Contextual information about the truck and the driver depends on sensor technologies installed in the road truck and other equipment carried onboard (handies, pads). Table 9.2 shows the various contexts obtained by onboard sensors.

9.1.2 Clouding

Data that is stable for some period could be obtained using the Internet services from the cloud (Svarnamugi & Chinnaiyang, 2017). In particular, metrological conditions, physical freight delivery conditions and other ones could be retrieved from several databases. In addition, data about traffic jumps, traffic lights could be obtained from street monitoring systems.

9.1.3 Supporting ITS

To have working services supporting ITS, a proper infrastructure should be foreseen. The services could be linked with many databases holding records about metrological conditions, parking slots, recent fuel or petrol prices, etc. This data is collected to a common database.

Table 9.2 Collect contextual data with sensors

Data	Initial raw data	Processed data
Truck and trailer data	Position of vehicle, capacity	Metrological conditions, physical truck conditions, timing about the route, current position
Driver data	People occurred on the street and matched the driving direction of the vehicle	Physical conditions of driver and driving skills.

9.1.4 Driver Application

Driver revises the collected and modelled data provided by the intelligent management system. Typically, drivers apply apps, such as smartphones or portable terminals, to enter data. In particular, a request to park a vehicle requires timing, space identification and is provided to the system by the driver.

9.2 Investigating Modelled Contextual Data

Under intelligent transport systems, the data is collected from different sensors having variations in the accuracy of the reported data. For the demonstration and modelling, it is necessary to know the characteristics and particularities of the data. In literature (Sezer et al., 2018; Bettini et al., 2010), there are different contextual modelling methods. The authors best explain the difference among the contextual modelling methods (Perera et al., 2014). Contextual modelling requirements for the intelligent transport systems and the applications of these methods are presented below.

Simplest Model Due to primitive formats and free text information, the data is not useful for merging and investigations. Due to these limitations, the records and not proper for the system of intelligent transport and could be applied for simply modelling.

Markup Model The markup model follows the structure of data in comparison to the simplest model. It applies XML or other formats of interfaces to collect data, keep it and use it for different implementations. Such helps to retrieve records, but even it is one step further than then the previous model, it is not guaranteeing efficient data validation. Several layers to process the data are used and associated with the use of ITS. For example, when forecasting a route, records from GPS sensors and pictures from cameras are transferred and analyzed to identify the best route. Using a markup model to manage records in a model that has several layers of data is not easy.

Graphical Model The implementation of ITS focuses on rapid data manipulation and solution taking. The graphical modelling is used in manipulating contextual data at run time and gathering relationships. The model presents contextual information with defined relations. A UML and other languages support the writing relationships under graphical modelling. Relational and other databases could be applied in the development of applications.

Entity-Oriented Model It identifies the classes of context and relationships. The model focus on the particularities of the paradigm related to the entity, such as expression, extraction, and reuse. Many applications of ITS development use computing, aiming that contextual data can be easily monitored as a hierarchy and integrated into the application.

Logic-Oriented Model Such modelling supports the understanding of contextual information following logical rules. It could help to identify the possibility of accidents and to detect incidents. For modelling reasons, the contextual data is grouped into several clusters based on its significance.

Taxonomy-Oriented Model Such modelling identifies relationships and helps data understanding. It organizes contextual information for the web environment and integrates knowledge from various domains, including operational ones.

Space-Oriented Modelling Modelling could follow facts, objects or ontology and identify suitability for the ITS application, which uses primary data about physical location. The location information could be geographical (presenting the coordinates reported by GPS) and street-oriented (providing the nearby address). This type of modelling is very important seeking to identify the environment of the tracked entity (Batty et al., 2012).

Mixed Modelling It links various techniques by taking best modelling particularities increasing representation of contextual information.

Summarizing Contextual Modelling Contextual information must be collected to identify the behaviour of the object. Meanwhile, the use of the ontological model is appreciated as it allows to represent relationships. The application of ITS, involving the data about object position, is revised under the space model. Table 9.2 provides an overview of the various models which are known and provided by researchers presenting studies in the area of context presentation in the last decades.

Contextual Understanding It is named as inference and could be stated as deriving from which is unknown to the understandable one. This could be described as the extraction of cognition from contextual data. The primary goal of contextual understanding in intellectual transport systems—the elimination of imperfections and ambiguities evident under the data sets. Some stages associated with contextual understanding should be taken

- Pre-processing of context information.
- Establishment of the connection to transfer data.
- Output of the process (Nurmi & Florin, 2004).

Under the first stage, the set of data is cleared to determine the appropriate contextual attributes. The process uses extraction methods such as separation of functions and downsizing to eliminate distortions and data revision. Under the second stage, the data from multiple sources are combined to produce data having higher accuracy, completeness, and reliability. Such stage is important under the usage of ITS because the big number of data sets is produced by the sensors. The third step is to infer context and to bring extra knowledge-based contextual data out of basic contextual information. Several methods of contextual understanding are presented by authors (Swarnamugi & Chinnayan, 2019):

Ontology Logic Method supports the description focusing on contextual understanding. One of the advantages of ontological understanding is that it could be linked with building models. The technique could be applied to combine information from multiple sources and to communicate the data sets (Mantyjärvi & Seppanen, 2002). The method also has disadvantages. In particular, it is not easy to link the model using ontologies with stages identifying which attributes are missed in data collection and communication.

Fuzzy Logic uses the logic of approximate understanding instead of a clear one. The method except 0 (out of the fuzzy set) or 1 (belongs to fuzzy set) also represents cases that are in between. Instead of referring to the minivan travelling at 140 km/h speed, we could express such speed as quite fast. Some cases of the method for representing contextual information are discussed by Mantyjärvi and Seppanen (2002).

Probabilistic Model This method of reasoning allows making decisions based on the likelihood of occurrence of incidents. It is applied under the application of the ITS helping for the prediction of traffic incidents and recognition of pedestrian actions, such as crossing streets, etc. The model incorporating Markov solution (Blunsom, 2004) is applied in ITS as the probabilistic method bridging gaps among collected basic contextual information (i.e. provided from GPS), and advanced contextual data, like the mean of transport or the purpose of the user.

Teaching Machines Method becomes popular in the last setup. The increasing data amounts have led to extensive use of machine learning techniques in various cases. Techniques for teaching machines are mainly used to gather extra knowledge stated under the basic code text. More classes of machine learning methods are used. Deep learning methods are used to gather knowledge from images. Both machine learning and deep learning have several learning approaches that are called learning with supervision, without supervision, or following reinforcement.

Learning with Supervision Machine learning method teaching how to estimate the exact outputs of tests and to classify objects. The authors Al-Shemarri et al. (2020) use the technique for supporting autonomous vehicles in semantic segmentation. The method is applied for objects classification nearby of roads' objects: motor and non-motor traffic participants, lanes, environmental issues, such as water, and so on. More algorithms are used under this methods group. Chen et al. (2019) propose an extreme learning method for reading truck number.

Learning Without Supervision The method of understanding which is applied to revise the information. The clustering method is used for contextual understanding. The method is applied to find similarities among attributes of contextual data. Laerhoven (2001) applies the method to classify data sets seeking to prepare them for context-sensitive usage in intellectual transport systems.

Learning Following Reinforcement Uncontrolled method group that are learned from the surroundings. The method helps the machine to revise signals and decide on better reasoning and behaviour. Reinforcement learning is applied in intellectual transport systems to reach their better performance (Abdus et al., 2011; Harini & Venkatesh, 2020; Santhoshkumar et al., 2020).

The methods are used in various cases of the application of intellectual transport systems, such as delivery forecasting, road light management, etc. Several methods could be applied together in a hybrid way seeking to minimize weaknesses. It is suggested to apply the method of ontological reasoning in combination with ontological modelling to get a higher understanding of contextual data. It has been discovered that in the latest years, scientists showed attention towards the adaptation of machine learning-oriented understanding methods to study contextual data.

The final stage in the contextual data lifecycle is the presentation of contextual information to decision makers. The user is getting contextual data and revises it. This means two-way communication. First, users form their requests based on their use. Later, the intellectual transport system outputs constantly signal in case certain conditions occur. Consequently, this means that the lifecycle of contextual data is cyclical. In cases when the advanced contextual data is derived from the basic contextual data in the understanding stage, the derived contextual information is provided to the users.

These sensors are constantly sensing and generating a big number of data sets, most of them have data that requires revision. The revision helps to add value to the data, it must be analyzed, interpreted and understood. Literature has proven that contextual computing is the best option for modelling, justifying and distributing the added value that was brought to the data. The development of intellectual transport systems is dependent on the Internet of Things and wireless technologies implemented in sensors. These technologies can communicate in various ways, such as wireless, radiofrequency, global positioning networks, etc., by using middleware for collecting, processing and analyzing context, ensuring consistency to transport conveniently, easily, affordably.

The learning approach helps to organize data distribution. Before the data transportation, it is revised to avoid issues that are common for intelligent transportation systems, such as removing noise, revising outliers, learning to commute, and optimizing data sets. The architecture of the Internet of Things and machine learning methods are designed to provide added value to data during its processing, clustering, noise detection, context aggregation, incident forecasting, etc. However, there are some issues for the future, such as:

- Automatic configuration used for the performance of various stages.
- Incorporation and integration of methods by indicating which method would be more appropriate.
- Supporting middleware seeking for the higher quality in service delivery (e.g. consistency, reliability, etc.).
- Ensuring trust—revise the data duplication and gaps of data in the intellectual transport system and middleware transportation layer.
- Building security between users—developing the data access procedure.

Intellectual transport systems are used for solving freight delivery issues and by supporting decision makers with data (Table 9.3). The revision of contextual

Table 9.3 Application of ITS

Year	References	Application of ITS	Contextual modelling method	Contextual understanding method
In 2015s	The study of Guo and co-authors (2014)	Predicting (providing vision about) delivery in urban areas	Knowledge-based modelling/ ontology	Hidden Markov method application
	The study De Brebisson and co-authors (2015)	Predicting arrival	Prediction using traffic density and status of the route	Controlled–Reverse Propagation
	The study Lv and co-authors (2015)	Forecasting flows	Knowledge-based method Historical traffic flow	Simulation under supervision used for ground vehicle system and subsystem
In 2016s	The study of Toke and co-authors (2016)	Volume forecasting	Forecasting future volumes	Models assuming various time horizons
	The study of Siripanpornchana and co-authors (2016)	Estimating timing of driving	Modelling following the data from previous trips	Under supervision
In 2017s	The study Bohmlander and co-authors (2017)	Detection of a possible collision and its prevention	Historical Base modelling techniques	Supervised machine learning
	The study of Xu and co-authors (2017)	Volume forecasting	Algorithm following historical data	Supervised machine learning for short and long time horizons
	The study of Jiang and co-authors (2017)	Identifying transport modes	Object detection algorithm	Under supervision
In 2018s	The study of Yao and co-authors (2018)	Volume forecasting	Knowledge, historical information	Supervised machine learning, adjustments
	The study of Jindal and co-authors (2018)	Driver service	Infrastructure revision model	Supervised machine learning, additional training
	The study of Nishi and co-authors (2018)	Traffic light control	–	Amplification teaching
	The study of Krishnakumari and co-authors (2018)	Traffic status	Algorithm measuring density following data of infrastructure capacity	Deep learning-based object status recognition
In 2019s	The study Chavhan and co-authors (2019)	Advanced ITS application	Probability for an event happening	Machine learning agents

(continued)

Table 9.3 (continued)

Year	References	Application of ITS	Contextual modelling method	Contextual understanding method
	The studies of Geng and co-authors (2019)	Forecasting the flow	Contextual data, outliers detection	Machine learning technique
	The study of Liang and co-authors (2019)	Control of street lights	Algorithm reacting to dynamic traffic information	Amplification teaching
In 2020s	The study of Dovis and co-authors (2020)	Integrated navigation systems for autonomous mobility in urban areas, which limits the visibility of satellites	Data fusion algorithms	Integration framework on extended Kalman filter when multiple reflections of the satellite signals are generated
	The study of Stepanova and co-authors (2020)	ITS road weather services	Temperature forecasting	Cellular network and hybrid models predicting route conditions
	The study of Dass and co-authors (2020)	The Internet of Things based ITS	Selecting optimal data for decision-making	Integer linear programming model
	The study of Choy and co-authors (2020)	Speed monitoring with RFID for ITS	The modular artificial neural network to increase vehicle detection	Speed calculation methodology
	The study of Faiyetole and co-authors (2020)	The intelligent transport pricing system	Pricing algorithm for shared-transport	Policy framework for intelligent transport systems, taking into account the needs of transport users as a context specificity
In 2021s	The study of Autili and co-authors (2021)	Cooperative-ITS application	Suggestions for eco-driving in respect to traffic	Multi-goal recognition technique
	The study of Mare and co-authors (2021)	V2V, I2V, V2I communication models	In GPS signal lost cases street light is integrated	Testing methodology
	The study of Ahmed and co-authors (2021)	Cooperative-ITS application	Collective behaviour (adaptive delivery window), Parallel computing	Statistical learning algorithms

(continued)

Table 9.3 (continued)

Year	References	Application of ITS	Contextual modelling method	Contextual understanding method
	The study of Rudskoy and co-authors (2021)	Digital twin model for roads traffic	Optimized planning and decision-making	Reference model of the services
	The study of Guillen-Perez and co-authors (2021)	Incorporation of artificial intelligence in ITS	Random early detection of traffic for vehicles	Evolutionary algorithms
	The study of Gao and co-authors (2021)	Automated vehicle support services	Evaluation of automatic parallel parking system	Genetic algorithm
In 2022s	The study of Tran and co-authors (2022)	ITS integration with smart cities	Avoiding traffic overcrowding in mega cities	Data analysis model for ITS using partial least squares structural equation
	The study of Shepelev and co-authors (2022)	Cooperative-ITS	Forecasting the queue of highly automated vehicles	Region-based heuristic, fuzzy logic methods
	The study of Li and co-authors (2022)	Energy-saving model for driving behaviour	Driver behaviour decision-making, modelling micro and macro effects	Recurrent Neural Network
	The study of Fourkiotis and co-authors (2022)	ITS which increase safety on the roads	Prediction of pedestrian behaviour	Deep learning techniques, deep neural network, logistic regression model
	The study of Altaf and co-authors (2022)	Autonomous vehicle technology used in ITS	Analysis of network, the active and passive driver behaviour	Approaches for Collision Prevention

information is the highest issue. The author analysed contextual understanding methods and summarized their application. The author revealed that increasing the number of records in the system of intelligent transport (ITS) is obtained from the sensor technologies and is used to transform raw data into investigations, using various machine learning methods and models. In addition, it becomes clear that learning methods have been indeed popular under the application of ITS at the stage of context understanding. The impact of the Internet of Things and various techniques using machine-oriented learning approach are supporting the extensions of services and applications of the system of intelligent transport (Rawal & Devadas, 2015).

9.3 Characteristics of the Transport System

Traffic forecasting and real-time response: mass data collection about traffic helps decision makers to revise traffic volume, bandwidth and the events that happened near to real-time, and traffic management systems should respond to accidents in real-time, allowing to immediately make solutions following the algorithms forecasting traffic and guiding traffic flows. The design of a contextual control system of traffic lights could reduce waiting times at intersections, to react to a rapid incident rescue system improving emergency response. Constant analysing and the system forecasting traffic by using extensive information about the traffic helps to reach efficiency.

By the way, the existing system for data collection, vehicular ad hoc networks (VANET) and the traditional model for traffic flow forecasting (Wan et al., 2016) are not sufficient to address the problems stated above. In case, VANET supports services and achieves success in some aspects, such as a collision avoidance system using a vehicle-to-vehicle or an infrared truck (Zheng et al., 2015). However, its specific shortcomings, such as imbalanced flow of traffic or lower throughput, affect the efficiency in the management of traffic (Li et al., 2016). Therefore, attempts have been made to urgently develop a higher efficiency system for traffic management.

Rapid development of emerging technologies, such as the fifth generation (5G), the network defined by software (SDN) (Hu et al., 2015) and mobile computing (MC) (Nunna et al., 2015), is expected to drive progress in traffic management. Having implemented technologies of 5G and SDN in the vehicular networks, such offers a communication service flowing high-bandwidth in a more flexible and programmable way (Ku et al., 2014), making the environment more mobile. Meanwhile, MEC transfers the calculation resource to the mobile network and takes critical action to respond following real-time data.

The transport system has big complexity, so different strategies can be analyzed to improve its efficiency. From the point of view of cost optimization in a two-stage (or two-tier) transport system, several sets of problems arise can be distinguished. The first set of problems is oriented to the optimization of costs for the movement of vehicles in the transport system, and the second set is the routing between locations.

Location-dispatching models have been used for the decisions focusing on customers' allocations, investigating costs of delivery of freight to customers (Gendron & Semet, 2009). The Location-dispatching model follows an estimated costs for vehicle fleets delivering freight from consignors to customers.

Lin and Lei (2009) formulated a 2S-LRP (Two-stage location routing problem) for investigating the choice of the best delivery route in each stage (Lin and Lei, 2009). Freight could be delivered directly from the consignor, forced to pass through a reloading warehouse or shipped to customers directly. The 2S-LRP has been used for defining the optimal route between the consignor and the consignee. It connects the two stages and manages their interdependence, i.e. the formulation can be considered as a special case of 2S-LRP (Nguyen et al., 2012). Moreover, 2S-LRP authors use the facility location problem (FLP) set to search for the best place to find

Table 9.4 Improving the efficiency in the transport system: a framework for assessment

Topics for investigations	Efficiency improvement criteria & strategies	Problem families	References
Nature of transport system	Multi-product	IRP	Shaabani and Kamalabadi (2016)
	Multi-period	IRP	De Arijit et al. (2020), Shaabani and Kamalabadi (2016)
	Number of stages	FTL	Lin et al. (2006)
	Clustering-based	VRP	Wang et al. (2018a, b)
Transport orders side-approaches	Vehicle usage	VRP	Nakhjirkan and Mokhatab Rafiei (2017)
	Number of loads	TTRP	Cuda et al. (2015), Marianov and Eiselt (2016)
	Volume per product	FTL	Lin et al. (2006
	Amount per client	LRP	Ambrosino and Scutella (2005)
Transportation type	Warehouse-to-warehouse delivery	LRP	Gonzalez-Feliu (2011)
	Direct delivery to the customer	LRP	Gonzalez-Feliu (2011)
	Reloading	IRP	Mirzapour Al-e-hashem and Rekik (2014)
	Transportation with consolidation	LRP	Gonzalez-Feliu (2011)
Delivery type	Round trip delivery	IRP, VRP	Gonzalez-Feliu (2011), Drezner et al. (2018)
	One-side delivery	IRP	De Arijit et al. (2020)
Distance type	Arch-based (Euclidean)	LDP	Gendron and Semet (2009), Aboolian et al. (2020)
	Path-based	LDP	Gendron and Semet (2009)

IRP routing of the inventory problem, *FTL* full truck load problem, *LRP* routing for location problem, *TTRP* routing truck and trailer problem, *VRP* routing vehicle problem, *LDP* location-distribution problem

a facility and use other problem sets to minimize delivery costs in light of constraints on a vehicle and its supply schedule.

The table below presents the main strategies and criteria important to setup an efficient system of transport and investigate specific problem sets (Table 9.4).

The authors listed in Table 9.5 focus on various time horizons. The authors show that the construction of service-oriented transport systems requires investigations on the time scale of the planning horizon (Timpe & Kallrath, 2000). They also suggest incorporating more elements.

The main elements provided in Table 9.5 are important for carriers. Works published by authors such as Kim et al. (2017), Ji and Sun (2017), Allen et al. (2018) also highlight their importance. Any activity requires delivery of freight to customers and deliveries from producers to customers if the product does not match their needs. Freight transport is a costly activity. Therefore, researchers analysing the

Table 9.5 Key elements for transport system revision

Elements	Lead-time	Transportation Time	Transportation Costs	Incomes for transport services
Studies	Lee and Kim (2002), Lim et al. (2006a, b)	Timpe and Kallrath (2000), Kallrath (2002)	Sakawa et al. (2001), Ryu et al. (2004), Oh and Karimi (2006)	Chen et al. (2003) Gupta and Maranas (2003), Perea-Lopez et al. (2003)
	Bilgen and Ozkarahan (2007), Romo et al. (2009)	Chen and Lee (2004), Rizk et al. (2006)	Selim and Ozkarahan (2008), Aliev et al. (2007)	Meijboom and Obel (2007), Jung et al. (2008)

costs structure propose to combine loads from manufacturers with deliveries of freights to clients to get lower service costs. The authors also identified that the integration of several streams minimizes transport service expenses. By following the research, an integration solves the problem when vehicles serving a set of routes fulfil many requests for transportation. Recently, Lahyani et al. (2015) have integrated additional constraints into the pickup and delivery. However, pickups and deliveries are usually treated as different destinations served by several modes of transport. Bergmann (2020) explored the impact of an integrated approach serving pickup and delivery on a single transport route (Manzini & Bindi, 2009). The effect of the one direction going from producers to customers and backwards from customers to producers can be combined, identifying the lowest total distance for the combined flows. Here transport costs going forward and backward can be illustrated by a cost function. This linear function was constructed first by Bergmann et al. (2020).

Designing the transport system following the recommendation by Bergmann et al. (2020) combines deliveries to clients with loads from manufacturers. However, these studies do not cover backward freight carriage (Bergmann et al., 2020; Linnainmaa et al., 1995), and they should be analyzed for specific cases, such as small-package delivery.

To fulfil the stages mentioned in Fig. 9.1 separate trucks are used. In various business cases, deliveries from producers could be made directly to business customers. Still, if a freight contains more than one product, more than one delivery could be required to the customer. Freights from different manufacturers are loaded and shipped to the warehouse and later on are separated to pick up the orders placed by customers (Cohen and Lee, 1985).

Based on Fig. 9.1, the transport system consists of stages. The majority of them involve a change of the vehicle in at least one intermediate facility. In such cases, freight is unloaded from an incoming vehicle and loaded onto another vehicle. Freight may be affected by package arrangement or change of a vehicle without any changes to the unit of measurement (i.e. when a full load does not change the load nature), or other important operations when reorganizing from big into small loads.

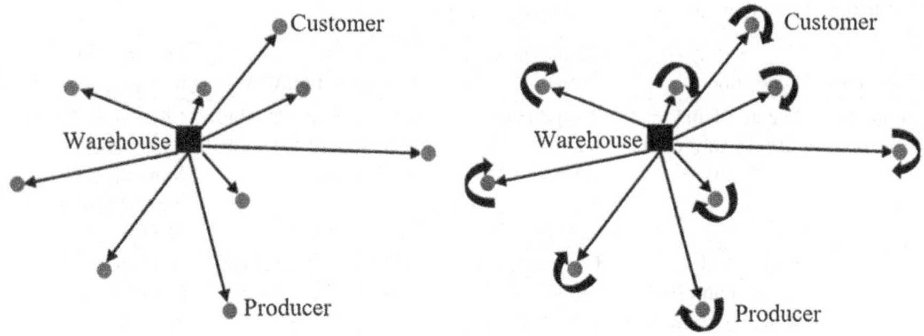

Fig. 9.1 Service-oriented transportation system (from the left side): delivery scenario; delivery and return with load scenario

The route is planned before its start and in the course of a route. The algorithm proposes the best possible transfers and has found a better transfer, changes are made. An optimization is possible for short distances (like last mile delivery) or when transfer points are available. Dynamic fleet management requires real-time data, monitoring of unforeseen events that could affect route execution.

This algorithm relying on dynamic data could be also installed in the construction of a new route and later matching it with real-time information using context-aware data for traffic forecasting and real-time response setup.

9.4 Routing Algorithm

Enterprises have many ways to solve routing problems, including route creation, two-phase, and route improvement algorithms (Cormier, 2005). Several classic methods will be introduced below.

The famous heuristic Clarke and Wright's method was developed in 1964 follows the savings of addressing two customers in succession on a combined trip, which is cheaper than going on several routes. In this algorithm, separate trips are combined if this gives a positive result on costs reduction. Another heuristics was developed 12 years later. The authors of this solution applied a sequential approach for clients between two customers in sequence, taking into account how far away clients are located from the warehouse (Mole and Jameson, 1976). Another author applied several stages of insertion heuristics which first has insertion sequential and after parallel, using them for scenario comparison (Christofides et al., 1979).

Two-phase methods have been provided by researchers to solve several problems: the construction of clusters and routes.

Savings algorithms are used for comparing the case when orders of two customers are combined in a single trip, including a case supporting individual transportation of different orders.

The researchers apply several stages approach for a constant amount of transport means (Bramel and Simchi-Levi, 1995). The technique helps to construct routes by reducing the total length between clients in the trip, considering the total volume of each client. In case no more customers are possible to assign to the vehicle, another route is initialized for it.

A list of vehicles is used for the construction of routes for the transport park, an intermediate decision is provided helping to form full trips.

A travelling salesman problem is also treated as an order pickup problem. Suppose an enterprise gets a notification about required freight delivery—goods required by the client. The capacity of a vehicle is assumed to be sufficient to pick up all the pallets listed in a transport order.

Ratliff and Rosenthal (1983) described a dynamic programming algorithm. Later, Carlier and Villon (1987) also analyzed this problem, calling it Steiner TSP.

The third category is route modification algorithms. Several techniques are applied to optimize pre-constructed routes using different heuristic methods. This begins by searching for a neighbourhood algorithm to get better costs. Having found an improvement, a new solution is used (Cordeau et al., 2007).

The standard algorithm includes Incapacitated and Capacitated truck routing problems. A truck routing problem with limited capacity (CVRP) is applied when the volume is constant. Under the papers, UVRP and CVRP solutions have been separated into ones that incorporate linear function and non-linear function. The model of the first type was developed for incapacitated and capacitated trucks (as resources). Some of the authors investigated the problem, where customers can be located anywhere, without limitation to predefined potential locations (Lopes et al., 2016).

In several models, authors analyze multi-product methods, considering solutions with cost variations, full loads discount and different transportation means. The authors defined shipment costs by following the volumes and concluded that they do not depend on the order size (Lin et al., 2006).

First, researchers use order constraints to revise capacity aspects; secondly, researchers think that demand is clear (particularly for CVRP); third, the construction of geographical clusters is still studied by the authors.

VRP could effectively deal with various restrictions and multi-use of trucks. Multi-tasking is referred to a minimization of the route distance when multiple tasks are allocated to a single-vehicle.

The new heuristic method includes the alignment of the trip with time window constraints. Recent heuristics are applied for solving freight shipment problems. After a task is placed, "real" trips are constructed taking into account determined time windows but no restriction on vehicle capacity. After the carrier leaves the warehouse, serves clients with the lowest costs, and travels back to the warehouse (Ribas et al., 2011).

An algorithm with time-window constraints is applicable also for carriers by sea seeking to avoid delays (Pace et al., 2015). VRP aims to assign trips for carriers travelling round trip and finishing routes at the warehouse so that only one vehicle arrives for each customer. The researchers assume that trailers are overloaded in planning their trips (UVRP applies to ships that carry a single load only between ports) (Yonglin & Awasthi, 2012).

Problems with time-window cases and varying trip duration due to delay must be accounted for in constructing routes and schedules for sea transport, were defined by Kong et al. (2015).

The algorithm is applied to solve the issue when the single carrier serves multiple trips with time windows constraints provided by different clients. To solve this problem, a method based on an elementary algorithm for finding the optimal way and estimating the constraint of resources is proposed. The algorithm is separated into several stages: creating all non-dominated valid routes and their selection and sequencing to organize workdays of vehicles (Azi et al., 2007).

The algorithm for serving several routes with one vehicle over the time horizon is provided by researchers (Azi et al., 2007).

The problem for routing inventory (IRP) is different from the vehicle routing problem (VRP). VRP occurs after the client sends the request and the person match deliveries to vehicle trips. IRP is the branch of VRP focusing on how to route inventory. In the case of the Vendor Managed Inventory (VMI), a supplier selects the size of delivery and time frame guaranteeing inventory availability in-full and on-time.

VRP deals with planning one-day horizon requiring that freights are shipped during the period. The algorithm for inventory routing is identified for the time frame, which is longer. Every period, suppliers make a what freights to ship and to which clients, evaluating shipment costs and time frames. To find the solution with IRP is not easy, but it uses some common elements as other algorithms do. These methods plan routes looking to short time horizons. Initial methods are used to consider a single day, while earlier ones took several days into account. There are several problems with that: the first one relates to modelling the effect for the longer time frame and the second with customers to be visited in the short period delivery schedule. The approach for a short time horizon plan with the lowest expenses could pre-plan deliveries for future periods. A solution must consider the added value to the client following not only expenses but also specified volume. Such an approach helps to avoid cases with immediate shipments to the client but rather deliver them earlier than needed (Jozefowiez et al., 2009).

When the plan for the short time horizon is used for a shorter period, the broadening of the vehicle routing problem could be used. The authors presented the applications showing that shipment volume fluctuates around the average fixed volume (Campbell & Savelsbergh, 2004). The mentioned researchers also discussed the generation of an optimized shipments time table, determining in which the clients are served, visit times, thus maximizing the amount delivered to customers per route. The use of the single day approach facilitates daily forecasts avoiding the difficulty to build forecasts for a longer time horizon.

In the initial stage, techniques with discrete time are applied for estimating the number of clients to be assigned to a route that will be fulfilled sooner, and the number of transport units for transportation is selected. Seeking to reduce expenses, LP helps in defining potential routes. It allows for determining delivery volumes to

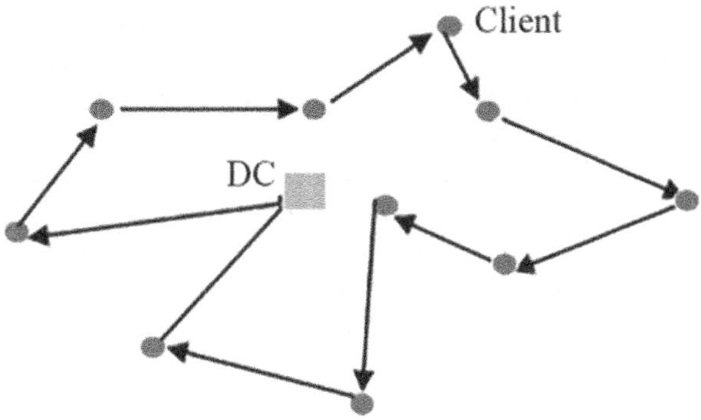

Fig. 9.2 Backhaulage route

customers, then assigning customers and vehicles to trips and planning when to start a route. The plan constructed for a short time horizon estimates product delivery volumes. In generating a long time horizon plan, the volume indicated by customers is applied, which means that transport units shipped to a client minimizes the volume left to be transferred in the near future. Shipment expenses for each time horizon are estimated by transport managers and are manually entered into the model.

The decisions depend on two characteristics: timing and routing.

The authors use two stages approaches (Campbell & Savelsbergh, 2004). The mentioned researchers determine visits to clients in the coming days and suggest shipment schedules to the same clients. During the next stage, the authors determine pre-planned shipment trips and volumes. This suggestion gives more precise planning and takes into consideration drivers' work time. Such helps to improve the use of vehicles. The rolling horizon framework can also be used.

The solution aims to determine the number of fully booked vehicles and deliveries. The amount of vehicles is determined at the initial stage, and the timetable per vehicle is created in the next step. The authors provided a new method that takes daily vehicle timetable with identified delivery time windows per client (Shen & Sessions, 1989; Ghilas et al., 2016).

The author suggested three-phase methods (Linnainmaa et al., 1995). The first phase is applied to figure out the distance for the trip and assess the volume for point allocation. The second phase is foreseen for the creation of one-week vehicles timetable, which is generated using developed algorithms. The last stage is applied to make modifications for single day routes.

The authors suggested a simulation method with a heuristic that follows the rule when a single truck load per time is applied using empty vehicles (Fig. 9.2) on the dynamic time horizon, and a one-day vehicle timetable is created (Weintraub et al., 1996). The researchers (such as Mendell et al., 2006) propose other methods using simulations to create trip timetables for vehicles.

The researchers have investigated dropping empty trailers at warehouses and picking new ones that are already filled [reducing time for trailer loading (McDonald et al., 2001a, b)]. Various rules have been proposed aiming to get solutions focusing on operations leading to higher efficiency. The researchers proposed an algorithm that deals with various complex aspects (Mendell et al., 2006). In particular, the vehicles are assigned by the algorithm to warehouses, and after routes are optimized, identifying which trucks will have routes from the warehouse the next day.

The authors suggested a three-stage algorithm for courier deliveries (Audy et al., 2011). The initial stage is applied, seeking to create possible trips. During the next stage, trips are allocated per vehicle to cover a one-week volume. At the last stage, the revision of the timetable is applied to get optimal solutions (Bell et al., 1983).

Authors Krommenacker et al. (2016) identified context perception at all levels (device independence, mobility, wireless support), the quality of service, the seamless discovery of services and content, better user control and effective delivery as important requirements.

The VRP algorithm could help to allocate trucks for clients aiming to minimize the transport costs. Afterwards, routes fulfilment is matched with real-time information which is retrieved by using context-aware services.

9.5 Clustering Applicable in Routing

The routing approach requires clustering. Grouping or classifying clients reflects a particular cluster. K-means clustering is the most popular method for implementing and attributing the solution to this problem, grouping customers into groups. A routing scheme following the destination context recommends taking different systems into account when forming groups, also considering the vehicle speed, vehicle destination, and delivery delays. Over time, new customers that need freight delivery services to appear, and customers are reclassified. For such cases, a learning-oriented system for routing vehicles is used, where the routing supporting system learns implicit provisions that planners have when manually creating route plans (or routes). The goal is to use these learnt subjective attitudes without an objective criterion of distance vehicle routing systems. It is an alternative way to practice the creation of distinctive customized trips for business that has their route construction requirements. Instead, planners think about previous vehicle routing decisions and compare them with similar customer groups. The learning decision support system is applied to make similar decisions (Murphy, 1995). The learning method follows the Markov model learning concept corresponding to the matrix of probabilistic distance transition rather than a deterministic one. It does, however, allow the use of existing arc route VRP software to create and optimize actual route selections, distances and priorities at the same time. Various schemes for learning are used by creating a probabilistic transition matrix that can evolve as preferences change.

Below is the presentation of a small clustering example, using the only preference—distance. When the first group of clients forms a cluster, the sequence of visiting them on the trip is revised. Several methods have been provided, helping to build clusters. The solution assigns a truck to delivery within the cluster.

In studies, demand points are clustered; and a delivery route is constructed within the cluster. Two paths are evident: one of them, first cluster customers and then plan the route. Herein, service points are allocated to clusters, and trips are constructed within the groups. The path starts with the nearest the warehouse service point, and other service points are assigned to the vehicle sequentially. Final trips for each cluster are constructed. Next path, the first route is constructed, and then customers are clustered. The first step—the construction of full load (FTL) routes.

Literature on clustering methods suggests the following for backward routes:

- Sequential clustering—one vehicle starts operating on a delivery tour. After the collection of deliveries in one area (zone) is finished, products are then collected from all other zones.
- Parallel clustering—all vehicles can work on deliveries at the same time. This synchronization of tasks is aimed at finishing any delivery tour faster.

Customers are assigned to a territory cluster focusing on the neighbourhood logic. In presence of a large variety of neighbourhoods, more routes will be created.

To review efficiency, customers are clustered. The clustering of demand points includes three steps. For clustering purposes, all service points are first associated with nearby cities; cities (towns) are then assigned into subclusters meaning regions. At the third stage, subclusters are combined into clusters, which are territorial units, the midpoints of which are located no closer than a given threshold value. After such hierarchical clustering and enabling distance parameters, cities (towns) are associated with clusters (initialization phase). In addition, a minimum distance threshold (termination step) is used to determine the number of clusters (aggregation step). During the initialization phase, cities are associated with subclusters [as specified in Eq. (9.1)]:

$$P'_a = (J_1, \ldots, J_z), \tag{9.1}$$

where: P' is subcluster, a—unique subcluster number, z—number of cities (towns).

Under the aggregative phase, two subclusters are grouped into a single cluster P [as presented in Eq. (9.2)]:

$$P_R = \left(P'_1, P'_2\right), \tag{9.2}$$

where: R—the unique number of cluster P.

Based on the results, the threshold value of the minimum distance is revised and subclusters are formed accordingly. Finally, a transport system based on clustering is built (Fig. 9.3).

Service points generate transport orders. In the development of the model, demand (service) points are assigned into clusters. Each demand point is assigned to the city by location place.

Parameters for clustering:

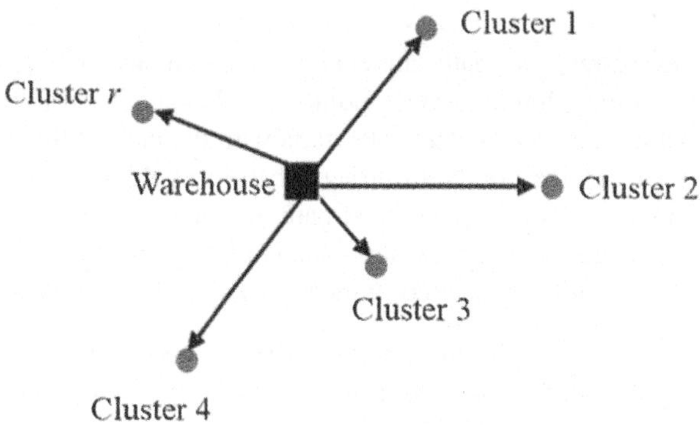

Fig. 9.3 Structure of transport system constructed from clusters

Table 9.6 Constructed clusters and geographical locations of their mid-points

Cluster No.	Cluster mid-points	Latitude (Lat) of mid-points	Longitude (Long) of mid-points
Cluster No 1	Marupe	56.8	24.1
Cluster No 2	Litene	57.2	27.0
Cluster No 3	Vanga	56.8	21.8
Cluster No 4	Pliavinios	56.6	25.6
Cluster No 5	Rozula	57.4	25.0

Table 9.7 Distances between mid-point of constructed clusters

Distances	Cluster No 1	Cluster No 2	Cluster No 3	Cluster No 4	Cluster No 5
Cluster No 1	0	186	135	98	89
Cluster No 2	186	0	317	109	123
Cluster No 3	135	317	0	233	202
Cluster No 4	98	109	233	0	97
Cluster No 5	89	123	202	97	0

- The threshold value is 85 km.
- Min variation of distances within-cluster is 34 km.
- Within-cluster max variation of distances: 52 km.

The mid-point represents the centre in the cluster. If mid-points are close to each other, a higher hierarchical cluster is constructed. In the case of Cluster No 1, there are two subclusters, and their mid-points is located 43 km away from each other (Ogre: Lat 56.87 and Long 24.17; Platuone: Lat 56.56 and Long 23.73). Having combined 10 and 3 cities, a cluster was formed with a mid-point in Marupe (Table 9.6).

The distances between mid-points are provided in Table 9.7.

Fig. 9.4 An example—Cluster No 4 with six cities and mid-point location

The different numbers of cities could be assigned to a cluster. There are clusters with more cities, and there are those with fewer. The minimum is two cities, and the maximum is 13 towns in the cluster.

Among any two points, we could draw straight lines. The intersection of these lines is identified as mid-point (Fig. 9.4)—the geographical point, which is defined by longitude and latitude values.

For modelling purposes, the demand among clusters could be weighted or unweighted (i.e. demand in one cluster is equal to demand in another cluster). The clustering option is selected following the results of the simulation representing 12% of the decrease in freight delivery expenses.

9.6 Development of Routes Using Intelligent Data Management Components

Main constraints are included in the developed algorithm (Fig. 9.5):

- Number of stops per route.
- Distances between service points.

9.6.1 The Construction of the Schedule

For the construction of the schedule, it is important to fix that each customer is visited only once. This option is used to achieve better costs. Having this fixed, the routes are created.

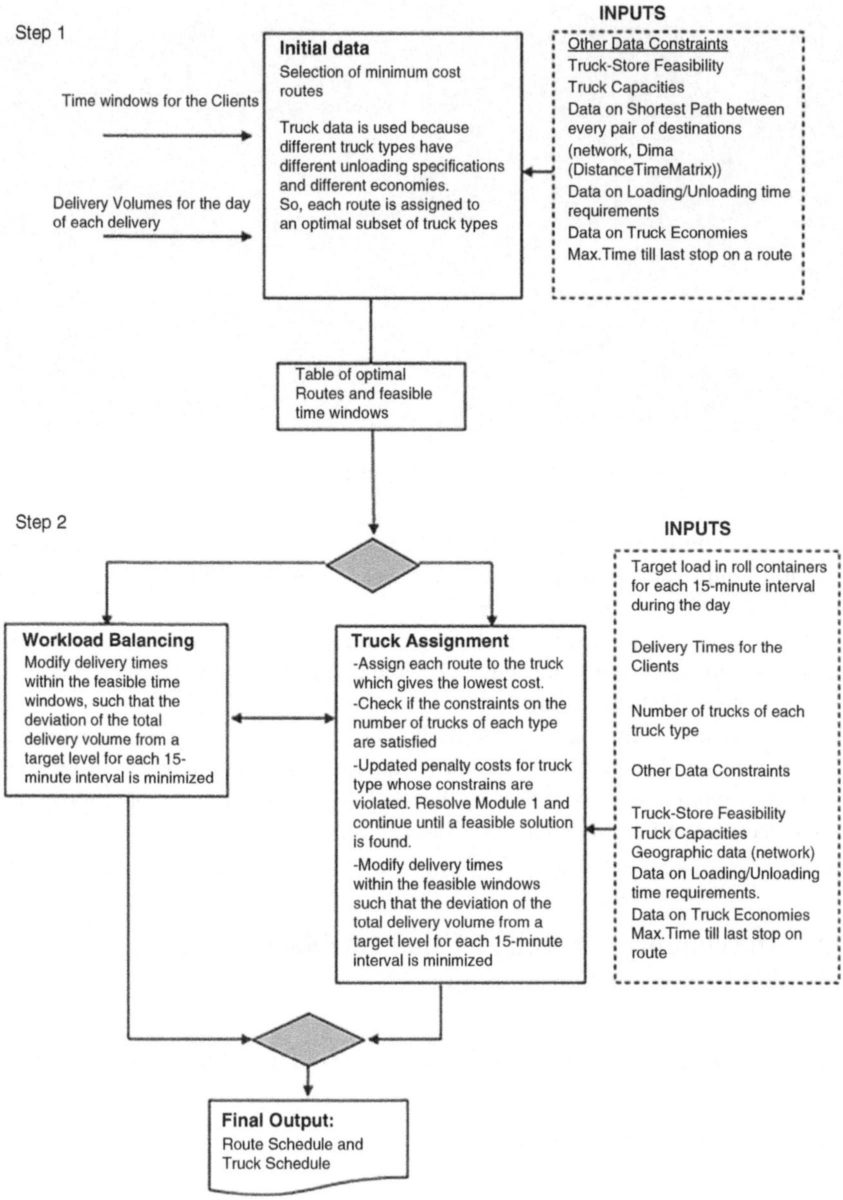

Fig. 9.5 Data flow and process flow for new route schedule

9.6.2 Assignment of Vehicle

The assignment of vehicle search for the possibility to reduce the number of trucks. Another important constrain is the number of stops per trip.

9.6.3 *Actual Routing*

Some key variables are important.

Variation of kilometres during the trip.

Freight shipment: time to pick the freight.

Freight delivery: time to drop the freight.

9.6.4 *Getting Started*

This is a brief description of the process. Information is needed for the route schedule include:

Customer position.

Characteristics for each trailer (the size and the tariff for service).

Distance to be covered under freight delivery service.

Customer data in most companies is received from a customer order entry system.

Having accepted the order, the customer is assigned to a cluster. So planners must keep that information.

Geographic information could be another problem for finding a network acceptable to carriers. Before finalizing the schedule, several possible schedules are compared and one is selected.

The customer orders is prioritized in such sequence:

- At first, customers whose location has the shortest distance parameter are selected from all candidate customers.
- Second, additional customers are added with intermediate distance parameter.
- In the last step, the last customers with the longest distance parameter are added. The revision in which clients' deliveries could be postponed for the next days is important (for such identification the review of agreements and longer lead-times are important).

Results show that customers with closest neighbour places are selected first (Fig. 9.6).

9.6.5 *Steps for Route Construction*

Several steps are taken during route construction (Fig. 9.7).

The author investigated cargo delivery distances from retailer warehouses to customers located in Baltic countries (see Table 9.8).

Almost one-half of all freight that arrive to customers in Baltic countries are delivered under short-distance delivery cases.

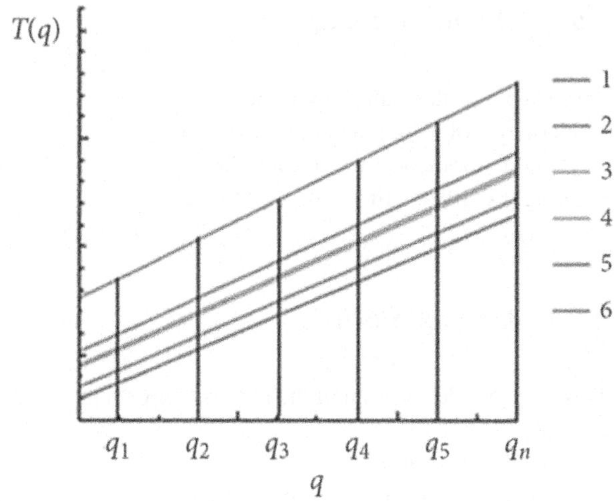

Fig. 9.6 Scenario analysis, herein: *q*—transport units

9.7 Costs Metrics Used to Calculate Transportation Service Costs

Intense competition in the market has forced the search for new ways of managing delivery services. Operating under such conditions may have additional benefits. Discounts on product or service costs or bonuses can be examples of such benefits. The task of optimizing multi-criteria of road transport services, which will ensure its efficient functioning TC, also ensuring that all contractual obligations to customers have been met, is interesting from a practical and theoretical point of view (Galkin, 2015). The aim is to attract new customers and retain their loyalty, at the same time reducing their costs for transportation services.

To compare solutions, service costs could be revised. Transport service costs can be calculated according to the formula Eq. (9.3):

$$T = w \sum_{R=1}^{Q} q_R d_R, \tag{9.3}$$

herein: *T*—transportation costs; *Q*—the number of *P* clusters; *q*—the average quantity of transported units; *w*—the average costs per transported unit per kilometre; *d*—the service distance in kilometres.

There are two basic equations used for transportation and transshipment [in Eq. (9.5)], where the inputs and variables are used to form Eq. (9.4):

p—Delivery of transport units from the shipper $\forall \in p,$

qR—Demand for transport units in cluster No R $\forall_R \in q,$

UR—Costs for sending transport unit from the shipper to cluster R (Eur/transport unit) $\forall_R$

xR—Flow of transport units from the shipper to cluster No R $\forall_R,$

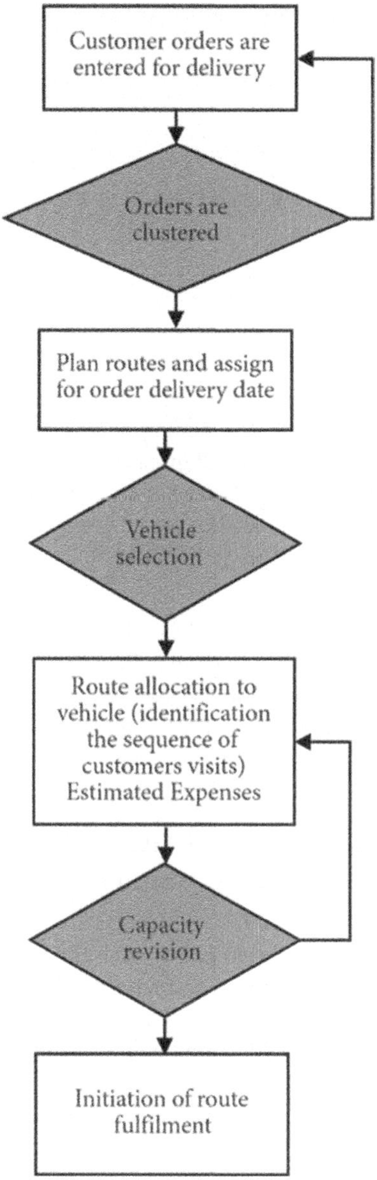

Fig. 9.7 Steps for route construction

Table 9.8 Freight delivery: case study results

Delivery	Share (%)	Distance type
0–149 km	47.4	Short-distance
150–299 km	11.7	Intermediate-distance
300–1000 km	40.8	Long-distance

Y—Minimum costs, euro

$$Y = \mathrm{Min}\left(\sum_R U_R x_R\right), \tag{9.4}$$

where are three constraints to be evaluated by Eq. (9.4):

$$\sum_R x_R \leq p \ \forall \in p,$$

$$\sum x_R \leq k_R \ \forall_R \in q,$$

$$x_R \geq 0 \ \forall_R.$$

In the reloading case, an extra condition was added for reloading at terminal s (Eq. 9.5). It states that the volume unloaded from delivering truck is equal to the volume loaded to the new truck.

$$\sum\nolimits_s x_{sR} - \sum\nolimits_R x_{Rs} = 0, \forall_R \notin p, \notin q. \tag{9.5}$$

The costs shown in the transportation and transshipment models may also include various operating costs. Next, a structure of costs elements is provided to increase the efficiency of transport system.

9.8 Distance Metrics Used for Calculating Transportation Costs

A decision support system (DSS) allows dispatchers-planners to interactively solve vehicle routing problems within a territory using time windows, appropriate calculation methods, and a customized knowledge base that includes traffic and spatial data. A DSS called map-route generates routes that meet time and vehicle capacity constraints. Its calculation engine is based on an efficient heuristic method to solve the main optimization problem, and its implementation is developed using the popular Geographic Information System (GIS) platform. Map-route offers highly efficient user-friendly solutions and can find answers to a variety of "what if" scenarios that can have high-cost implications.

Most commonly in studies used the types of distances: Euclidean one and path one.

Authors analyzing the location of loading place, which increases the efficiency of the transport system, follow Euclidean distance logic Zhao and Batta (1999), Elshaikh et al. (2016), Erkut et al. (1994), You et al. (2019), Al-Loughani (1997), Murray and Tong 2007

Table 9.9 Measuring distance for freight delivery

Service points	J_1 (lat$_1$, long$_1$)	...	J_n (lat$_n$, long$_n$)
I_1 (lat$_1$, long$_1$)	–	...	
...		...	
I_n (lat$_n$, long$_i$)		...	–

To adjust distances of freight delivery from suppliers to the delivery point, applying the Euclidean distance calculation method, which helps to identify the length of deliveries is possible.

For Euclidean distance estimation, the values of Latitudes (Lat) and Longitudes (Long) of two locations are inserted into Eq. (9.6) and presented below (in Table 9.9):

$$\begin{aligned} d_{IJ} &= \text{ACOS}(\text{COS}(\text{RADIANS}(90 - \text{lat\_I})) \times \text{COS}(\text{RADIANS}(90\text{-lat\_J}))+ \\ &\text{SIN}(\text{RADIANS}(90\text{-lat\_I})) \times \text{SIN}(\text{RADIANS}(90\text{-lat\_J}))\times \\ &\text{COS}(\text{RADIANS}(\text{long\_J-long\_I}))) \times 6371, \end{aligned} \tag{9.6}$$

where: d–Euclidean distance (kilometres).

The shortest Euclidean distances L_j for all locations I are summarized and computed in Eq. (9.7):

$$L_j = \sum_{i=1}^{I} d_{ij}. \tag{9.7}$$

A path linking producers, terminals and customer locations follows distance and is covered by land transport. To choose the loading place, the path-based shortest service distance approach could be used the same way as specified in Eq. (9.7). The sum of path-based shortest distances L_j' to all locations is calculated in Eq. (9.8):

$$L_j' = \sum_{i=1}^{I} d_{ij}', \tag{9.8}$$

where:d_{ij}'—the path-based distance from the analyzed loading location j to location i; I—the group of points.

Euclidean and path-based distances (forward and back) are included in Eqs. (9.9) and (9.10):

$$L_{jr} = \sum_{i=1}^{I} d_{ij} + \sum_{i=1}^{I} d_{ijr}, \tag{9.9}$$

$$L'_{jr} = \sum_{i=1}^{I} d'_{ij} + \sum_{i=1}^{I} d'_{ijr}, \tag{9.10}$$

where: d_{ijr}—Euclidean distance for back delivery from manufacturer j to customer i; d'_{ijr}—path-based distance for back delivery from manufacturer j to customer i; r—back deliveries.

All these scenarios impacts the costs for transportation services.

9.8.1 Summary of Distance in Backhauling Case

Companies organize freight transportation following return orders placed by customers. Having delivered freight to clients, carriers drive to suppliers to ship freight, however, in the situation of extra loads, trucks must operate additional routes, which means more kilometres and higher expenses.

The author estimated the impact of returns on service distance. These results are presented graphically in Fig. 9.8.

The figure illustrates the impact on distances, which grows exponentially when more freights are returned. The impact of increased returns is the highest in the path distance cases and has effect on transportation costs.

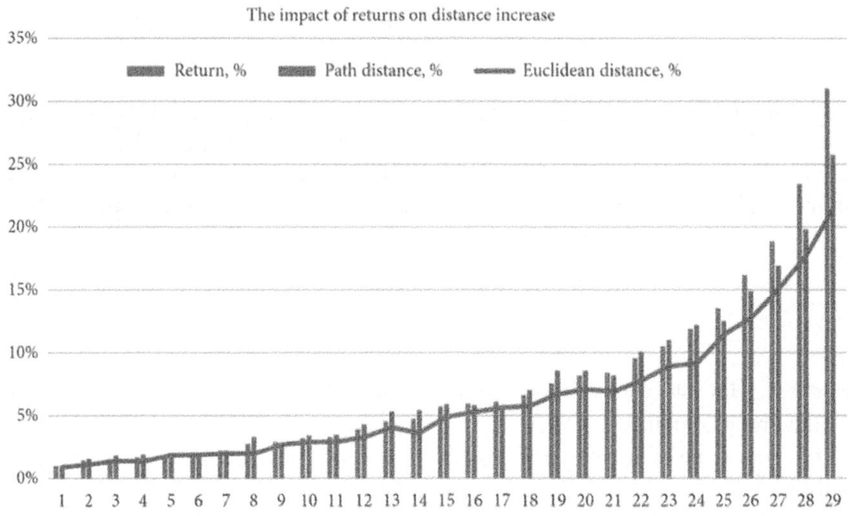

Fig. 9.8 The effect of returns on distance

9.9 Timetable Metrics for Scheduling Services

In general, there are not so many optimization algorithms that focus on scheduled transportation. Other models use synchronized timetables which notify the transfer time and stops between supply and destination points, where minor schedule changes are possible. Unforeseen stops hurt the total scheduled delivery time, also gaps between expected and real-time occur, which could mean penalties for unnecessary waiting times.

A timetabling problem is a particular problem that occurs in certain events (pickup, delivery, etc.) that must be divided into several time slots in light of various constraints. Powerful methods to solve the principal scheduling problem are obvious and simple in the application, for example, assigning visits to customers into time slots is the possible schedules of candidates, which will later render some optimality and change in light of constraints.

The problem of periodic planning of several resources was first approached by Liu in 1969. Mathematical periodic activity planning models were used. Among them, a model for scheduling recurring events with certain time constraints was proposed. This is a problem that could be seen as an extension to periodic routine planning phenomena with priority constraints. Periodic algorithms are optimal for independent periodic real-time tasks performed using a multi-resource system, but they incur significant planning costs, making planning decisions in each time unit which would be proportional to the progress of each task.

The scheduling of preventive vehicle maintenance follows flexible periods between service interventions. Models developed on these bases eliminate time-step and use less complicated algorithms than those used to solve periodic scheduling problems.

The problem of quadratic semi-assignment is usually difficult to solve optimally, although many trials have been made. Branches and associated algorithms are the most efficient solution methods, but they can only solve small-scale cases. Nevertheless, many effective heuristics procedures have been proposed, and they offer solutions that are very close to optimal.

A schedule of events means focusing on events, i.e. moments when a status changes.

The target function of mixed-integer programming (Fig. 9.9) is to minimize costs, i.e. transportation costs.

Various factors, such as the transportation distance and the unit distance cost, the weight of the carried goods, and even more importantly, the way goods are directed to their destination, are taken into consideration.

The timetabling problem is important for many transportation companies. They transport cargo based on contracts and seek to generate extra revenues from the transportation of emergency freights. The introduction of split cargo has lifted this restriction, and each freight can be transported by two ships. Authors propose extensive neighbourhood search heuristics to address the problem of sea transport mode routing and planning related to cargo splitting.

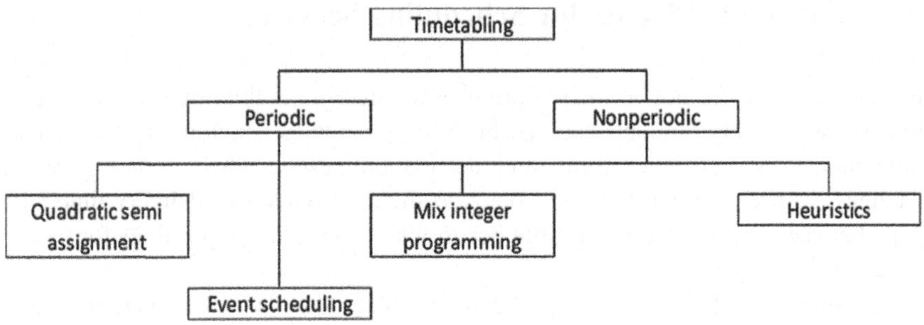

Fig. 9.9 Timetabling models for scheduled transportation/Scheduling vehicle trips

Table 9.10 Studies focusing on freight unloading

Vehicle path selection	Gate assignment	Transportation to storage
Beasley (1983)	Lim et al. (2006a, b)	Lee and Ng (1997)
Sinriech and Tanchoco (1991)	Yu and Egbelu (2008)	Ruben and Jacobs (1999)
Schouwenaars et al. (2001)	Miao et al. (2009)	Heragu et al. (2005)
Velenis and Tsiotras (2005)	Zhu et al. (2009)	Kovacs (2011)
Xu et al. (2007)	Agustina et al. (2010)	Gyulai et al. (2013)
Chunyan (2008)	Bierwirth and Meisel (2010)	Ming-Huang Chiang et al. (2014)
Wei (2010)	Boysen et al. (2010)	Burinskiene (2015)
Edison and Shima (2011)	Berghman et al. (2014)	
De Filippis and Guglieri 2012	Guignard et al. (2012)	Li et al. (2016)
Chen et al. (2014)	Vahdani et al. (2013)	Burinskiene et al. (2018)
Xiao and Chen (2015)	Ehm and Freitag 2016	Ladosz et al. (2018)
Moon et al. (2016)	Deng et al. (2017)	Xie et al. (2018)
Evans (2017)	Wang et al. (2018a, b)	Anoop et al. (2018)
Hadikurniawati et al. (2018)	Shadman et al. (2017)	Dua and Sinha (2019)
Croce et al. (2020)	Fügenschuh et al. (2018)	Tavasszy et al. (2020)

Timetabling also deals with cases of gate assignment at the destination point. The driver gets information about the delivery gate which is assigned to the cargo arrival time-window and other rules.

The gate assignment and re-assignment is important for all types of cargo (full load, part load) deliveries to storage facilities.

Table 9.10 presents studies are grouped to represent several directions.

The analysis of gate assignment reveals that there are many constraints and several researches conducted in the area. The author focus on the reduction of the total travelling distance. Some optimization approaches helping to minimize distance could be named: metaheuristics, gate selection solutions, etc.

All pallets are routed from inbound to storage. Some genetic algorithms have been developed for such routing and helped to construct a master schedule for carriers upfront with the application of optimization techniques.

9.9.1 The Model for Optimisation

Every day a large number of orders (freights) reach consignees. Now receivers ask shippers to indicate the delivery of freight time.

When the delivery plan for freights is constructed, suppliers can find information on delivery details (gate number and inbound time, giving 10 min per freight and 3 min per transport unit. The created schedule follows the shortest path distance (which is a vector distance in this case) estimation method. The model below discusses horizontal freight movements between inner-transportation points at the warehouse.

The model includes two components (Fig. 9.10):

- The composition of order.
- Physical gate allocation.

The created strategic freight inbound plan can be updated each day. In the strategic plan, the number of pallets and delivery days are indicated. In the case, couple of cargos arrives at the same time, the unloading of freight is prioritized which stock level in the storage area is the lowest.

Figure 9.10 illustrates different trajectories. The first example shows a case when freight is received straight in front of gate 12 and a wide-aisle storage area (D) in the shortest way. The next example shows the ways for freight delivery from the same gate 12 to all other storage areas than D, but the distance for such shipment becomes much longer.

This stimulates to plan and update deliveries of freights to the gates and use proper gate assignment for these freights.

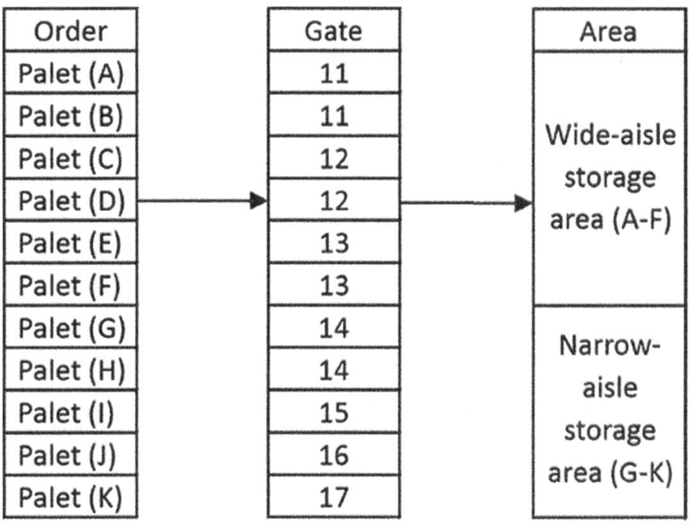

Fig. 9.10 Concept of the freight delivery model

In summary, there are many FRP versions available, allowing to make routes in different ways: to visit clients by regions, deliver and pick up on the same routes, organize freights delivery in travelling salesman way, combine and separate large routes into the small ones and so on.

Most of these plans are not adjusted by following the availability of B2C customers, real-time information. In many cases, the freight pick up from the customers' routes separated from the freight delivery to customers routes.

9.10 Conclusions

Transport systems must be integrated into the supply chain development to achieve the indivisible development of the relevant functional logistics area. A review of the scientific literature was conducted to expand knowledge about the transport system and its functional areas.

Some key aspects have to be revised to achieve high efficiency in freight delivery management.

Real-time comprehension of traffic: high-resolution road and onboard sensors need to be installed to monitor all delivery conditions, such as the speed of the vehicle, the direction of the trip, the location of the position, the permeability of the road, climate conditions (including temperature and humidity), etc.

The communication of low latency and the storage of bulk data: sensors that are able intermittently to create the petabytes of raw data. Given the differences in types of data, dimensions and extensive volume, bandwidth, the capacities of storage and the speed of data processing are required.

The majority of methods search how to solve the various problem which is rising in complex cases and to reach optimization of resources following time, capacity, distance, number of stops restrictions and products specifics. Most algorithms aim to offer the lowest possible transport costs for freight delivery service. Practical implementation requires real-time data collection, data processing, traffic and unloading forecasting and real-time response to unforeseen situations.

References

Abdus, M. Mozayani, N., & Bazzan, A. L. C. Traffic light management in non-stationary environments based on multi-agent Q-learning. *14th IEEE International Conference on Intelligent Transportation Systems*, Washington, DC, October 5–7, 2011.

Aboolian, R., Berman, O., & Krass, D. (2020). Optimizing facility location and design. *European Journal of Operational Research, 5*, 34–46. https://doi.org/10.1016/j.ejor.2020.06.044

Abowd, G. D., Dey, A. K., Orr, R., & Brotherton, J. (1997). Context-awareness in wearable and ubiquitous computing. *1st International Symposium on Wearable Computing* (pp. 179–180).

Agustina, D., Lee, C. K. M., & Piplani, R. (2010). A review: Mathematical models for cross docking planning. *International Journal of Engineering Business Management, 2*(2), 47–54.

Ahmed, U., Srivastava, G., Djenouri, Y., & Lin, J. C. W. (2021). Deviation point curriculum learning for trajectory outlier detection in cooperative intelligent transport systems. *IEEE Transactions on Intelligent Transportation Systems*. https://doi.org/10.1109/tits.2021.3131793

Alhamdia, W., Shakshukia, E., & Sheltamiba, T. E. (2012). Context-sensitive driver assistance system. *International Conference on Mobile Web Information Systems, Procedia Computer Science* (pp. 785–794).

Aliev, R. A., Fazlollahi, B., Guirimov, B. G., & Aliev, R. R. (2007). Fuzzy-genetic approach to aggregate production–distribution planning in supply chain management. *Information Sciences, 177*, 4241–4255. https://doi.org/10.1016/j.ins.2007.04.012

Allen, J., Piecyk, M., Piotrowska, M., McLeod, F., Cherrett, T., Ghali, K., Nguyenc, T., Bektasc, T., Batesd, O., Fridayd, A., et al. (2018). Understanding the impact of e-commerce on last-mile light goods vehicle activity in urban areas: The case of London. *Transportation Research Part D: Transport and Environment, 61*, 325–338. https://doi.org/10.1016/j.trd.2017.07.020

Al-Loughani, I. M. (1997). *Algorithmic approaches for solving the Euclidean distance location and location-allocation problems*. Virginia Tech.

Al-Shemarri, M. S., Lee, Y., & Abdullah, S. (2020). An effective texture descriptor for detecting license plates from vehicle images in challenging environments. *IEEE Transactions on Intelligent Transportation Systems, 21*(2), 553–564.

Altaf, I., & Kaul, A. (2022). A survey on autonomous vehicles in the field of intelligent transport system. In *Applications of networks, sensors and autonomous systems analytics* (pp. 11–31). Springer.

Ambrosino, D., & Scutella, M. G. (2005). Distribution network design: New problems and related models. *European Journal of Operational Research, 165*(3), 610–624. https://doi.org/10.1016/j.ejor.2003.04.009

Anoop, K. P., Panicker, V. V., Narayanan, M., & Sunil Kumar, C. T. (2018). A mathematical model and solution methods for rail freight transportation planning in an Indian food grain supply chain. *Sādhanā, 43*(12), 1–20.

Antoine, G., Mikeka, C., Bajpai, G., Valko, A., & Jayavel, K. (2022). Towards a framework for context-aware intelligent transportation system: Case of Kigali. In *ICT systems and sustainability* (pp. 591–599). Springer.

Audy, J. F., El Hachemi, N., Michel, L., & Rousseau, L. M. (2011). Solving a combined routing and scheduling problem in forestry. *Proceedings of International Conference on Industrial Engineering and Systems Management* (pp. 50–59), ISBN 978-2-9600532-3-4, Metz, May 2011, International Institute for Innovation, Industrial Engineering and Enterpreneurship, Hornu.

Autili, M., Chen, L., Englund, C., Pompilio, C., & Tivoli, M. (2021). Cooperative intelligent transport systems: Choreography-based urban traffic coordination. *IEEE Transactions on Intelligent Transportation Systems, 22*(4), 2088–2099.

Azi, N., Gendreau, M., & Potvin, J.-Y. (2007). An exact algorithm for a single vehicle routing problem with time windows and multiple routes. *European Journal of Operational Research, 178*, 755–766. ISSN 0377-2217.

Batty, M., Crooks, A. T., See, L. M., & Heppenstall, A. J. (2012). Perspectives on agent-based models and geographical systems. In *Agent-based models of geographical systems* (pp. 1–15). Springer.

Beasley, J. E. (1983). Route first—cluster second methods for vehicle routing. *Omega, 11*(4), 403–408.

Bell, W., Dalberto, L., Fisher, M. L., Greenfield, A., Jaikumar, R., Kedia, P., Mack, R., & Prutzman, P. (1983). Improving the distribution of industrial gases with an on-line computerized routing and scheduling optimizer. *Interfaces, 13*(6), 4–23. ISSN 0092-2102.

Berghman, L., Leus, R., & Spieksma, F. C. (2014). Optimal solutions for a dock assignment problem with trailer transportation. *Annals of Operations Research, 213*, 3–25.

Bergmann, F. M., Wagner, S. M., & Winkenbach, M. (2020). Integrating first-mile pick-up and last-mile delivery on shared vehicle routes for efficient urban e-commerce distribution. *Transportation Research: Part B Methodological, 131*, 26–62. https://doi.org/10.1016/j.trb.2019.09.013

Bettini, C., Indulska, J., Nicklas, D., Ranganathan, A., & Roboni, D. (2010). An overview of contextual modeling and reasoning methods. *Pervasive and Mobile Computing*, 161–180.

Bierwirth, C., & Meisel, F. (2010). A survey of berth allocation and quay crane scheduling problems in container terminals. *European Journal of Operational Research, 202*, 615–627.

Bilgen, B., & Ozkarahan, I. (2007). A mixed-integer linear programming model for bulk grain blending and shipping. *International Journal of Production Economics, 107*, 555–571. https://doi.org/10.1016/j.ijpe.2006.11.008

Blunsom, P. (2004). *Hidden Markov models*. University of Melbourne.

Bohmlander, D., Dirndorfer, T., Al Bayatti, A. H., & Brandmeier, T. (2017). Context-sensitive system for pre-starting irreversible vehicle safety drives. *Accident Analysis and Prevention*.

Boysen, N., Fliedner, M., & Scholl, A. (2010). Scheduling inbound and outbound trucks at cross docking terminals. *OR-Spektrum, 32*, 135–161.

Bramel, J., & Simchi-Levi, D. (1995). A location-based heuristic for general routing problems. *Operations Research, 43*, 649–660. ISSN 1109-2858.

Brown, P. J., Bovey, J. D., & Chen, X. (1997). Context-sensitive applications: From the lab to the market. *IEEE Personal Communications, 4*(5), 58–64.

Brum-Bastos, V., Łoś, M., Long, J. A., Nelson, T., & Demšar, U. (2022). Context-aware movement analysis in ecology: A systematic review. *International Journal of Geographical Information Science, 36*, 405–427.

Burinskiene, A. (2015). Optimising forklift activities in wide-aisle reference warehouse. *International Journal of Simulation Modelling, 14*(4), 621–632.

Burinskiene, A., Lorenc, A., & Lerher, T. (2018). A simulation study for the sustainability and reduction of waste in warehouse logistics. *International Journal of Simulation Modelling, 17*(3), 485–497.

Campbell, A. M., & Savelsbergh, M. W. P. (2004). A decomposition approach for the inventory-routing problem. *Transportation Science, 38*, 488–502. ISSN 0041-1655.

Carlier, J., & Villon, P. (1987). *A well solved case of the traveling salesman problem*, Research Report, No. 8, July 1987. Université de Technologie de Compiègne.

Chang, H., Tai, Y., & Hsu, J. Y.-j. (2010). Context-sensitive taxi demand forecast. *International Journal of Business Intelligence and Data Mining, 5*(1), 3–18.

Chavhan, S., Gupta, D., Chandana, B. N., Khanna, A., & Rodrigues, J. J. (2019). IoT-based context-aware intelligent public transport system in a metropolitan area. *IEEE Internet of Things Journal, 7*(7), 6023–6034.

Chen, C. L., & Lee, W. C. (2004). Multi-objective optimization of multi-echelon supply chain networks with uncertain product demands and prices. *Computers and Chemical Engineering, 28*, 1131–1144. https://doi.org/10.1016/j.compchemeng.2003.09.014

Chen, C. L., Wang, B. W., & Lee, W. C. (2003). Multi-objective optimization for a multienterprise supply chain network. *Industrial and Engineering Chemistry Research, 42*, 1879–1889. https://doi.org/10.1021/ie0206148

Chen, Y. Z., Shen, S. F., Chen, T., & Yang, R. (2014). Path optimization study for vehicles evacuation based on Dijkstra algorithm. *Procedia Engineering, 71*, 159–165.

Chen, B., Gong, C., & Yang, J. (2019). Semantic segmentation with importance for autonomous vehicles. *IEEE Transactions on Intelligent Transportation Systems, 20*(1), 137–148.

Chinnaiyan, R., & Swarnamugi, M. (2021). *Modelling and reasoning techniques for context–aware computing in intelligent transportation system.*

Choy, J. L. C., Wu, J., Long, C., & Lin, Y. B. (2020). Ubiquitous and low power vehicles speed monitoring for intelligent transport systems. *IEEE Sensors Journal, 20*(11), 5656–5665.

Christofides, N., Mingozzi, A., & Toth, P. (1979). The vehicle routing problem. In N. Christofides, A. Mingozzi, P. Toth, & C. Sandi (Eds.), *Combinatorial optimization* (pp. 315–338). Wiley. ISBN 9780471997498.

Chunyan, D. (2008). A compare between the two algorithm for shortest path. *Journal Artificial Intelligence and Identification Technology, 11*, 511–513.

Clarke, G., & Wright, J. W. (1964). Scheduling of vehicles from a central depot to a number of delivery points. *Operations Research, 12*, 568–581. ISSN 1109-2858.

Cohen, M., & Lee, H. (1985). Equilibrium analysis of disaggregate facility choice system subject to congestion-elastic demand. *Operations Research, 33*(2), 293–311.

Cordeau, J. F., Laporte, G., Savelsbergh, M. W. P., & Vigo, D. (2007). Vehicle routing. In C. Barnhart & G. Laporte (Eds.), *Operations research and management science* (pp. 367–428). Elsevier. ISBN-9780444513465.

Cormier, G. (2005). Operational research methods for efficient warehousing. In A. Langevin & D. Riopel (Eds.), *Logistics systems: Design and optimization* (pp. 93–122). Springer. ISBN 978-0-387-24971-1.

Croce, A. I., Musolino, G., Rindone, C., & Vitetta, A. (2020). Route and path choices of freight vehicles: A case study with floating car data. *Sustainability, 12*(20), 8557.

Cuda, R., Guastaroba, G., & Speranza, M. G. (2015). A survey on two-echelon routing problems. *Computers and Operations Research, 55*, 185–199. https://doi.org/10.1016/j.cor.2014.06.008

Dass, P., Misra, S., & Roy, C. (2020). T-safe: Trustworthy service provisioning for IoT-based intelligent transport systems. *IEEE Transactions on Vehicular Technology, 69*(9), 9509–9517.

Day, A. K. (1998). Context-aware computing: The cyber desk project. *AAAI Spring Symposium on Intelligent Environments, Technical Report*, SS-98-02 (pp. 51–54).

De Arijit, A., Mogale, D. G., Zhang, M., Pratap, S., Kumar, S. K., & Huang, G. Q. (2020). Multi-period multi-echelon inventory transportation problem considering stakeholders behav-ioural tendencies. *International Journal of Production Economics, 225*, 107566. https://doi.org/10.1016/j.ijpe.2019.107566

De Brebisson, A., Simon, E., Auvolat, A., Vincent, P., & Bengio, Y. (2015). Artificial neural networks applied to taxi destination prediction. In *Proceedings of the International Conference on ECML PKDD Discovery Challenge* (vol. 1526, pp. 40–51).

De Filippis, L., & Guglieri, G. (2012). Advanced graph search algorithms for path planning of flight vehicles. In: *Proceedings of Recent Advances in Aircraft Technology* (pp. 56–64).

de Jiang, X., Souza, E. N. Pesarangader, A., Silver, B., Hu, D. L., & Matwin, C. (2017). TrajectoryNet: A embedded GPS trajectory representation for point-based classification using recurrent neural networks. In *International Conference on Computer Science Software Engineering* (pp. 192–200).

Deng, W., Li, B., & Zhao, H. (2017). Study on an airport gate reassignment method and its application. *Symmetry, 9*(11), 258.

Deveaux, D., Higuchi, T., Uçar, S., Härri, J., & Altintas, O. (2022). *Role of context in determining transfer of risk knowledge in roundabouts*. Preprint.

Dinh, L. T. N., Karmakar, G., & Kamruzzaman, J. (2020). A survey on context awareness in big data analytics for business applications. *Knowledge and Information Systems, 62*(9), 3387–3415. https://doi.org/10.1016/j.cor.2008.02.009

Doush, I. A., Damaj, I., Al-Betar, M. A., Awadallah, M. A., Al-khatib, R. E. M., Alchalabi, A. E., & Bolaji, A. L. (2020). A survey on accessible context-aware systems. In *Technological trends in improved mobility of the visually impaired* (pp. 29–63). Springer.

Dovis, F., Ruotsalainen, L., Toledo-Moreo, R., Kassas, Z. Z. M., & Gikas, V. (2020). Recent advancement on the use of global navigation satellite system-based positioning for intelligent transport systems [guest editorial]. *IEEE Intelligent Transportation Systems Magazine, 12*(3), 6–9.

Drezner, T., Drezner, Z., & Schöbel, A. (2018). The Weber obnoxious facility location model: A big arc small arc approach. *Computers and Operations Research, 98*, 240–250. https://doi.org/10.1016/j.cor.2018.06.006

Dua, A., & Sinha, D. (2019). Quality of multimodal freight transportation: A systematic literature review. *World Review of Intermodal Transportation Research, 8*(2), 167–194.

Edison, E., & Shima, T. (2011). Integrated task assignment and path optimization for cooperating uninhabited aerial vehicles using genetic algorithms. *Computers and Operations Research, 38*(1), 340–356.

Ehm, J., & Freitag, M. (2016). The benefit of integrating production and transport scheduling. *Procedia CIRP, 41*, 585–590.

Elshaikh, A., Salhi, S., Brimberg, J., Mladenović, N., Callaghan, B., & Nagy, G. (2016). An adaptive perturbation-based heuristic: An application to the continuous p-centre problem. *Computers and Operations Research, 75*, 1–11.

Erkut, E., Ülküsal, Y., & Yeniçerioğlu, O. (1994). A comparison of p-dispersion heuristics. *Computers and Operations Research, 21*(10), 1103–1113.

Erl, T. (2005). *Service-oriented architecture: Concepts, technology, and design*. Prentice Hall PTR.

Evans, J. (2017). *Optimization algorithms for networks and graphs* (pp. 1–153). Routledge.

Faiyetole, A. A., & Fulani, E. O. (2020). Consideration of users' needs in regard to planning of intelligent transport solutions in an African Metropolis. *Transportation in Developing Economies, 6*(1), 1–12.

Fan, Y., Wolfson, J., Adomavicius, G., Das, K. V., Khandelwal, Y., & Kang, J. (2014). *SmarTrAC: A smartphone solution for context-aware travel and activity capture*. Research Paper, University of Minnesota.

Fourkiotis, M., Kazaklari, C., Kopsacheilis, A., & Politis, I. (2022). Applying deep learning techniques for the prediction of pedestrian behaviour on crossings with countdown signal timers. *Transportation Research Procedia, 60*, 536–543.

Franklin, D., & Flaschbart, J. (1998). All gadgets and lack of representation make Jack a boring environment. In *AAAI 1998 Spring Symposium on Intelligent Environments, Technical Report SS-98-02* (pp. 155–160).

Fuchs, S., Rass, S., & Kyamakya, K. (2008). A constraint-based and context-aware overtaking assistant with fuzzy-probabilistic risk classification. In *IADIS International Conference Wireless Applications and Computing* (pp. 93–100).

Fügenschuh, A., Homfeld, H., Johann, M., Schülldorf, H., & Stieber, A. (2018). Use of optimization tools for routing in rail freight transport. In *Handbook of optimization in the railway industry* (pp. 161–179). Springer, .

Galkin, A. (2015). Distribution channels management in different regions. *Management, 1*(2), 92–96.

Gao, F., Zhang, Q., Han, Z., & Yang, Y. (2021). Evolution test by improved genetic algorithm with application to performance limit evaluation of automatic parallel parking system. *IET Intelligent Transport Systems, 15*(6), 754–764.

Gendron, B., & Semet, F. (2009). Formulations and relaxations for a multi-echelon capacitated location–distribution problem. *Computers and Operations Research, 36*(5), 1335–1355.

Geng, X. Li, Y. Wang, L. Zhang, L. Yang, Q. Ye, J., & Liu, Y. (2019). Spatiotemporal multi-graph convolution network for ride-hailing demand forecasting. In *Proceedings of 32rd AAAI Conference on Artificial Intelligence* (pp. 1–8).

Ghilas, V., Demir, E., & Van Woensel, T. (2016). An adaptive large neighborhood search heuristic for the pickup and delivery problem with time windows and scheduled lines. *Computers and Operations Research, 72*, 12–30. https://doi.org/10.1016/j.cor.2016.01.018

Gonzalez-Feliu, J. (2011). Two-echelon freight transport optimisation: Unifying concepts via a systematic review. *Working Papers on Operations Management, 2*, 18–30. https://doi.org/10.1016/j.cor.2008.02.009

Guignard, M., Hahn, P. M., Pessoa, A. A., & da Silva, D. C. (2012). Algorithms for the cross- dock door assignment problem. In *Proceedings of the 4th International Workshop on Model-Based Metaheuristics* (pp. 17–20). Rio de Janeiro.

Guillen-Perez, A., & Cano, M. D. (2021). Intelligent IoT systems for traffic management: A practical application. *IET Intelligent Transport Systems, 15*(2), 273–285.

Guo, C., Meguro, J., Kojima, Y., & Naito, T. (2014). A multimodal ADAS system for unmarked urban scenarios based on road context understanding. *IEEE Transactions on Intelligent Transportation Systems, 16*(4), 1690–1704.

Gupta, A., & Maranas, C. D. (2003). Managing demand uncertainty in supply chain planning. *Computers and Chemical Engineering, 27*, 1219–1227. https://doi.org/10.1016/S0098-1354(03)00048-6

Gyulai, D., Pfeiffer, A., Sobottka, T., & Váncza, J. (2013). Milkrun vehicle routing approach for shop-floor logistics. *Procedia CIRP, 7*, 127–132.

Hadikurniawati, W., Winarno, E., Hernawan, A., & Abdullah, D. (2018). Optimization of ISP service maintenance router using Dijkstra and Flyod-Warshall Algorithm. *Journal of Physics, 1114*(1), 121–131.

Harini, U., & Venkatesh, V. (2020). A context aware framework for collision avoidance with deep reinforcement learning in VANET. *International Journal of Advanced Science and Technology, 29*(6c), 4383–4391.

Heragu, S. S., Du, L., Mantel, R. J., & Schuur, P. C. (2005). Mathematical model for warehouse design and product allocation. *International Journal of Production Research, 43*(2), 327–338.

Hu, L., Qiu, M., Song, J., Hossain, M. S., & Ghoneim, A. (2015). Software defined Healthcare Networks. *IEEE Wireless Communications, 22*(6), 67–75.

Hull, R., Neaves, P., & Bedford-Roberts, J. (1997). On the way to located computing. In *1st International Symposium on Wearable Computers* (pp. 146–153).

Iqbal, R., Butt, T. A., Shafiq, M. O., Talib, M. W. A., & Umar, T. (2018). Context-aware data-driven intelligent framework for fog infrastructures in internet of vehicles. *IEEE Access, 6*, 58182–58194.

Ji, S., & Sun, Q. (2017). Low-carbon planning and design in B&R logistics service: A case study of an e-commerce big data platform in China. *Sustainability, 9*, 20–52. https://doi.org/10.3390/su9112052

Jindal, Y., Qin, Z. T., Chen, X., Noclebi, M., & Ye, J. D. (December 2018). Optimizing taxi policy through reinforcement learning and spatio-temporal mining. In *Proceedings of the IEEE International Conference on Big Data* (pp. 1417–1426).

Jnr, B. A., Petersen, S. A., Ahlers, D., & Krogstie, J. (2020). Big data driven multi-tier architecture for electric mobility as a service in smart cities: A design science approach. *International Journal of Energy Sector Management*. https://doi.org/10.1108/IJESM-08-2019-0001

Jozefowiez, N., Semet, F., & Talbi, E. G. (2009). An evolutionary algorithm for the vehicle routing problem with route balancing. *European Journal of Operational Research, 195*(3), 761–769.

Jung, H., Jeong, B., & Lee, C. G. (2008). An order quantity negotiation model for distributor-driven supply chains. *International Journal of Production Economics, 111*, 147–158.

Kachousangi, F. T., Araghi, Y., van Oort, N., & Hoogendoorn, S. (2022). *Passengers preferences for using emerging modes as first/last mile transport to and from a multimodal hub Case study Delft Campus railway station. Case Studies on Transport Policy.*

Kalatian, A., & Farooq, B. (2022). A context-aware pedestrian trajectory prediction framework for automated vehicles. *Transportation Research Part C: Emerging Technologies, 134*, 103453.

Kallrath, J. (2002). Combined strategic and operational planning–An MILP success story in chemical industry. *OR Spectrum, 24*, 315–341. https://doi.org/10.1007/s00291-002-0102-6

Kembro, J. H., & Norrman, A. (2020). Which future path to pick? A contingency approach to omnichannel warehouse configuration. *International Journal of Physical Distribution and Logistics Management*. https://doi.org/10.1108/IJPDLM-08-2019-0264

Kim, T. Y., Dekker, R., & Heij, C. (2017). Cross-border electronic commerce: Distance effects and express delivery in European Union markets. *International Journal of Electronic Commerce, 21*, 184–218. https://doi.org/10.1080/10864415.2016.1234283

Kong, X. T., Chen, J., Luo, H., & Huang, G. Q. (2015). Scheduling at an auction logistics centre with physical internet. *International Journal of Production Research*, 1–21.

Kovacs, A. (2011). Optimizing the storage assignment in a warehouse served by milkrun logistics. *International Journal of Production Economics, 133*(1), 312–318.

Krishnakumari, P., Perotti, A., Pinto, V., Cats, O., & van Lint, X. (November 2018). Understanding network traffic conditions using transfer training. In *Proceedings of the 21st International Conference on Intelligent Transport Systems* (pp. 1396–1401).

Krommenacker, N., Charpentier, P., Berger, T., & Sallez, Y. (2016). On the usage of wireless sensor networks to facilitate composition/decomposition of physical internet containers. In T. Borangiu, D. Trentesaux, A. Thomas, & D. McFarlane (Eds.), *Service orientation in holonic and multi-agent manufacturing* (pp. 81–90). Springer International Publishing.

Ku, I., Lu, Y., Gerla, M., Gomes, R. L., Ongaro, F., & Cerqueira, E. (2014). Towards software-defined VANET: Architecture and services. In *Proceedings of the 13th Annual IEEE Mediterranean Ad floc Networking Workshop* (pp. 103–110).

Ladosz, P., Banjo, O., De Guido, S., & Zalasiewicz, M. (2018). A genetic algorithm optimiser for dynamic product routing in agile manufacturing environment. In *Proceedings of 16th International Conference on Industrial Informatics* (pp. 1079–1084). IEEE.

Laerhoven, T. I. (2001). A combination of self-organizing map and K-mean clustering for online classification of sensor data. In *Proceedings of the International Conference on Artificial Neural Networks* (pp. 464–469). Springer.

Lahyani, R., Khemakhem, M., & Semet, F. (2015). Rich vehicle routing problems: From a taxonomy to a definition. *European Journal of Operational Research, 241*, 1–14. https://doi.org/10.1016/j.ejor.2014.07.048

Lee, Y. H., & Kim, S. H. (2002). Production–distribution planning in supply chain considering capacity constraints. *Computers and Industrial Engineering, 43*, 169–190. https://doi.org/10.1016/S0360-8352(02)00063-3

Lee, H. L., & Ng, S. M. (1997). Introduction to the special issue on global supply chain management. *Production and Operations Management, 6*(3), 191–192.

Li, J., Moghaddam, M., & Nof, S. Y. (2016). Dynamic storage assignment with product affinity and ABC classification – A case study. *International Journal of Advanced Manufacturing Technology, 84*(9), 2179–2194.

Li, Y., Yu, B., Chen, Y., & Hu, Z. (2022). A new theory of driver vision pressure energy field and its application in driver behaviour decision-making model. *IET Intelligent Transport Systems, 16*(1), 1–12.

Liang, H., Doo, X., Wang, G., & Khan, Z. (2019). A deep reinforcement learning network for traffic light cycle management. *IEEE Transaction on Vehicular Technologies, 68*(2), 1243–1253.

Lim, A., Ma, H., & Miao, Z. (2006a). Truck dock assignment problem with operational time constraint within crossdocks. *Advances in Applied Artificial Intelligence, 4031*, 262–271.

Lim, S. J., Jeong, S. J., Kim, K. S., & Park, M. W. (2006b). A simulation approach for production–distribution planning with consideration given to re-plenishment policies. *International Journal of Advanced Manufacturing Technology, 27*, 593–603. https://doi.org/10.1007/s00170-004-2208-2

Lin, J., & Lei, H. (2009). Distribution systems design with two-level routing considerations. *Annals of Operations Research, 172*(1), 329–347. https://doi.org/10.1007/s10479-009-0628-y

Lin, J. R., Nozick, L. K., & Turnquist, M. A. (2006). Strategic design of distribution systems with economies of scale in transportation. *Annals of Operations Research, 144*(1), 161–180. https://doi.org/10.1007/s10479-006-0004-0

Linnainmaa, S., Savola, J., & Jokinen, O. (1995). EPO: A knowledge-based system for wood procurement management. In *Proceedings of the 7th Conference on Innovation Applications of Artificial Intelligence* (pp. 117–113). Montreal, 20–23 August 1995. AAAI Press.

Liu, Y., Singleton, A., & Arribas-Bel, D. (2020). Considering context and dynamics: A classification of transit-orientated development for New York City. *Journal of Transport Geography, 85*, 102711.

Liu, R., Wang, S., Pi, Y., & Qin, Q. (2021a). An effective heuristic with evolutionary algorithm for the coordinated capacitated dynamic lot-size and delivery problem. *Computers and Industrial Engineering, 153*, 107051.

Liu, Y., Singleton, A., Arribas-Bel, D., & Chen, M. (2021b). Identifying and understanding road-constrained areas of interest (AOIs) through spatiotemporal taxi GPS data: A case study in New York City. *Computers, Environment and Urban Systems, 86*, 101592.

Lopes, R. B., Ferreira, C., & Santos, B. S. (2016). A simple and effective evolutionary algorithm for the capacitated location–routing problem. *Computers and Operations Research, 70*, 155–162.

Love, P. E., & Ika, L. A. (2021). The 'context' of transport project cost performance: Insights from contract award to final construction costs. *Research in Transportation Economics, 90*, 101062.

Lv, Y., Duan, Y., Kahn, W., Li, Z., & Wang, F.-Y. (2015). Predicting transport flows with big data: An approach to deep learning. *IEEE Transactions on Intelligent Transportation Systems, 16*(2), 865–873.

Mantyjärvi, J., & Seppanen, T. (2002). Adapting applications in mobile terminals using fuzzy context information. In *Proceedings of the 4th International Symposium on Mobile Human-Computer Interaction* (pp. 95–107). Springer.

Manzini, R., & Bindi, F. (2009). Strategic design and operational management optimization of a multi stage physical distribution system. *Transportation Research Part E: Logistics and Transportation Review, 45*, 915–936.

Mare, R. M., Marte, C. L., Cugnasca, C. E., Sobrinho, O. G., & dos Santos, A. S. (2021). Feasibility of a testing methodology for visible light communication systems applied to intelligent transport systems. *IEEE Latin America Transactions, 19*(3), 515–523.

Marianov, V., & Eiselt, H. A. (2016). On agglomeration in competitive location models. *Annals of Operations Research, 246*(1–2), 31–55. https://doi.org/10.1007/s10479-014-1704-5

McDonald, T., Taylor, S., & Valenzuela, J. (2001a). Information needs for increasing log transport efficiency. In: *Proceedings of 1st International Precision Forestry Cooperative Symposium* (pp. 181–184), Seattle, June 17–20, ISBN 978-80-228-2056-1, College of Forest Resources.

McDonald, T., Taylor, S., & Valenzuela, J. (2001b). Potential for shared log transport services. In: 24th *Annual COFE Meeting* (pp. 115–120), Snowshoe, US, July 2001. Council on Forest Engineering.

Meijboom, B., & Obel, B. (2007). Tactical coordination in a multi-location and multi-stage operations structure: A model and a pharmaceutical company case. *OMEGA: International Journal of Management Science, 35*, 258–273. https://doi.org/10.1016/j.omega.2005.06.003

Mendell, B. C., Haber, J. A., & Sydor, T. (2006). Evaluating the potential for shared log truck resources in middle Georgia. *Southern Journal of Applied Forestry, 30*(2), 86–91. ISSN 0148-4419.

Meyer, R., Harrington, A., & Cahill, W. (September 2006). Towards providing context-sensitive transportation services to users. *IEEE Intelligent Transportation Systems Conference* (pp. 369–376).

Miao, Z., Lim, A., & Ma, H. (2009). Truck dock assignment problem with operational time constraint within crossdocks. *European Journal of Operational Research, 192*, 105–115.

Ming-Huang Chiang, D., Lin, C. P., & Chen, M. C. (2014). Data mining based storage assignment heuristics for travel distance reduction. *Expert Systems, 31*(1), 81–90.

Mirzapour Al-e-hashem, S. M. J., & Rekik, Y. (2014). Multi-product multi-period Inventory Routing Problem with a transshipment option: A green approach. *International Journal of Production Economics, 157*, 80–88. https://doi.org/10.1016/j.ijpe.2013.09.005

Mole, R. H., & Jameson, S. R. (1976). A sequential route-building algorithm employing a generalized savings criterion. *Operational Research Quarterly, 27*, 503–511. ISSN 0030-3623.

Moon, J. H., Ruy, W. S., & Cha, J. H. (2016). Comparison of optimal path algorithms and implementation of block transporter planning system. *Journal of the Society of Naval Architects of Korea, 53*(2), 115–126.

Murphy, L. D. (1995). Geographic information systems: Are they decision support systems? Paper presented at the *System Sciences, 1995. Proceedings of the Twenty-Eighth Hawaii International Conference.*

Murray, A. T., & Tong, D. (2007). Coverage optimization in continuous space facility siting. *International Journal of Geographical Information Science., 21*(7), 757–776.

Nakhjirkan, S., & Mokhatab Rafiei, F. (2017). An integrated multi-echelon supply chain network design considering stochastic demand: A genetic algorithm based solution. *Promet-Traffic and Transportation, 29*(4), 391–400. https://doi.org/10.7307/ptt.v29i4.2193

Nguyen, V. P., Prins, C., & Prodhon, C. (2012). A multi-start iterated local search with tabu list and path relinking for the two-echelon location-routing problem. *Engineering Applications of Artificial Intelligence, 25*(1), 56–71. https://doi.org/10.1016/j.engappai.2011.09.012

Nishi, T., Otaki, K., Hayakawa, K., & Yoshimura, T. (November 2018). Traffic light control based on reinforcement learning with graph convolutional neural networks. In *Proceedings of the 21st International Conference on Intelligent Transportation Systems* (pp. 877–883).

Nunna, S., Kousaridas, A., Ibrahim, M., Dillinger, M., Thuemmler, C., Feussner, H., & Schneider, A. (2015). Enabling real-time context-aware collaboration through 5G and mobile edge computing. In *Proceedings of the 12th IEEE International Conference on Information Technology-New Generations* (pp. 601–605).

Nurmi, P., & Florin, P. (2004). Reasoning in context-dependent systems. *Helsinki Institute of Information Technology, 1*, 1–6.

Oh, H. C., & Karimi, I. A. (2006). Global multi-product production–distribution planning with duty drawbacks. *AICHE Journal, 52*, 595–610. https://doi.org/10.1002/aic.10669

Otebolaku, A., & Lee, G. M. (2018). A framework for using the internet of things for context-sensitive, personalized services based on trust. *Mobile Information Systems*, Hindui.

Pace, S., Turky, A., Moser, I., & Aleti, A. (2015). Distributing fibre boards: A practical application of the heterogeneous fleet vehicle routing problem with time windows and three-dimensional loading constraints. In *ICCS, 2257–2266*. ICCS.

Parola, F., Satta, G., Notteboom, T., & Persico, L. (2021). Revisiting traffic forecasting by port authorities in the context of port planning and development. *Maritime Economics and Logistics, 23*(3), 444–494.

Perea-Lopez, E., Ydstie, B. E., & Grossmann, I. E. (2003). A model predictive control strategy for supply chain optimization. *Computers and Chemical Engineering, 27*, 1201–1218. https://doi.org/10.1016/S0098-1354(03)00047-4

Perego, A., Perotti, S., & Mangiaracina, R. (2011). ICT for logistics and freight transportation: A literature review and research agenda. *International Journal of Physical Distribution and Logistics Management, 41*(5), 457–483. https://doi.org/10.1108/09600031111138826

Perera, C., Zaslavsky, A., Christen, P., & Georgakopoulos, D. (2014). Context aware computing for the internet of things: A survey. *IEEE Communications Surveys and Tutorials, 16*(1), 414–454.

Ramesh, M. V., Vidya, P. T., & Pradeep, P. (October 2013). Context aware wireless sensor system integrated with cooperative sensing for real time traffic accident detection. *IEEE International Conference on Wireless and Optical Communications Networks (WOCN)* (pp. 1–5), Bhopal.

Ratliff, H. D., & Rosenthal, A. S. (1983). Order-picking in a rectangular warehouse: A solvable case of the traveling salesman problem. *Operations Research, 31*(3), 507–521.

Rawal, T., & Devadas, V. (2015). Intelligent transportation system in India. *Journal of Development Management and Communication, 2*(3), 299–308.

Ribas, S., Subramanian, A., Coelho, I. M., Ochi, L. S., & Souza, M. J. F. (2011). A hybrid algorithm for the Vehicle Routing Problem with Time Windows. *International Journal on Artificial Intelligence Tools, 10*(3), 431–449. ISSN 0218-2130.

Rizk, N., Martel, A., & D'amours, S. (2006). Multi-item dynamic production–distribution planning in process industries with divergent finishing stages. *Computers and Operations Research, 33*, 3600–3623. https://doi.org/10.1016/j.cor.2005.02.047

Romo, F., Tomasgard, A., Hellemo, L., Fodstad, M., Eidesen, B. H., & Pedersen, B. (2009). Optimising the Norwegian natural gas production and transport. *Interfaces, 39*, 46–56. https://doi.org/10.1287/inte.1080.0414

Ruben, R. A., & Jacobs, F. R. (1999). Batch construction heuristics and storage assignment strategies for walk/ride and pick systems. *Management Science, 45*(4), 575–596.

Rudskoy, A., Ilin, I., & Prokhorov, A. (2021). Digital twins in the intelligent transport systems. *Transportation Research Procedia, 54*, 927–935.

Ryan, N., Pascoe, J., & Morse, D. (April 1997). Enhanced reality fieldwork: The context-aware archaeological assistant. In W. Gaffney, M. Van Leysen, & S. Exxon (Eds.), *Computer applications and quantitative methods in archaeology. Proceedings of the 25th Anniversary Conference* (pp. 269–274), University of Birmingham.

Ryu, J. H., Dua, V., & Pistikopoulos, E. N. (2004). A bilevel programming framework for enterprise-wide process networks under uncertainty. *Computers and Chemical Engineering, 28*, 1121–1129. https://doi.org/10.1016/j.compchemeng.2003.09.021

Sakawa, M., Nishizaki, I., & Uemura, Y. (2001). Fuzzy programming and profit and cost allocation for a production and transportation problem. *European Journal of Operational Research, 131*, 1–15. https://doi.org/10.1016/S0377-2217(00)00104-1

Salber, D., Dey, A. K., & Abowd, G. D. (1998). Ubiquitous computing: Defining a HCI research agenda for a emerging interaction paradigm. *Technical Report GVU Georgia Tech GVU GIT-GVU-98-01*.

Santhoshkumar, S., Gayathri, A., Hemalatha, R., Indhumathi, A., & Kavishree, C. (2020). Intelligent traffic light cycle control using machine learning. *International Journal of Advanced Science and Technology, 29*(6s), 2644–2655.

Schilit, B., Adams, N., & Want, R. (1994a). Contextual computer programs. In *Mobile Computer Systems and Applications, 1994 MCSA 1994 First Seminar* (pp. 85–90). IEEE.

Schilit, B. N., Teimer, M., & Hrsg, G. (1994b). Dissemination of active map information for mobile hosting devices. *IEEE Network, 8*(5), 22–32.

Schouwenaars, T., De Moor, B., Feron, E., & How, J. (2001). Mixed integer programming for multi-vehicle path planning. In *Proceedings of Control Conference, 2603–2608*. IEEE.

Selim, H., & Ozkarahan, I. (2008). A supply chain distribution network design model: an interactive fuzzy goal programming-based solution approach. *The International Journal of Advanced Manufacturing Technology, 36*(3–4), 401–418.

Sezer, O. B., Dogdu, E., & Ozbayoglu, A. M. (2018). context-sensitive computing, learning and big data in the internet of things: A survey. *IEEE Internet of Things Journal, 5*(1), 1–27.

Shaabani, H., & Kamalabadi, I. N. (2016). An efficient population-based simulated anneal-ing algorithm for the multi-product multi-retailer perishable inventory routing problem. *Computers and Industrial Engineering, 99*, 189–201. https://doi.org/10.1016/j.cie.2016.07.022

Shadman, A., Bozorgi-Amiri, A., & Rahmani, D. (2017). A mathematical model for vehicle routing and scheduling problem with cross-docking by considering risk. *International Journal of Industrial Engineering and Production Research, 28*(2), 189–199.

Shen, Z., & Sessions, J. (1989). Log truck scheduling by network programming. *Forest Products Journal, 39*(10), 47–50. ISSN 0015-7473.

Shepelev, V., Zhankaziev, S., Aliukov, S., Varkentin, V., Marusin, A., Marusin, A., & Gritsenko, A. (2022). Forecasting the passage time of the queue of highly automated vehicles based on neural networks in the services of cooperative intelligent transport systems. *Mathematics, 10*(2), 282.

Sinriech, D., & Tanchoco, J. M. A. (1991). Intersection graph method for AGV flow path design. *International Journal of Production Research, 29*(9), 1725–1732.

Siripanpornchana, S., Panihpapibun, S., & Chaovalit, P. (November 2016). Time-determination of deep learning travel. In *Proceedings of IEEE Region Conference* (pp. 1859–1862).

Stepanova, D., Sukuvaara, T., & Karsisto, V. (2020, May). Intelligent Transport Systems–Road weather information and forecast system for vehicles. In *2020 IEEE 91st Vehicular Technology Conference (VTC2020-Spring)* (pp. 1–5). IEEE.

Sun, J., Wu, Z., & Pan, G. (2009). Context-sensitive smart car: Model to prototype. *Journal of Zhejiang University, 10*(7), 1049–1059.

Svarnamugi, M., & Chinnaiyang, R. (2017). Cloud and fog computing models for the internet of things. *International Journal of Applied Sciences and Engineering Technology Research*.

Svarnamugi, M., & Chinnaiyang, R. (February 2018). IoT Hybrid Computing Model for Intelligent Transportation System (ITS). In *IEEE Second International Conference on Computational Methodologies and Communications* (pp. 802–806).

Swarnamugi, M., & Chinnayan, R. (2019). Context-aware intelligent reliable service model for an intelligent transport system based on ontology. *Lecture Notes on Electrical Engineering* (Vol. 597, pp. 22–30). Springer.

Tavasszy, L., de Bok, M., Alimoradi, Z., & Rezaei, J. (2020). Logistics decisions in descriptive freight transportation models: A review. *Journal of Supply Chain Management Science, 1*(3–4), 74–86.

Timpe, C. H., & Kallrath, J. (2000). Optimal planning in large multi-site production networks. *European Journal of Operational Research, 126*, 422–435. https://doi.org/10.1016/S0377-2217(99)00301-X

Toke, F., Kom, E., Al-Mahrsi, M. T., & Oukhellou, L. (November 2016). Predicting dynamic patterns of origin and destination of public transport with long-term recurrent neural networks with short-term memory. In *IEEE 19th International Conference on Intelligent Transportation Systems* (pp. 1071–1076).

Tran, C. N., Tat, T. T. H., Tam, V. W., & Tran, D. H. (2022). Factors affecting intelligent transport systems towards a smart city: A critical review. *International Journal of Construction Management*, 1–17.

Vahdani, B., Tavakkoli-Moghaddam, R., & Mousavi, S. M. (2013). Scheduling of trucks in cross-docking systems: A hybrid meta-heuristic algorithm. *Lecture Notes in Management Science, 5*, 125–132.

Vahdat-Nejad, H., Ramazani, A., Mohammadi, T., & Mansour, V. (2016). Review of context-aware applications of automotive networks. *Automotive Communications* (pp. 1–15). Elsevier.

Velenis, E., & Tsiotras, P. (2005). Minimum time vs maximum exit velocity path optimization during cornering. In *Proceedings of 2005 IEEE International Symposium on Industrial Electronics* (pp. 355–360). IEEE.

Veres, M., & Moussa, M. (2019). Deep learning for intelligent transportation systems: An overview of emerging trends. *IEEE Transactions on Intelligent Transport Systems* (pp. 1–17).

Vieira, V., Caldas, L. R., & Salgado, A. C. (2011, July). Towards an ubiquitous and context sensitive public transportation system. In *2011 Fourth International Conference on Ubi-Media Computing* (pp. 174–179). IEEE.

Wan, J., Liu, J., Shao, Z., Vasilakos, A. V., Imran, M., & Zhou, K. (2016). Mobile crowd sensing for traffic prediction in internet of vehicles. *Sensors, 16*(1), 88.

Wang, Y., Zhang, S., Assogba, K., Fan, J., Xu, M., & Wang, Y. (2018a). Economic and environmental evaluations in the two-echelon collaborative multiple centers vehicle routing optimization. *Journal of Cleaner Production, 197*, 443–461. https://doi.org/10.1016/j.jclepro.2018.06.208

Wang, Z., Lu, Y., Zhao, L., & Cao, N. (2018b). Improved harmony search algorithm for truck scheduling problem in multiple-door cross-docking systems. *Discrete Dynamics in Nature and Society*, 20–38.

Ward, A., Jones, A., & Hopper, A. (1997). A new location technique for an active office. *IEEE Personal Communications, 4*(5), 42–47.

Wei, D. (2010). An optimized Floyd algorithm for the shortest path problem. *Journal of Networks, 5*(12), 1496.

Weintraub, A., Epstein, R., Morales, R., Seron, J., & Traverso, P. (1996). A truck scheduling system improves efficiency in the forest industries. *Interfaces, 26*(4), 1–12. ISSN 0092-2102.

Woerndl, W., Schueller, C., & Wojtech, R. (2007, April). A hybrid recommender system for context-aware recommendations of mobile applications. In *2007 IEEE 23rd International Conference on Data Engineering Workshop* (pp. 871–878). IEEE.

Worndl, W., Hubner, J., Bader, R. & Gallego-Vico, D. (2011). A model for proactivity in mobile, context-aware recommender systems. In *Conference on Recommender systems* (pp. 273–276), Chicago, IL.

Xiao, Z., & Chen, J. (2015). Research on path optimization based on improved adaptive genetic algorithm. *Intelligent Human-Machine Systems and Cybernetics, 1*, 207–209.

Xie, J., Mei, Y., Ernst, A. T., Li, X., & Song, A. (2018). A bi-level optimization model for grouping constrained storage location assignment problems. *IEEE Transactions on Cybernetics, 48*(1), 385–398.

Xu, M., Liu, Y., Huang, Q., Zhang, Y., & Luan, G. (2007). An improved Dijkstra shortest path algorithm for sparse network. *Applied Mathematics and Computation, 185*, 247–254.

Xu, D., Rahmatizade, R., Bölöni, L., & Turgut, D. (2017). Real-time prediction of taxi demand using recurrent neural networks. *IEEE Transaction on Intelligent Transportation Systems., 19*(8), 2572–2581.

Yao, H., Wu, F., Ke, J., Tang, X., Jia, Y., Lu, S., Gong, P., Ye, J., & Li, Z. (2018). *Deep multi-view space-time network for forecasting taxi demand*. In *Proceedings of the 32nd AAAI Conference on Artificial Intelligence* (pp. 1–8).

Yonglin, R., & Awasthi, A. (2012). Location allocation planning of logistics depots using genetic algorithm. *Research in Logistics and Production, 2*, 247–257.

You, M., Xiao, Y., Zhang, S., Yang, P., & Zhou, S. (2019). Optimal mathematical programming for the warehouse location problem with Euclidean distance linearization. *Computers and Industrial Engineering, 136*, 70–79.

Younes, M. B., Boukerche, A., & Mammeri, A. (2016, April). Context-aware traffic light self-scheduling algorithm for intelligent transportation systems. In *2016 IEEE Wireless Communications and Networking Conference* (pp. 1–6). IEEE.

Yu, W., & Egbelu, P. J. (2008). Scheduling of inbound and outbound trucks in cross docking systems with temporary Storage. *European Journal of Operational Research, 184*, 377–396.

Zampou, E., Mourtos, I., Pramatari, K., & Seidel, S. (2022). A design theory for energy and carbon management systems in the supply chain. *Journal of the Association for Information Systems, 23*(1), 329–371.

Zhao, P., & Batta, R. (1999). Analysis of centroid aggregation for the Euclidean distance p-median problem. *European Journal of Operational Research., 113*(1), 147–168.

Zheng, K., Zheng, Q., Yang, H., Zhao, L., Hou, L., & Chatzimisios, P. (2015). Reliable and efficient autonomous driving: The need for heterogeneous vehicular networks. *IEEE Communications Magazine, 53*(12), 72–79.

Zhu, Y. R., Hahn, P. M., Liu, Y., & Guignard, M. (2009). New approach for the cross-dock door assignment problem. In *Anais do XLI Simp'osio Brasileiro de Pesquisa Operacional* (pp. 30–52). SOBRAPO.

Zhu, L., Yu, F. R., Wang, Y., Ning, B., & Tang, T. (2019). Big data analytics in intelligent transportation systems: A survey. *IEEE Transactions on Intelligent Transportation Systems, 20*(1).

Chapter 10
Cargo Delivery Management Process: Risks, Solutions, and Optimization

Aurelija Burinskienė

Abstract Researchers pay more attention to the process of managing the delivery of goods. An overview of the latest scientific articles is presented to identify optimization methods applied in the process of cargo delivery management and analyse the risks covered. The chapter also gives a review of the application of different decision support techniques including several attributes in studies on the transportation of freights. A recent analysis of cargo delivery management includes some metaheuristic algorithms. However, the research potential on the topic is quite high. Research in the literature shows that research in mathematics and science involves a variety of methods. Next, the methods involved in the process are explained and a set of metaheuristic techniques capable to improve these processes is proposed. In the chapter, the application of algorithms: genetic, evolutionary, and other which are used to optimize transportation processes is demonstrated.

Keywords Micro-risks · Cause-and-effect modelling · Multi-attribute methods of decision support · The process of cargo delivery management · Metaheuristics for optimization

10.1 Cargo Delivery Process and Risk Management in the Application of Automatic Risk Assessment Tools

From the point of view of the design of service-oriented transport systems, the application of common methods of reducing or mitigating risks with identified risks is listed in order of increased severity of risk:

- Strengthening the flow of data on the delivery of goods (environmental events and other failures).

A. Burinskienė (✉)
Vilnius Gediminas Technical University, Vilnius, Lithuania
e-mail: aurelija.burinskiene@vilniustech.lt

D. Dzemydienė et al., *Development of Smart Context-Aware Services for Cargo Transportation*, International Series in Operations Research & Management Science 330, https://doi.org/10.1007/978-3-031-07199-7_10

- Identification of potential risks and losses (comparison of the actual delivery with the plan).
- Exchange of data between stakeholders, receipt of data from institutions and its transfer to institutions.

Automated risk assessment systems incorporate various data sources that view, evaluate, and distribute information in an automatical way. From an automated risk management perspective, most of these multiple groups of regulating data are important. Control over the content of the data, the organization, and the timing of data. Customizable data provides consistent, specific, and easier to analyse results than other types of data that do not provide the same level of detail or have lower coverage across all organizations and transactions.

Automated risk assessment systems incorporate the following functions: they provide visibility, integrate data about objects to a particular shipment or transit to be able to comprehensively see its context and history; analyse transaction-based information; user interfaces integrating and merging it with other data. They include interfaces for creating custom risk management rules; rules for identifying specific types of threats; request functions, which help to update the system and build risks revision promptly, as well as to update qualitative (official intuition) and quantitative (statistical revision) sets of rules; the ability to adapt to new methodologies and approaches focusing on the increased ability to identify risks (Tomlin, 2006).

To do this, the following tools are available: (1) business intelligence tools are designed to manage historical analysis, forecasting, and predictive analysis using a variety of mathematical methods and (2) visualization tools are designed to identify anomalies or patterns using mapping connectors and relationships between data.

In the transport system, there are many risks of unexpected events that disrupt the uninterrupted flow of goods at the national and international levels. The thematic of risk assessment is analysed in different areas of study: economic, strategic, and transport management.

Lavastre et al. (2012) identified some elements for determining risks: the appearance of lost events (the number of lost events), their importance (the volume of losses), and the probability of their occurrence (the uncertainty of cases with lost events). The probability of losses and their significance can be investigated from several points of view: (1) from the perspective of the enterprise and (2) from the position of the client.

The literature distinguishes between macro-risks and micro-risks. Klibi et al. (2010) analysed the sources of destruction and discussed environmental risks. Blome and Schoenherr (2011) pointed out that transportation systems have become more complex, resulting in higher supply chain vulnerabilities. Juttner (2005) emphasized security, and Wakolbinger and Cruz (2011) identified risk-sharing contracts. They argue that in the last years, macroeconomic disruptions had been happened due to fuel protests and terrorist attack events having a significant impact on the system's vulnerability. Tan et al. (2006) said this was due to uncertainty in unstable market demand and price pressures. Hendricks and Singhal (2005) described that the focus on efficiency in the latest years (i.e. cost reductions) has

Table 10.1 The interest of researchers on micro-risks

Micro-risks	Drivers	Authors
Delay	Accidents	Christopher and Peck (2004), Özdemir et al. (2018)
	Equipment and labour issues	Kleindorfer and Saad (2005), Satta et al. (2019)
	Performance	Hendricks and Singhal (2005), Ouhader (2020)
	Unreliable transport system	Christopher and Peck (2004), Halse et al. (2019)
	Unstable packaging	Atilgan and McCullen (2007)
	Unreliable provider	Chopra et al. (2007), Talley and Ng (2020)
Other issues	Obsolescence of goods	Tummala and Schoenherr (2011)
	Damage during transportation	Sawik (2013), Batarlienė (2018), Zaikin et al. (2020)
	Human error	Atilgan and McCullen (2007), Singh and Singh (2021)
	Mismatch of documents	Atilgan and McCullen (2007), Christopher and Peck (2004), Popescu (2018)
	Quality issues	Sawik (2013), Wagner and Bode (2008), Dua and Sinha (2019)
	Lack of coordination	Schmitt and Singh (2009), Raimbekov et al. (2018)

led to an increase in supply disruptions, and management companies have not made sufficient efforts to address them.

Some examples of micro-risks are shown in Table 10.1.

According to Tang and Tomlin (2016), the measurement of risk has two dimensions: (1) the probability of occurrence and (2) the impact of failures. Baghalian et al. (2013) divided risks into two categories: (1) systematic risks associated with environmental factors that companies cannot control and (2) unsystematic risks associated with factors controlled by companies. Sadghiani et al. (2015) noted that mitigating unsystematic risks can improve a company's reputation. Simchi-Levi et al. (2014) suggested revising common unsystematic disruptions, namely ones of unreliable carriers and unreliable transport systems. Scholten et al. (2014) proposed that the creation of a sustainable transport system can lead to the prevention of exposure (vulnerability). Juttner et al. (2003) highlighted ownership of risks assessment systems and the risks of obsolescence of freight, low level of response, which leads to unforeseen circumstances.

10.1.1 Methods Applicable to Risk Analysis in a Service-Oriented Transport System

Bode and Wagner (2015) use normal accident theory to explain failures. Literature analyses the supply disruption process from a single company's perspective. Most models focus on single-period cases, and only a few deals with multi-period periods, the ability of companies to adjust capacity in multi-period conditions (Pettit et al., 2013).

The documents presenting the optimal strategy deal with supply disruption, which is the opposite of demand uncertainty, deal with random quantities, and occur more frequently (for each order), but less seriously than supply uncertainty (Snyder & Shen, 2006). In most models, supply disruption means failure to secure freight (Snyder et al., 2016).

Below are the three main methods used in risk analysis:

- Probabilistic analysis,
- Cause-and-effect modelling,
- Analysis of system vulnerabilities.

These methods are described below.

Probabilistic analysis. Different failure scenarios are modelled for different levels of probability and severity (Deleris & Erhun, 2005). Some studies attempt to identify lost sales in multi-product or multi-supply source simulation scenarios. Other authors have simply introduced case analysis. For example, Jaberidoost et al. (2015) used probabilistic analysis to analyse regulatory risks of operations management, quality management, and supply and supplier issues, etc. They estimated the probability of each type of risk and created a probability-hazard chart. The probability component was also included in economic models. One such model is the Schawell model, where probability is considered as a violation over some time and as an investment function to reduce the likelihood of disruption or related losses. The Schawell model is the simplest useful framework in risk analysis, resulting in more efficient costs for risk reduction and the number of accident cases minimization (Kleindorfer & Saad, 2005).

The literature presents a variety of quantitative models (Ho et al., 2015). Tang (2006) revised about 200 articles that rely on quantitative models. Halpern and Pearl proposed accurate modelling of many important cause-and-effect relationships (Eiter & Lukasiewicz, 2012). Eiter and Lukasiewicz (2012) presented cause-and-effect models with weak and actual causes and proposed to reduce weak causes. In these analyses, more than 50 internal and environmental characteristics were used to prove the truth of the causal hypothesis put forward by the authors. The analysis is presented based on descriptive statistics.

One well-known model is called the Rubin causal model, developed in 1974. The TSLS (two-stage least square) method can be used to estimate the average causal effect when variability occurs. The average causal effect is different on freight. It starts from probability assumptions along with the period required to respond to a well-defined cause-and-effect relationship (Angrist & Imbens, 1995). It is also noted that causal modelling is partially covered by the ISO 31000 risk management standard (Bharathy & McShane, 2014).

Multiple attribute assessment methods were used in the analysis of system vulnerabilities. Various types of methodologies were proposed to assess the vulnerability risks of the system. There are problems of decision-making on several attributes when less risky alternatives are selected based on a set of assessed risks

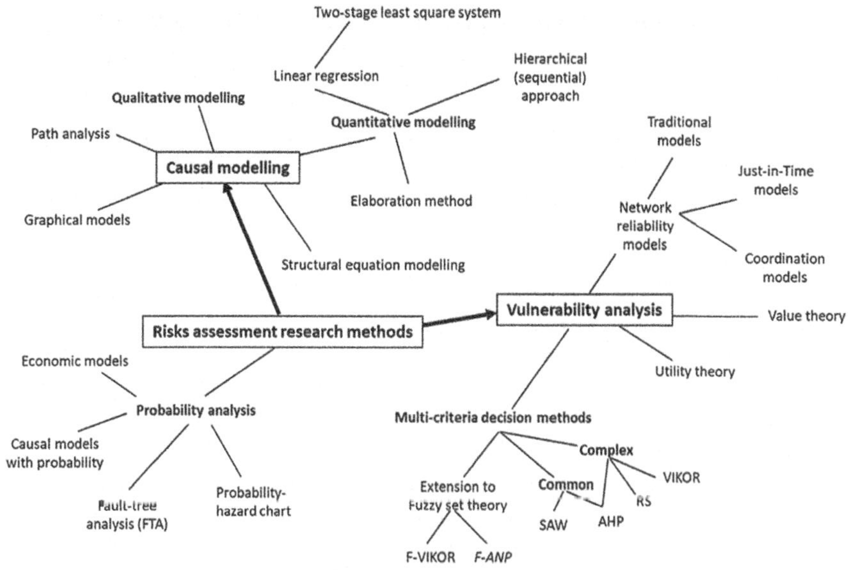

Fig. 10.1 Analysis of methods used in risk assessment studies

(Samvedi et al., 2013). The authors consider the identified integrated approach and structure of classification of risk indices:

- Ensure that decision-makers follow the "rational" behaviour of the system;
- Development of a system of risk assessment indicators;
- Find out preferred solution.

Numerous multi-criteria decision support approaches are used in risk assessment of freight transport systems, including AHP (Ilbahar et al., 2018), RS, SAW (Jaberidoost et al., 2015), ANP (Yüksel & Dagdeviren, 2007), F-ANP (Moeinzadeh & Hajfathaliha, 2009), VIKOR (Sanayei et al., 2009), F-VIKOR (Moeinzadeh & Hajfathaliha, 2009), etc.

Risks assessment research methods are separated into two groups: vulnerability analysis and causal modelling. Vulnerability analysis includes network reliability models. The network reliability models take into account the design of the network. Causal models could be used to analyse the improvement of the reliability of the system and focus on cost-effectiveness.

Figure 10.1 shows a summary of all the methods presented above.

Probability models assume that every element in the network always works according to plan, which in practice does not always happen.

10.2 Application of Multi-attribute Methods of Decision Support in the Process of Cargo Delivery Management

For a long time, the authors have used multi-attribute decision-making and decision support (MADM) techniques to solve transportation issues (Table 10.2). In numerous situations, it is necessary to make decisions for solving many economic, managerial, and other problems. Several methods have been created to analyse many attributes.

There are COPRAS (Comprehensive Proportional Evaluation of Alternatives), TOPSIS (Method of Ordering Preference Systems by Analogy with the Ideal Solution), VIKOR (Vise Criterion Optimization and Compromise Solution), AHP (Analytic Hierarchy Process), and others (Inti & Tandon, 2017). An overview of an evolution of various multi-attribute decision-making and decision support (MADM) techniques is presented in Table 10.3.

The COPRAS multi-attribution method is one of the most important and most commonly used methods, which assumes a direct and proportional relationship. For the situations with the uncertainty aspects, COPRAS-G, fuzzy TOPSIS, or game theory methods can be used. The advantages of the VIKOR method were manifested in the solution of multi-attribute optimization.

Table 10.2 Overview of the evolution of various MADM methods

Types of MADM	1970	1980	1990	2000	2010
MADM method	VIKOR (1973)	TOPSIS (1981)	COPRAS (1996)	MURA (2006)	ARAS (2010); WASPAS (2012)
MADM method + method to determine the importance of an attribute					SWARA-COPRAS (2015); BWM-COPRAS (2019)
MADM method + fuzzy sets, grey numbers				COPRAS-G (2008); Fuzzy TOPSIS (2007)	Fuzzy COPRAS (2018)
MADM + another method				DEA TOPSIS (2009)	AHP & COPRAS-G (2011); AHP & TOPSIS (2013); TOPSIS FADR (2018)
MADM + method that determines the importance of the attribute + other method					SAW, TOPSIS & GAME (2016)

SWARA stepwise weight assessment ratio analysis, *ARAS* additive ratio estimation method, *MOORA* multi-purpose optimization based on ratio analysis, *WASPAS* weighted aggregate product estimation sum, *DEA* enveloping data analysis, *FADR* modified fuzzy method for preference order similarity to the ideal solution, *COPRAS-G* complex proportional evaluation of grey ratio alternatives, *SAW* simple additive weighting, *BWM* is the best worst method

Table 10.3 Overview of applications of various MADM in the field of transport research

Areas of application of MADM in the transport sector	MADM Method	Authors of such studies
Selection of transport indicators	SAW	Castillo and Pitfield (2010)
Assessment of road design	COPRAS	Zavadskas et al. (2007)
Comparison of container terminal technologies	COPRAS-G	Barysiene (2012)
Strategic transport infrastructure planning	AHP	López and Monzón (2010)
Evaluating transport projects	AHP	Cadena and Magro (2015)
Identification of location of forest roads	AHP & COPRAS-G	Hashemkhani et al. (2011)
Selection of road transport for transportation	BVM-COPRAS	Pamučar and Savin (2020)
Assessment of critical infrastructure risk	Fuzzy COPRAS	Yazdani et al. (2011)
Assessment of railway transport operation in regards to safety	TOPSIS	Yan et al. (2020)
Assessment of the quality of services of the railway transit system	Fuzzy TOPSIS	Li et al. (2019)
Evaluation of transport systems	Fuzzy TOPSIS	Awasthi et al. (2011)
Evaluation of life cycle of heavy cargo vehicle	DEA TOPSIS	Wang et al. (2019)
Determination of location for radar used to control air traffic	TOPSIS	Petrovic and Kankaras (2020)
Selection of the station for transport mean sharing	WASPAS and TOPSIS	Deveci et al. (2018)
Ranking of alternatives to bridge design	TOPSIS FADR	Keshavarz-Ghorabaee et al. (2018)
Solving other transport issues	ARAS & TOPSIS, WASPAS, COPRAS, VIKOR, MOORA	Petrović et al. (2018), Awasthi and Balezentis (2016)
Selection of intermodal transport terminal location	DVIKOR	Zečević et al. (2017)

DVIKOR fuzzy Delphi-based fuzzy VIKOR

By identifying that the transport system is important for countries, Barić et al. (2016) propose the application of the AHP approach when ranking traffic-related urban projects. Barić et al. (2016) pointed out serious work in the application of the AHP technique, which has a lot of input data, so it is difficult to evaluate output. One of the solutions is to increase the stability of transport using various alternative fuels and propulsion systems of vehicles. Thus, the stability index of Mitropoulos and Prevedouros (2016) was used to assess the characteristics of the vehicle.

The identified indicators were divided into five sustainability categories: environment, technology, energy, economy, and consumers, and then summarized using the weighted sum (WSM) method. Safaei Mohamadabadi et al. (2009) also selected vehicle engines for fuel based on three key aspects of sustainability. Preference

Ranking for Enrichment Assessment (PROMETHEE) was used to break down alternatives by preference ranking method. The authors integrated decision-making methods using several attributes into the proposed approach.

For the revision of sustainability of transport systems in selected European countries, Bojković et al. (2010) presented economic, environmental, and social indicators. The technique for ranking transport-related urban projects was proposed by Jones et al. (2013). The selection of alternatives follows a localized sustainability assessment index calculated by applying the AHP technique (Awasthi & Chauhan, 2012). To assess the sustainability of various transport projects, there is a necessity for ranking the modes of transport. For example, a study by Pamučar et al. (2020) proposes a joint technique (BWM with COPRAS) for selecting the appropriate road vehicles (Pamučar & Savin, 2020).

Although, many studies have been given to cargo transportation thematic, all these suggested techniques and named problems do not focus on dynamic context data in decision-making but in operation freight management where such decisions are necessary.

10.2.1 Description of Attributes in Decision Models for the Selection of Alternatives in Freight Transport

An overview of the usage of various attributes in decision-making models is presented in Table 10.4 and provides some ideas for choosing the best alternatives in freight transport management. Determining the specific characteristic of accidents in various studies is a complex procedure since ideal safety in transport processes and their management depends on many modes of transport and shipping factors.

The review shows important issues in the application of multi-attribute decision support systems. It involves the approval of MADM results that are obtained by:

- Comparison of results with other multi-attribute (MADM) models;
- Application of stability analysis with changing attribute weights.

The design of transport systems follows the approach of analytical modelling. Authors researching transport systems use analytical techniques such as fuzzy, stochastic, linear and nonlinear programming with single-objective and multi-objective, metaheuristics, and hybrid programming methods. When designing transport systems, McDonald and Karimi (1997) presented an integer model of LP to solve the problem in multistage cases. Park (2005) suggested applying a model for the construction of transportation systems using the linear programming method. Goetschalckx et al. (2002) proposed the same method that involves revising seasonal demand in placing transport orders.

Lababidi et al. (2004) proposed an NLP model to optimize the transportation system of the chemical business. Chern and Hsieh (2007) and Selim et al. (2008) developed MOLP for a basic transport planning approach. Aliev et al. (2007)

Table 10.4 Results of an overview of attributes in freight transport decision-making models

No	Attribute name	MADM method	Authors
1	Transportation costs	IT2FSs	Samanta et al. (2020)
2	Duration of transportation	ELECTRE and AHP	Anton et al. (2004)
3	Corridor reliability	ELECTRE and AHP	Alazzawi and Zak (2020)
4	Safety and security in transport	Fuzzy TOPSIS ELECTRE and AHP	Pak et al. (2015); Alazzawi and Zak (2020)
5	Quality indicators in transport system	Fuzzy TOPSIS	Pak et al. (2015)
6	Efficiency of customs processes	AHP	Srisawat et al. (2017)
7	Level of transport competitiveness	ELECTRA	Zak and Węgliński (2014)
8	Running a mode-related service	EATVOS	Iyigun (2019)
9	Vehicle capacity	IT2FSs	Samanta et al. (2020)
10	Energy consumption (environmental friendliness)	ELECTRE and AHP	Alazzawi and Zak (2020)
11	Infrastructure sustainability (environmental sustainability)	Enveloping data analysis	Rahimi-Golkhandan et al. (2019)
12	Number of possible mode ratings	Enveloping data analysis	Rahimi-Golkhandan et al. (2019)
13	Availability of a communication network	AHP and TOPSIS	Wang et al. (2018)
14	Number of accidents at work	AHP	Özdemir et al. (2018)
15	Performance of service provider	AHP	Wang et al. (2020)

IT2FS interval type-2 of fuzzy sets, *EATWOS* a method consisting of an analysis of entropy and efficiency with the satisfaction of the output

proposed an FMP to create a transport system that is analysed over several periods. Roghanian et al. (2007) presented a MADM structure applicable to transport system planning. Park (2005) provided metaheuristic with the several-stage applications for the architecture of the transportation system.

10.3 Application of Metaheuristic Methods Designed for Optimizing the Performance of Process of Cargo Delivery Management

There are many optimization programmes. Every process can be optimized. Indeed, many complex applications in science and industry can be called optimization tasks. Optimization occurs by reducing time, costs, and risks or by increasing profits, quality, and efficiency. For example, there are many possible ways to build a network to optimize costs and improve service quality. There are many ways to plan production, task optimization, and so on. Metaheuristics solves problems that are generally considered serious and solutions can be found in these cases (Table 10.5). Metaheuristics serves for solving problems faster, solving basic

Table 10.5 Application of metaheuristic methods in the process of managing the delivery of goods

	Cargo delivery management process				
Method	Transportation order management	Order delivery management	Return management	Routing management	Cargo shipping management
Genetic algorithms (GA)	Kuo (2001), Lawrence and Pasternack (2002), Ossipov (2005), Zhang and Qinghe (2014), Jun and Qing (2015), Devika et al. (2016), Moghimi and Beheshtinia (2021)	Liu et al. (2014), Pierre and Zakaria (2014), Mohr (2014), Kumar et al. (2014), Zhang and Li (2014), Psychas et al. (2015), Cai et al. (2015), Karakatic and Podgorelec (2015), Ahmadizar et al. (2015), Ladosz et al. (2018)	Rogers et al. (2002), Min et al. (2006), Lieckens and Vandaele (2007), Trappey et al. (2010), Zhou and Min (2011), Boudhar et al. (2013), Diabat et al. (2013), Li et al. (2013), Ghezavati and Nia (2014), Guo and Li (2014), Park et al. (2021)	Altiparmak et al. (2006), Baker and Ayechew (2003), Nair and Sooda (2012), Dirgayusari et al. (2021), Yetis and Karakose (2021)	Moura (2008), Jiang et al. (2009), Huang et al. (2009), Galindo-Muro et al. (2020)
Evolutionary algorithms (EA)	Bandyopadhyay and Bhattacharya (2014), Liu et al. (2021)	Tan et al. (2006, 2007), Erbao and Mingyong (2009), Liu et al. (2021)	Kumar et al. (2017)	Prins (2004), Jozefowiez et al. (2009), Leyva Lopez et al. (2016), Lopes et al. (2016), Peng et al. (2020), Ghannadpour et al. (2021)	
Differential evolution (DE) algorithms	Wang et al. (2012)	Wang et al. (2010), Moonsri et al. (2022)		Sethanan and Jamrus (2020)	Qu et al. (2020)
Optimization of the particle swarm (PSO)	Coello and Lechuga (2002)	Ai & Kachitvichyanukul (2009), Marinakis and Marinaki (2010), Singh and Singh (2021)		Venkatesan & Kumanan (2010), Islam et al. (2021)	Zhiyue and Maoqi (2020)

problems, and getting reliable algorithms. What is more, they can be easily designed and implemented, and are highly flexible.

Many metaheuristics in various fields have been developed by communities over 20 years. There are three common metaheuristic design problems:

- Presentation: Triple (e.g. linear/nonlinear, direct/indirect) or specific coding can be used to present solutions to problems. Coding plays an important role in the efficiency and effectiveness of any metaheuristic and is an important step in the development of metaheuristic. The representation must have several desired characteristics, such as completeness, coherence, and efficiency. Coding must be appropriate and appropriate for the optimization task. In addition, the accuracy of mapping is also related to the location search functionality that is used for this representation (e.g. neighbourhood generation, solution recombination).
- Objective function: A meaningful element of the metaheurism design. It directs the search into the "good" search solutions.
- Constraint management: Bridge constraint management strategies affect the presentation of decisions or the function of a goal (e.g. reject, penalize, correct, decode, and save strategies) and address constraints.

There are freight delivery processes in which metaheuristics can be applied. The author selected five freight delivery processes.

The order delivery management process includes constraints important for the customer. The size of customer order and their placement frequency could be analysed as multi-purpose, including various factors important for the number of orders forecasting. In such cases, genetic and other algorithms can be used to find optimal solutions.

The process for freight delivery management process concentrates on planning deliveries and solving planning problems. Costs allocated for each customer order delivery are quite high, so these situations require finding optimal values which could be reached by applying GA and other metaheuristics.

The return shipping management process includes the management of cargo flows. These methods include mathematical optimization when methods require to reach the optimal solution.

Several studies have been carried out also on the management of loading and unloading of goods.

Modelling the process of managing transport orders helps to analyse the links. Metaheuristic algorithms help to solve problems with higher complexity where traditional methods do not help.

The metaheuristic algorithms unite genetic algorithms (GA), evolutionary algorithms (EA), differential algorithms of evolution (DE), and optimization of the particle swarm (PSO). The author indicates the algorithms that were used in the above processes and the number of peer-reviewed publications.

The more precise analysis is given in Table 10.5. According to the information presented in Table 10.5, GA was most widely used by the authors in the process of managing return shipping.

The fewest publications on evolutionary algorithms have been produced, but they dominate studies covering all five functional processes which are applied for the management of freight delivery.

Following the results of this literature analysis, the use of metaheuristic methods for revising the study of service-oriented processes of the transport system is recommended. A literature review of the five transportation processes shows that the researchers mainly apply genetic and evolutionary algorithms for the optimization studies focusing on the three main functional processes (ordering, delivering, and routing) in cargo delivery.

10.4 Conclusions

The application of various methods for the freight delivery process is quite complex. The author presents micro-risks management, decision-making, and optimization oriented metaheuristics methods. The study presents the evolution of multi-decision support methods from early 1970 and their application for solving various problems evident in the freight delivery process.

The revision of papers on main cargo delivery management stages suggests that studies focus on the usage of both algorithms which are called genetic and evolutionary and revise their application in optimization studies.

The application demonstrates that above-mentioned metaheuristics are recommended for studies in the field of cargo delivery management, especially in several functional processes.

References

Ahmadizar, F., Zeynivand, M., & Arkat, J. (2015). Two-level vehicle routing with cross-docking in a three-echelon supply chain: A genetic algorithm approach. *Applied Mathematical Modelling, 39*(22), 7065–7081.

Ai, T. J., & Kachitvichyanukul, V. (2009). Particle swarm optimisation and two solution representations for solving the capacitated vehicle routing problem. *Computers and Industrial Engineering, 56*(1), 380–387.

Alazzawi, A., & Zak, J. (2020). MADM/A based design of sustainable logistics corridors combined with suppliers selection. The case study of freight movement to Iraq. *Transportation Research Procedia, 47*, 577–584. https://doi.org/10.1016/j.trpro.2020.03.134

Aliev, R. A., Fazlollahi, B., Guirimov, B. G., & Aliev, R. R. (2007). Fuzzy-genetic approach to aggregate production–distribution planning in supply chain management. *Information Sciences, 177*, 4241–4255. https://doi.org/10.1016/j.ins.2007.04.012

Altiparmak, F., Gen, M., Lin, L., & Paksoy, T. (2006). A genetic algorithm approach for multi-objective optimisation of supply chain networks. *Computers & Industrial Engineering, 51*(1), 196–215.

Angrist, J. D., & Imbens, G. W. (1995). Two-stage least squares estimation of average causal effects in models with variable treatment intensity. *Journal of the American Statistical Association, 90*(430), 431–442.

Anton, J. M., Grau, J. B., & Andina, D. (2004). ELECTRE and AHP MADM methods versus CP method and the official choice applied to high-speed railway layout alternative election. *WSEAS Transactions on Business and Economics, 1*, 64–69.

Atilgan, C., & McCullen, P. (2007). Using "quick scan" at a medium-sized food company: a change management perspective. *Developing Sustainable Collaborative Supply Chains, 43*(1), 215.

Awasthi, A., & Balezentis, T. (2016). A hybrid approach based on BOCR and fuzzy MULTIMOORA for logistics service provider selection. *International Journal of Logistics Systems and Management, 27*(3), 261–282.

Awasthi, A., & Chauhan, S. S. (2012). A hybrid approach integrating affinity diagram, AHP and fuzzy TOPSIS for sustainable city logistics planning. *Applied Mathematical Modelling, 36*(2), 573–584.

Awasthi, A., Chauhan, S. S., & Omrani, H. (2011). Application of fuzzy TOPSIS in evaluating sustainable transportation systems. *Expert Systems with Applications, 38*, 12270–12280. https://doi.org/10.1016/j.eswa.2011.04.005

Baghalian, A., Rezapour, S., & Farahani, R. Z. (2013). Robust supply chain network design with service level against disruptions and demand uncertainties: A real-life case. *European Journal of Operational Research, 227*(1), 199–215.

Baker, B. M., & Ayechew, M. A. (2003). A genetic algorithm for the vehicle routing problem. *Computers and Operations Research, 30*(5), 787–800.

Bandyopadhyay, S., & Bhattacharya, R. (2014). Solving a tri-objective supply chain problem with modified NSGA-II algorithm. *Journal of Manufacturing Systems, 33*(1), 41–50.

Barić, D., Pilko, H., & Strujić, J. (2016). An analytic hierarchy process model to evaluate road section design. *Transport, 31*, 312–321. https://doi.org/10.3846/16484142.2016.1157830

Barysiene, J. (2012). A multi-criteria evaluation of container terminal technologies applying the COPRAS-G method. *Transport, 27*, 364–372. https://doi.org/10.3846/16484142.2012.750624

Batarlienė, N. (2018). Risk and damage assessment for transportation of dangerous freight. *Transport and Telecommunication Journal, 19*(4), 356–363.

Bharathy, G. K., & McShane, M. K. (2014). Applying a systems model to enterprise risk management. *Engineering Management Journal, 26*(4), 38–46.

Blome, C., & Schoenherr, T. (2011). Supply chain risk management in financial crises—A multiple case-study approach. *International Journal of Production Economics, 134*(1), 43–57.

Bode, C., & Wagner, S. M. (2015). Structural drivers of upstream supply chain complexity and the frequency of supply chain disruptions. *Journal of Operations Management, 36*, 215–228.

Bojković, N., Anić, I., & Pejčić-Tarle, S. (2010). One solution for cross-country transport-sustainability evaluation using a modified ELECTRE method. *Ecological Economics, 69*, 1176–1186. https://doi.org/10.1016/j.ecolecon.2010.01.006

Boudhar, H., Dahane, M., & Rezg, N. (2013). Spare part returns in stochastic deteriorating manufacturing system under a condition-based maintenance policy: Simulation-based genetic algorithm approach. *Manufacturing Modelling and Management Control, 7*(1), 1399–1404.

Cadena, P. C. B., & Magro, J. M. V. (2015). Setting the weights of sustainability criteria for the appraisal of transport projects. *Transport, 30*, 298–306. https://doi.org/10.3846/16484142.2015.1086890

Cai, Y. G., Tang, Y. L., & Yang, Q. J. (2015). An improved genetic algorithm for multi-depot heterogeneous vehicle routing problem with simultaneous pickup and delivery time windows. *Applied Mechanical Material, 738*, 361–365.

Castillo, H., & Pitfield, D. E. (2010). ELASTIC – A methodological framework for identifying and selecting sustainable transport indicators. *Transportation Researcg Part D: Transport and Environment, 15*, 179–188. https://doi.org/10.1016/j.trd.2009.09.002

Chern, C. C., & Hsieh, J. S. (2007). A heuristic algorithm for master planning that satisfies multiple objectives. *Computers and Operations Research., 34*, 3491–3513. https://doi.org/10.1016/j.cor.2006.02.022

Chopra, S., Reinhardt, G., & Mohan, U. (2007). The importance of decoupling recurrent and disruption risks in a supply chain. *Naval Research Logistics, 54*(5), 544–555.

Christopher, M., & Peck, H. (2004). Building the resilient supply chain. *The International Journal of Logistics Management, 15*(2), 1–14.

Coello, C. C., & Lechuga, M. S. (2002). MOPSO: A proposal for multiple objective particle swarm optimisation. In *Proceedings of the 2002 Congress on Evolutionary Computation, 5*(2), 1051–1056. IEEE.

Deleris, L. A., & Erhun, F. (2005). Risk management in supply networks using Monte-Carlo simulation. In M. E. Kuhl, N. M. Steiger, F. B. Armstrong, & J. A. Joines (Eds.), *Proceedings of the 2005 winter simulation conference* (pp. 19–34). IEEE.

Deveci, M., Canıtez, F., & Gökaşar, I. (2018). WASPAS and TOPSIS based interval type-2 fuzzy MADM method for a selection of a car sharing station. *Sustainable Cities and Society, 41*, 777–791. https://doi.org/10.1016/j.scs.2018.05.034

Devika, K., Jafarian, A., Hassanzadeh, A., & Khodaverdi, R. (2016). Optimising of bullwhip effect and net stock amplification in three-echelon supply chains using evolutionary multi-objective metaheuristics. *Annals of Operations Research, 242*(2), 457–487.

Diabat, A., Kannan, D., Kaliyan, M., & Svetinovic, D. (2013). An optimisation model for product returns using genetic algorithms and artificial immune system. *Resources, Conservation and Recycling., 74*, 156–169.

Dirgayusari, A. M., Asana, I. M. D. P., Sudipa, I. G. I., Komalasari, R., & Dwipayani, S. A. (2021). Application for determining shipping routes at cargo companies with genetic algorithms. *Jurnal Mantik, 5*(2), 1034–1044.

Dua, A., & Sinha, D. (2019). Quality of multimodal freight transportation: A systematic literature review. *World Review of Intermodal Transportation Research, 8*(2), 167–194.

Eiter, T., & Lukasiewicz, T. (2012). Causes and explanations in the structural-model approach: Tractable cases. *Arxiv*, 55–66.

Erbao, C., & Mingyong, L. (2009). A hybrid differential evolution algorithm to vehicle routing problem with fuzzy demands. *Journal of Computational and Applied Mathematics, 231*(1), 302–310.

Galindo-Muro, A. B., Mora-Vargas, J., Cedillo-Campos, M. G., & Regis-Hernández, F. (2020, October). A genetic algorithm approach for a truck and trailer routing problem in a loading/ Unloading bays application. In *Mexican International Conference on Artificial Intelligence* (pp. 296–310). Springer.

Ghannadpour, S. F., Zandieh, F., & Esmaeili, F. (2021). Optimizing triple bottom-line objectives for sustainable health-care waste collection and routing by a self-adaptive evolutionary algorithm: A case study from Tehran province in Iran. *Journal of Cleaner Production, 287*, 125010.

Ghezavati, V., & Nia, N. S. (2014). Development of an optimisation model for product returns using genetic algorithms and simulated annealing. *Soft Computer, 19*, 3055–3069. https://doi.org/10.1007/s00500-014-1465-8

Goetschalckx, M., Vidal, C. J., & Dogan, K. (2002). Modeling and design of global logistics systems: A review of integrated strategic and tactical models and design algorithms. *European Journal of Operational Research., 143*, 1–18. https://doi.org/10.1016/S0377-2217(02)00142-X

Guo, H. & Li, Y. (2014). Multiobjective location-inventory routing problem taking returns into consideration. In: *Proceedings of the Ninth International Symposium on Linear Drives for Industry Applications*, 2, 19–26. Springer 1016.

Halse, A. H., Killi, M., & Landmark, A. D. (2019). Train unreliability and demand for rail freight transport in Norway. *Journal of Transport Economics and Policy, 53*(1), 1–18.

Hashemkhani, Z. S., Rezaeiniya, N., Kazimieras Zavadskas, E., & Turskis, Z. (2011). *Forest roads locating based on AHP and COPRAS-G methods: An empirical study based on Iran.*

Hendricks, K. B., & Singhal, V. R. (2005). An empirical analysis of the effect of supply chain disruptions on long-run stock price performance and equity risk of the firm. *Production and Operations Management, 14*(1), 35–52.

Ho, W., Zheng, T., Yildiz, H., & Talluri, S. (2015). Supply chain risk management: A literature review. *International Journal of Production Research, 53*(16), 5031–5069.

Huang, Y., Liang, C., & Yang, Y. (2009). The optimum route problem by genetic algorithm for loading/unloading of yard crane. *Computers and Industrial Engineering, 56*(3), 993–1001.

Ilbahar, E., Karaşan, A., Cebi, S., & Kahraman, C. (2018). A novel approach to risk assessment for occupational health and safety using Pythagorean fuzzy AHP & fuzzy inference system. *Safety Science, 103*, 124–136.

Inti, S., & Tandon, V. (2017). Application of fuzzy preference-analytic hierarchy process logic in evaluating sustainability of transportation infrastructure requiring multicriteria decision making. *Journal of Infrastructure Systems, 23*, 04017014. https://doi.org/10.1061/(ASCE)IS.1943-555X.0000373

Islam, M. A., Gajpal, Y., & ElMekkawy, T. Y. (2021). Hybrid particle swarm optimization algorithm for solving the clustered vehicle routing problem. *Applied Soft Computing, 110*, 107655.

Iyigun, I. (2019). Evaluation of efficiency of rail transportation of black sea countries by using an integrated MADM approach. *Economy and Business Journal, 13*, 305–323.

Jaberidoost, M., Olfat, L., Hosseini, A., Kebriaeezadeh, A., Abdollahi, M., Alaeddini, M., & Dinarvand, R. (2015). Pharmaceutical supply chain risk assessment in Iran using analytic hierarchy process (AHP) and simple additive weighting (SAW) methods. *Journal of Pharmaceutical Policy and Practice, 8*(1), 1–10.

Jiang, Y., Xu, L., Wang, H., & Wang, H. (2009). Influencing factors for predicting financial performance based on genetic algorithms. *Systems Research and Behavioral Science: The Official Journal of the International Federation for Systems Research, 26*(6), 661–673.

Jones, S., Tefe, M., & Appiah-Opoku, S. (2013). Proposed framework for sustainability screening of urban transport projects in developing countries: A case study of Accra, Ghana. *Transportation Research Part A: Policy and Practice, 49*, 21–34. https://doi.org/10.1016/j.tra.2013.01.003

Jozefowiez, N., Semet, F., & Talbi, E. G. (2009). An evolutionary algorithm for the vehicle routing problem with route balancing. *European Journal of Operational Research, 195*(3), 761–769.

Jun, T. & Qing, W. (2015). Optimise BP neural network structure on car sales forecasts based on genetic algorithm. In: *International Industrial Informatics and Computer Engineering Conference* (pp. 75–79). Atlantis Press.

Juttner, U. (2005). Supply chain risk management: Understanding the business requirements from a practitioner perspective. *The International Journal of Logistics Management, 16*(1), 120–141.

Juttner, U., Peck, H., & Christopher, M. (2003). Supply chain risk management: Outlining an agenda for future research. *International Journal of Logistics: Research and Applications, 6*(4), 197–210.

Karakatic, S., & Podgorelec, V. (2015). A survey of genetic algorithms for solving multi depot vehicle routing problem. *Applied Soft Computer, 27*, 519–532.

Keshavarz-Ghorabaee, M., Amiri, M., Zavadskas, E. K., Turskis, Z., & Antuchevičienė, J. (2018). Ranking of bridge design alternatives: A TOPSIS-FADR method. *Baltic Journal of Road and Bridge Engineering, 13*, 209–237. https://doi.org/10.7250/bjrbe.2018-13.413

Kleindorfer, P. R., & Saad, G. H. (2005). Managing disruption risks in supply chains. *Production and Operations Management, 14*(1), 53–68.

Klibi, W., Martel, A., & Guitouni, A. (2010). The design of robust value-creating supply chain networks: A critical review. *European Journal of Operational Research, 203*(2), 283–293.

Kumar, V. S., Thansekhar, M. R., Saravanan, R., & Amali, S. M. J. (2014). Solving multi-objective vehicle routing problem with time windows by FAGA. *Proceedings Engineering, 97*, 2176–2185.

Kumar, V. N. S. A., Kumar, V., Brady, M., Garza-Reyes, J. A., & Simpson, M. (2017). Resolving forward-reverse logistics multi-period model using evolutionary algorithms. *International Journal of Production Economics, 183*, 458–469.

Kuo, R. J. (2001). A sales forecasting system based on fuzzy neural network with initial weights generated by genetic algorithm. *European Journal of Operational Research, 129*(3), 496–517.

Lababidi, H. M. S., Ahmed, M. A., Alatiqi, I. M., & Al-enzi, A. F. (2004). Optimizing the supply chain of a petrochemical company under uncertain operating and economic conditions. *Industrial and Engineering Chemistry Research.*, *43*, 63–73. https://doi.org/10.1021/ie030555d

Ladosz, P., Banjo, O., De Guido, S., & Zalasiewicz, M. (2018). A genetic algorithm optimiser for dynamic product routing in agile manufacturing environment. In *Proceedings of 16th International Conference on Industrial Informatics* (pp. 1079–1084). IEEE.

Lavastre, O., Gunasekaran, A., & Spalanzani, A. (2012). Supply chain risk management in French companies. *Decision Support Systems, 52*(4), 828–838.

Lawrence, J. A., & Pasternack, B. A. (2002). *Applied management science: Modelling, spreadsheet analysis, & communication for decision making* (2nd ed.). Wiley.

Leyva Lopez, J. C., Solano Noriega, J. J., Garcia Alcaraz, J. L., & Gastelum Chavira, D. A. (2016). Exploitation of a medium-sized fuzzy outranking relation based on multi-objective evolutionary algorithms to derive a ranking. *International Journal of Computational Intelligence Systems, 9*(4), 745–764.

Li, Y., Guo, H., Wang, L., & Fu, J. (2013). A hybrid genetic simulated annealing algorithm for the location-inventory routing problem considering returns under E-supply chain environment. *Science World Journal, 2013*, 125893. https://doi.org/10.1155/2013/125893

Li, J., Xu, X., Yao, Z., & Lu, Y. (2019). Improving service quality with the fuzzy TOPSIS method: A case study of the Beijing rail transit system. *IEEE Access, 7*, 114271–114284. https://doi.org/10.1109/ACCESS.2019.2932779

Lieckens, K., & Vandaele, N. (2007). Reverse logistics network design with stochastic lead-times. *Computer Operational Research, 34*(2), 395–416.

Liu, R., Jiang, Z., & Geng, N. (2014). A hybrid genetic algorithm for the multi-depot open vehicle routing problem. *OR Spectrum, 36*(2), 401–421.

Liu, R., Wang, S., Pi, Y., & Qin, Q. (2021). An effective heuristic with evolutionary algorithm for the coordinated capacitated dynamic lot-size and delivery problem. *Computers and Industrial Engineering, 153*, 107051.

Lopes, R. B., Ferreira, C., & Santos, B. S. (2016). A simple and effective evolutionary algorithm for the capacitated location–routing problem. *Computers and Operations Research, 70*, 155–162.

López, E., & Monzón, A. (2010). Integration of sustainability issues in strategic transportation planning: A multi-criteria model for the assessment of transport infrastructure plans. *Computer-Aided Civil and Infrastructure Engineering, 25*, 440–451. https://doi.org/10.1111/j.1467-8667.2010.00652.x

Marinakis, Y., & Marinaki, M. (2010). A hybrid multi-swarm particle swarm optimization algorithm for the probabilistic traveling salesman problem. *Computers and Operations Research, 37*(3), 432–442.

McDonald, C. M., & Karimi, I. A. (1997). Planning and scheduling of parallel semicontinuous processes. 1. Production planning. *Industrial and Engineering Chemistry Research, 36*, 2691–2700. https://doi.org/10.1021/ie960901

Min, H., Ko, H. J., & Ko, C. S. (2006). A genetic algorithm approach to developing the multi-echelon reverse logistics network for product returns. *Omega, 34*(1), 56–69.

Mitropoulos, L. K., & Prevedouros, P. D. (2016). Incorporating sustainability assessment in transportation planning: An urban transportation vehicle-based approach. *Transportation Planning and Technology, 39*, 439–463. https://doi.org/10.1080/03081060.2016.1174363

Moeinzadeh, P., & Hajfathaliha, A. (2009). A combined fuzzy decision-making approach to supply chain risk assessment. *International Scholarly and Scientific Research and Innovation, 3*(12), 1631–1647.

Moghimi, M., & Beheshtinia, M. A. (2021). Optimization of delay time and environmental pollution in scheduling of production and transportation system: A novel multi-society genetic algorithm approach. *Management Research Review*.

Mohr, C. M. (2014). Optimisation of warehouse order-picking routes using vehicle routing model and genetic algorithm. Doctoral dissertation, State University of New York at Binghamton.

Moonsri, K., Sethanan, K., Worasan, K., & Nitisiri, K. (2022). A hybrid and self-adaptive differential evolution algorithm for the multi-depot vehicle routing problem in egg distribution. *Applied Sciences, 12*(1), 35.

Moura, A. (2008). A multi-objective genetic algorithm for the vehicle routing with time windows and loading problem. In *Intelligent decision support* (pp. 187–201). Gabler.

Nair, T. G., & Sooda, K. (2012). An intelligent routing approach using genetic algorithms for quality graded network. *International Journal of Intelligent Systems Technologies and Applications, 11*(3–4), 196–211.

Ossipov, P. (2005). Heuristic optimisation of sequence of customer orders. *Applied Mathematical Computation, 162*(3), 1303–1313.

Ouhader, H. (2020). Assessing the economic and environmental benefits of horizontal cooperation in delivery: Performance and scenario analysis. *Uncertain Supply Chain Management, 8*(2), 303–320.

Özdemir, Ü., Altinpinar, İ., & Demirel, F. B. (2018). A MADM Approach with Fuzzy AHP method for occupational accidents on board. *TransNav International Journal on Marine Navigation and Safety of Sea Transportation, 12*, 93–98. https://doi.org/10.12716/1001.12.01.10

Pak, J. Y., Thai, V. V., & Yeo, G. T. (2015). Fuzzy MADM approach for evaluating intangible resources affecting port service quality. *The Asian Journal of Shipping and Logistics, 31*, 459–468. https://doi.org/10.1016/j.ajsl.2016.01.004

Pamučar, D. S., & Savin, L. M. (2020). Multiple-criteria model for optimal off-road vehicle selection for passenger transportation: BWM-COPRAS Model. *Vojnotehnički Glasnik, 68*, 28–64. https://doi.org/10.5937/vojtehg68-22916

Pamučar, D., Ecer, F., Cirovic, G., & Arlasheedi, M. A. (2020). Application of improved Best Worst Method (BWM) in real-world problems. *Mathematics, 8*. https://doi.org/10.3390/math8081342

Park, Y. B. (2005). An integrated approach for production and distribution planning in supply chain management. *International Journal of Production Research, 43*, 1205–1224. https://doi.org/10.1080/00207540412331327718

Park, H., Son, D., Koo, B., & Jeong, B. (2021). Waiting strategy for the vehicle routing problem with simultaneous pickup and delivery using genetic algorithm. *Expert Systems with Applications, 165*, 113959.

Peng, B., Wu, L., Yi, Y., & Chen, X. (2020). Solving the multi-depot green vehicle routing problem by a hybrid evolutionary algorithm. *Sustainability, 12*(5), 2127.

Petrovic, I., & Kankaras, M. (2020). A hybridized IT2FS-DEMATEL-AHP-TOPSIS multicriteria decision making approach: Case study of selection and evaluation of criteria for determination of air traffic control radar position. *Decision Making: Applications in Management and Engineering, 3*, 146–164. https://doi.org/10.31181/dmame2003134p

Petrović, G. S., Madić, M., & Antucheviciene, J. (2018). An approach for robust decision making rule generation: Solving transport and logistics decision making problems. *Expert Systems with Applications, 106*, 263–276. https://doi.org/10.1016/j.eswa.2018.03.065

Pettit, T. J., Croxton, K. L., & Fiksel, J. (2013). Ensuring supply chain resilience: Development and implementation of an assessment tool. *Journal of Business Logistics, 34*(1), 46–76.

Pierre, D. M. & Zakaria, N. (2014). Partially optimised cyclic shift crossover for multi-objective genetic algorithms for the multi-objective vehicle routing problem with time windows. In: *2014 IEEE Symposium on Computational Intelligence in Multi-Criteria Decision-Making* (pp. 106–115).

Popescu, M. C. (2018). Eliminating transportation waste using the transportation value stream map. *Proceedings of the International Conference on Business Excellence, 12*(1), 793–803.

Prins, C. (2004). A simple and effective evolutionary algorithm for the vehicle routing problem. *Computers & Operations Research, 31*(12), 1985–2002.

Psychas, I. D., Marinaki, M. & Marinakis, Y. (2015). A parallel multi-start NSGA II algorithm for multiobjective energy reduction vehicle routing problem. In: *International Conference on*

Evolutionary Multi-criterion Optimization (pp. 336–350). Springer International Publishing. doi:https://doi.org/10.1007/978-3-319-15934-8_23.

Qu, H., Ai, X. Y., & Wang, L. (2020). Optimizing an integrated inventory-routing system for multi-item joint replenishment and coordinated outbound delivery using differential evolution algorithm. *Applied Soft Computing, 86*, 105863.

Rahimi-Golkhandan, A., Garvin, M. J., & Brown, B. L. (2019). Characterizing and measuring transportation infrastructure diversity through linkages with ecological stability theory. *Transportation Research Part A: Policy and Practice, 128*, 114–130. https://doi.org/10.1016/j.tra.2019.07.013

Raimbekov, Z., Syzdykbayeva, B., & Sharipbekova, K. (2018). Economic aspects of freight transportation along the east-west routes through the transport and logistics system of Kazakhstan. In *Transport systems and delivery of cargo on East–West routes* (pp. 205–230). Springer.

Rogers, D. S., Lambert, D. M., Croxton, K. L., & Garcia-Dastugue, S. J. (2002). The returns management process. *International Journal Logistics Management, 13*(2), 1–18.

Roghanian, E., Sadjadi, S. J., & Aryanezhad, M. B. (2007). A probabilistic bi-level linear multi-objective programming problem to supply chain planning. *Applied Mathematics and Computation., 188*, 786–800. https://doi.org/10.1016/j.amc.2006.10.032

Sadghiani, N. S., Torabi, S. A., & Sahebjamnia, N. (2015). Retail supply chain network design under operational and disruption risks. *Transportation Research Part E: Logistics and Transportation Review, 75*, 95–114.

Safaei Mohamadabadi, H., Tichkowsky, G., & Kumar, A. (2009). Development of a multi-criteria assessment model for ranking of renewable and non-renewable transportation fuel vehicles. *Energy, 34*, 112–125. https://doi.org/10.1016/j.energy.2008.09.004

Samanta, S., Jana, D. K., Panigrahi, G., & Maiti, M. (2020). Novel multi-objective, multi-item and four-dimensional transportation problem with vehicle speed in LR-type intuitionistic fuzzy environment. *Neural Computing and Applications, 32*, 11937–11955. https://doi.org/10.1007/s00521-019-04675-y

Samvedi, A., Jain, V., & Chan, F. T. S. (2013). Quantifying risks in a supply chain through integration of fuzzy AHP and fuzzy TOPSIS. *International Journal of Production Research, 51*(8), 2433–2442. https://doi.org/10.1080/00207543.2012.741330

Sanayei, A., Mousavi, S. F., & Yazdankhah, A. (2009). Group decision making process for supplier selection with VIKOR under fuzzy environment. *Expert Systems with Applications, 11*, 20–29.

Satta, G., Maugeri, S., Panetti, E., & Ferretti, M. (2019). Port labour, competitiveness and drivers of change in the Mediterranean Sea: a conceptual framework. *Production Planning and Control, 30*(13), 1102–1117.

Sawik, T. (2013). Integrated selection of suppliers and scheduling of customer orders in the presence of supply chain disruption risks. *International Journal of Production Research, 51*(23–24), 7006–7022.

Schmitt, A. J., & Singh, M. 2009. Quantifying supply chain disruption risk using Monte Carlo and discrete-event simulation. In *Proceedings of the 2009 winter simulation conference* (pp. 1237–1248). IEEE.

Scholten, K., Sharkey, S. P., & Fynes, B. (2014). Mitigation processes – Antecedents for building supply chain resilience. *Supply Chain Management: An International Journal, 19*(2), 211–228.

Selim, H., Am, C., & Ozkarahan, I. (2008). Collaborative production–distribution planning in supply chain: A fuzzy goal programming approach. *Transportation Research Part E: Logistics and Transportation Review, 44*, 396–419. https://doi.org/10.1016/j.tre.2006.11.001

Sethanan, K., & Jamrus, T. (2020). Hybrid differential evolution algorithm and genetic operator for multi-trip vehicle routing problem with backhauls and heterogeneous fleet in the beverage logistics industry. *Computers and Industrial Engineering, 146*, 106571.

Simchi-Levi, D., Schmidt, W., & Wei, Y. (2014). From superstorms to factory fires: Managing unpredictable supply chain disruptions. *Harvard Business Review, 92*(1–2), 96–101.

Singh, G., & Singh, A. (2021). Extension of particle swarm optimization algorithm for solving transportation problem in fuzzy environment. *Applied Soft Computing, 110*, 107619.

Snyder, L. V., & Shen, Z. J. M. (2006). *Supply and demand uncertainty in multi-echelon supply chains* (pp. 20–32). Lehigh University.

Snyder, L. V., Atan, Z., Peng, P., Rong, Y., Schmitt, A. J., & Sinsoysal, B. (2016). OR/MS models for supply chain disruptions: A review. *IEEE Transactions, 48*(2), 89–109.

Srisawat, P., Kronprasert, N., & Arunotayanun, K. (2017). Development of decision support system for evaluating spatial efficiency of regional transport logistics. *Transportation Research Procedia, 25*, 4832–4851. https://doi.org/10.1016/j.trpro.2017.05.493

Talley, W. K., & Ng, M. (2020). Note: Determinants of cargo port, hinterland cargo transport and port hinterland cargo transport service chain choices by service providers. *Transportation Research Part E: Logistics and Transportation Review, 137*, 101921.

Tan, K. C., Chew, Y. H., & Lee, L. H. (2006). A hybrid multi-objective evolutionary algorithm for solving vehicle routing problem with time windows. *Computational Optimization and Applications, 34*(1), 115.

Tan, K. C., Cheong, C. Y., & Goh, C. K. (2007). Solving multi-objective vehicle routing problem with stochastic demand via evolutionary computation. *European Journal of Operational Research, 177*(2), 813–839.

Tang, C. S. (2006). Robust strategies for mitigating supply chain disruptions. *International Journal of Logistics: Research and Applications, 9*(1), 33–45.

Tang, C., & Tomlin, B. (2016). The power of flexibility for mitigating supply chain risks. In *Developments in logistics and supply chain management* (pp. 80–89). Palgrave Macmillan.

Tomlin, B. (2006). On the value of mitigation and contingency strategies for managing supply chain disruption risks. *Management Science, 52*(5), 639–657.

Trappey, A. J., Trappey, C. V., & Wu, C. R. (2010). Genetic algorithm dynamic performance evaluation for RFID reverse logistic management. *Expert Systems Applied, 37*(11), 7329–7335.

Tummala, R., & Schoenherr, T. (2011). Assessing and managing risks using the supply chain risk management process. *Supply Chain Management: An International Journal, 16*(6), 474–483.

Venkatesan, S. P., & Kumanan, S. (2010). Bi-criteria allocation of customers to warehouses using a particle swarm optimisation. *International Journal of Operational Research, 9*(1), 65–81.

Wagner, S. M., & Bode, C. (2008). An empirical examination of supply chain performance along several dimensions of risk. *Journal of Business Logistics, 29*(1), 307–325.

Wakolbinger, T., & Cruz, J. M. (2011). Supply chain disruption risk management through strategic information acquisition and sharing and risk-sharing contracts. *International Journal of Production Research, 49*(13), 4063–4084. https://doi.org/10.1080/00207543.2010.501550

Wang, Y., Li, B., & Weise, T. (2010). Estimation of distribution and differential evolution cooperation for large scale economic load dispatch optimisation of power systems. *Information Sciences, 180*(12), 2405–2420.

Wang, L., He, J., Wu, D., & Zeng, Y. R. (2012). A novel differential evolution algorithm for joint replenishment problem under interdependence and its application. *International Journal of Production Economics, 135*(1), 190–198.

Wang, X., Li, D., Zhang, X., & Cao, Y. (2018). MCDM-ECP: Multi Criteria Decision Making Method for Emergency Communication Protocol in Disaster Area Wireless Network. *Applied Sciences, 8*, 1165. https://doi.org/10.3390/app8071165

Wang, Z., Hao, H., Gao, F., Zhang, Q., Zhang, J., & Zhou, Y. (2019). Multi-attribute decision making on reverse logistics based on DEA-TOPSIS: A study of the shanghai end-of-life vehicles industry. *Journal of Cleaner Production, 214*, 730–737. https://doi.org/10.1016/j.jclepro.2018.12.329

Wang, G., Hu, X., Li, X., Zhang, Y., Feng, S., & Yang, A. (2020). Multi-objective decisions for provider selection and order allocation considering the position of the CODP in a logistics service supply chain. *Computers and Industrial Engineering, 140*, 106216.

Yan, X., Peng, Q., Yin, Y., Zhang, Y., & Zhong, Q. (2020). Evaluating railway operation safety situation in China based on an improved TOPSIS method: A regional perspective. *Journal of Advanced Transportation, 2020*, 1796132. https://doi.org/10.1155/2020/1796132

Yazdani, M., Alidoosti, A., & Zavadskas, E. K. (2011). Risk analysis of critical infrastructures using fuzzy copras. *Economic Research-Ekonomska Istraživanja, 24*, 27–40. https://doi.org/10.1080/1331677X.2011.11517478

Yetis, H., & Karakose, M. (2021, February). Collaborative truck-drone routing optimization using quantum-inspired genetic algorithms. In *2021 25th International Conference on Information Technology (IT)* (pp. 1–4). IEEE.

Yüksel, I., & Dagdeviren, M. (2007). Using the analytic network process (ANP) in a SWOT analysis. *Information Sciences, 177*(16), 3364–3382.

Zaikin, R., Filatov, S., Komkova, D., & Efimenko, D. (2020, April). Remote identification of damages of freight by means of transport telematics. In *IOP Conference Series: Materials Science and Engineering* (Vol. 832, No. 1, p. 012018). IOP Publishing.

Zak, J., & Węgliński, S. (2014). The selection of the logistics center location based on MADM/A methodology. *Transportation Research Procedia, 3*, 555–564. https://doi.org/10.1016/j.trpro.2014.10.034

Zavadskas, E. K., Kaklauskas, A., Peldschus, F., & Turskis, Z. (2007). Multi-Attribute Assessment of Road Design Solutions by Using the Copras Method. *The Baltic Journal of Road and Bridge Engineering, 2*, 195–203.

Zečević, S., Tadić, S., & Krstić, M. (2017). Intermodal transport terminal location selection using a novel hybrid MADM model. *International Journal of Uncertainty, Fuzziness and Knowledge-Based Systems, 25*, 853–876. https://doi.org/10.1142/S0218488517500362

Zhang, J. & Li, J. (2014). A hybrid genetic algorithm to the vehicle routing problem with fuzzy cost coefficients. In: *2014 IEEE 11th International Conference on Fuzzy Systems and Knowledge Discovery* (pp. 147–152).

Zhang, S. & Qinghe, H. (2014). Application of GA optimising neural network to coal sales forecasts. In: *2nd International Conference on Information, Electronics and Computer*. Atlantis Press.

Zhiyue, Z. H. A. O., & Maoqi, C. U. I. (2020). Optimization of intelligent logistics distribution route of integrated loading and unloading based on improved particle swarm optimization. *Academic Journal of Manufacturing Engineering, 18*(1), 137–143.

Zhou, G., & Min, H. (2011). Designing a closed-loop supply chain with stochastic product returns: A Genetic Algorithm approach. *International Journal Logistics Systems Management, 9*(4), 397–418.

Chapter 11
Methods of Data Collection, Aggregation, and Dissemination and Their Application

Aurelija Burinskienė

Abstract The flows of data and methods for their revision are different and depend on the needs of the transport business. Various parameters are communicated and regularly transmitted during cargo delivery. The set of data is compiled for each point in time and includes all the information and review of big data for making decisions about the transportation of goods.

The author divided the description into two parts. The beginning of the chapter is devoted to the collection of data from various sources, and the second part is devoted to the revision of data sets. Finally, the chapter ends with issues of trust and confidentiality in cases of data analysis.

Keywords Data collection · Data analysis · Grubb test · Big data · Operational data revision · Conflicts resolution · Decision hierarchy · Trust and privacy data review

11.1 Collecting Data from Data Sources

The information flow is modelled to meet operational freight transportation needs:

- information flows provided to report traffic jams;
- data set reported in cases of traffic accident;
- data set for monitoring vehicle (Table 11.1).

Data concerning the status in a traffic network, parking slots, other conditions are collected through real-time messaging and can be included in different identifiers:

The following are the individual components of the information flow, and these identifiers help to revise the data.

The most important information is the freight status (its place, date, time, physical conditions), the status of resources (the truck and the driver), and the purpose of the trip (delivery, backhauling, etc.).

A. Burinskienė (✉)
Vilnius Gediminas Technical University, Vilnius, Lithuania
e-mail: aurelija.burinskiene@vilniustech.lt

D. Dzemydienė et al., *Development of Smart Context-Aware Services for Cargo Transportation*, International Series in Operations Research & Management Science 330, https://doi.org/10.1007/978-3-031-07199-7_11

Table 11.1 Overview of data objects

Locations	
Service points	• Suppliers • Warehouses • Terminals
Resources	
Trucks/trailers	• Types: rigid, semi-trailer (tractor + trailer), double (rigid + trailer) • Temperature zones: single-temperature and double-temperature trucks • Ownership of trucks: own trucks, subcontracted trucks.
Drivers	• Number of drivers: one per truck, two per truck • Driver shift interval—8 h
Freight	
Products	• Loading unit (pallets, roll cages, etc.) • Freight temperature. Freight temperature combinations: • A freezer should be delivered separately or with a double-temperature truck. • Freezer goods without or with temperature-controlled boxes

Active context-aware data includes (Table 11.2):

- Monitoring of the on-time delivery ratio for all service points according to the truck location and incident date/time.
- Calculation of actual kilometres travelled by each truck and their comparison to the planned kilometres by revision according to odometer readings or checking route distance data using the map distance calculation functionality.
- Providing information about a completed delivery (proof of delivery) according to the door status, truck position, and incident date/time.

Passive context-aware data includes:

- Review of past incidents in case of contingencies or emergencies,
- Review of the odometer, engine hours,
- Review of seal numbers, load on-axis,
- Review of fuel consumption, fuel level, etc.

11.1.1 Information Flows in Traffic Jams

For traffic management systems connection to an information network with large amounts of data is required (Evangelista, 2017). To support real-time data reporting, some critical hardware has been required, such as RFID, GPS, sensors, radars, etc. The control of traffic lights on streets can be simulated using a representation of a multi-optimization task. These systems use sensors that are single-circuit and/or dual-circuit able to identify vehicles at intersections.

For example,

Table 11.2 Monitoring of contextual information for road transport mode

	Field name	Type	Comment	Form
1	Truck Nr	Text	Reg no. of truck	PL
2	Driver1 card No	Text	Number of inserted driver card into tachograph	PL
	Driver2 card No	Text	Number of inserted driver card into tachograph	PL
3	Engine on/off	Num	Engine on 1, engine off 0	ST
4	Truck door open/close	Num	Truck door open 1, truck door close 0	PL
5	Trailer door open/close	Num	Trailer door open 1, trailer door close 0	PL
6	Date/time	Num	Incident date and time	PL
7	Truck position, X	Num	Coordinate X	PL
8	Truck position, Y	Num	Coordinate Y	PL
9	Truck speed	Num	Speed of truck	PL
10	Truck seal no.	Text	Truck seal status and number	ST
11	Trailer seal no.	Text	Trailer seal status and number	ST
12	Odometer (km)	Num	Odometer of truck	PL/ ST
13	Engine hours	Num	Engine hours metre of truck	ST
14	Load on first axle	Num	Load on first axle of truck	ST
15	Load on 2nd + third axle	Num	Load on 2nd + third axles of truck	ST
16	Load of trailer	Num	Total load on trailer axles	ST
17	Consumed fuel of truck	Num	Consumed fuel litres of truck	ST
18	Fuel level in tank	Num	Fuel level in truck tank	ST
19	Truck temperature 1	Num	Truck temperature 1	PL
20	Truck temperature 2	Num	Truck temperature 2	PL
21	Trailer temperature 1	Num	Trailer temperature 1	PL

PL Execution/Monitoring/Control (Active), *ST* Statistical (Passive)

- UTCS—Urban traffic control system
- REHODES—Hierarchical optimized distributed efficient real-time system
- SCOOT—Split-cycle bias optimization method
- SCATS—Sydney coordinated adaptive motion system

The information from sensors is used to revise the flow of traffic. By having information on how many vehicles are waiting in the lanes, it is possible to improve the efficiency of traffic flow. Typically, the strategies for traffic management and optimization targets use strategic and operational data of truck stops, latency, and travelling. Monitoring data is represented in big data repositories, recording data such as queue length, traffic intensity, vehicle speed, CO_2 emissions, and geographic location (Hazen et al., 2018).

The studies on the geography of transport network began in the 1960s but still was not able to provide the mechanisms for network design. Optimization studies have been evident since about 1970. Models aim to minimize the costs for physical network flows. Models that solve the problems of minimum cost are aimed at searching how to transport freight through the transport network. Optimization models take into account current and predicted volumes. There has been uncomplicated user behaviour when users choose a route between points of departure and destination, which minimizes their freight transport costs and leads to an increase in traffic intensity.

The changes in traffic effects traffic on streets and intersections. Congestion occurs when automotive density is higher and their speed—too low, this means that physical flow on the network approaches maximum capacity. The primary analysis of traffic flows began from static modelling, which estimates congestion costs. However, it is not dynamic and is not able to revise congestion and its changes over the period. The simulation is used to construct a timetable. The collected traffic flow data is applied for the analysis using optimization techniques. Collected results after the analysis are applied for the control of traffic lights, which gives ideas on how to minimize congestions.

11.1.2 Data Set Reported in Cases of Traffic Accident

The authors present behavioural applications where drivers of heavy cargo vehicles choose the optimal speed to avoid accidents, which could be minor or major, i.e. blocking the road for quite a long time (Gyulyev and Dolia, 2017). Such an analysis describes the time lost in the cases of an accident.

The authors still searching for how to report the cases of accidents. Information about identified traffic incidents can be provided with GPS and other means.

Other researches focus on identifying risky accident-related places on the traffic network. To report this, the most important is the collection of data and the identification of coordinates. Many accidents are reported on social media. New methods for collecting data use satellite imagery reported by the Application Programming Interface of Google Map when social media posts appear with an identified section of a street that is blocked due to an accident (Dadashova et al., 2020).

The last type of research focuses on the time to react to the number of accidents minimization. These studies use the results of previous studies but incorporate traffic control features helping emergency cars to cross streets with traffic lights faster (Ordu et al., 2019). And when emergency cars are coming up to help people who are in an accident, it is a system that reduces the time of roadblocks.

The author analyses cases of accidents involving heavy trucks in the countries of the European Union. Most of them (i.e. 88.2%) were registered outside urban areas. Table 11.3 lists the accidents.

The fewest accidents occurred in Croatia, Denmark, Estonia, Finland, Greece, Latvia, Slovakia, and Slovenia; some countries (Bulgaria, Lithuania, Luxembourg) had not registered them at all, and the highest number of accidents were in Germany, Italy, and Poland.

Such unplanned events evident in cargo transportation can lead to disruptions in freight delivery due to loss of time and damage to the truck or cargo.

Table 11.3 Number of accidents met by cargo carriers (>3.5 tonnes) on the EU level (European Commission, 2021)

Countries	Number of accidents, 2019	Countries	Number of accidents, 2019
Austria	2	Ireland	
Belgium	11	Latvia	3
Bulgaria	0	Lithuania	–
Croatia Cyprus	1	Luxembourg Malta	0
Czech Republic	17	Netherlands	5
Denmark	2	Poland	26
Estonia	5	Portugal	
Finland	3	Romania	5
France	36	Slovakia	1
Germany	80	Slovenia	2
Greece	4	Spain	55
Hungary	12	Sweden	3
Italy	67	United Kingdom	
The total number of EU-28 is 353			

11.1.3 Data Set for Monitoring Vehicle and Freights

Monitoring the positions of freights is especially important for buyers. Several tracking possibilities exists such as: (a) monitoring for the cargo vehicle or/and (b) monitoring for the goods.

Studies in the literature cover a variety of tracking tools. For the monitoring of cargo vehicles, the most of authors focus on GPS and radio frequency identification (RFID) tracking possibilities. The studies focus on data retrieval, storage, archiving, clustering, and a frequency of data retrieval. By using monitoring technologies, the status of infrastructure could be reported and road conditions according to real-time information could be determined. The main advantage is the possibility to reach the information that is effectively processed to address traffic congestion (Pérez and Dafonte, 2017).

11.1.4 Tracking of Road Vehicles

For tracking the vehicle we could use GIAI-96 standard. GS1 standard has components (as presented in Table 11.4) that help to track Vehicle ID number. The technical realization of the standard is implemented at heavy vehicle production factories. The first number follow WMI World Manufacturer Identification.

By following GIAI-96 standard described in 24 characters, the VIN can be specified starting from partition (i.e. 2–6 numbers), when it holds a company prefix

Table 11.4 The implementation of GS1 standard

Partition	Company prefix	Reference to asset	Partition	Company prefix
	Numbers	Bytes		Numbers
2	10	34	2	10
3	9	30	3	9
4	8	27	4	8
5	7	24	5	7
6	6	20	6	6

Table 11.5 RFID tags

Tags	Costs	Metal areas	Position to reader	Temperature
Active	High	Friendly	Shorter	Upper level
Semi-passive	Average	Un-friendly	Larger	Under level
Passive	Low	Mid-friendly	Larger	Under level

Table 11.6 The application of radio frequency identification scenarios

Type of connection	Communication with a shorter reading range	Communication with greater reading distance
Tag-to-reader	Toplink and bottom-up link reader tag	–
Multiple tags to read	–	Reading top-down link and bottom-up link tags

and reference to assets. The application of the standard helps to track delivery, i.e. the tag of RFID is scanned on a truck when it passes RFID readers till it arrives at its destination (Table 11.5).

We use passive tags for tracking and 10 metres distance. In addition to RFID/EPC tags and GIAI, the data can be also transmitted to storage (which follows ISO/IEC 15424 standard). The system identifies cargo vehicles, which are traversing the tag reader.

The communication between tags and readers has been organized accordingly (Hui and Kan 2018), as indicated in Table 11.6.

There is a collision in the communication between the reader and the readers (R2R) in case the conflicting readers are unable to process.

11.1.5 Cargo Tracking

The literature on tracing of goods indicates some advantages: (a) each shipper can control the freight delivery and (b) the receiver can track freight position (Nyiendo, 2018).

GSIN Global Shipment Identification Number is required to analyse shipment tracking (group of transport units), which is routed by the shipper to the consignee on a single consignment note. The GSIN can be indicated in the consignment note or dispatch notice; pointing to the number of the shipment. All this information is transferred via EDI to the freight shipper and receiver.

According to WCO World Customs Organization standards, the unique GSIN is issued for 10 years and during this period can be used once. However, for the domestic transport cases, the reuse of GSIN is locally regulated (Tengler and Pekná, 2020).

11.2 Methods Used to Analyse Datasets

There are several methods:

- Validate data using the outlier discovery method.
- Analysing Big Data data sets.
- Revision of operational data for the delivery of goods and their scheduling.

11.2.1 Validate Data Using the Outlier Discovery Method

After collecting the contextual datasets, the next step is to validate the data. Data validation is applied to avoid the negative effect of an event. A literature review highlights some papers which detect outliers. Many data validation studies have been conducted (Carrion et al., 2017; Lin et al., 2013; Jackson and Mazzei, 2012; Greaves and Figliozzi, 2008; Broach et al., 2012; Kuppam et al., 2014) to develop techniques helping quickly indicate the outliers (Shen et al., 2015).

It includes the detection of outliers by reviewing data with time-series patterns and by applying the Grubbs method for the identification of confidence.

Other data validation methods are also available, such as the Dixon method (1953), the David method (1961), the Barnett and Lewis method (1994), and the Rosner method (2011).

The Grubbs test is important for the revision of historical data and its adjustment. It is applied with a confidence level—of 99%. The application of the Grubbs method (1950) is widely used for the detection of transport accidents. It is applied for the identification of emissions exceeding the boundaries of the standard distribution (Djenouri et al., 2018). This results in probability showing the number of data points with outlier detected from all data points (Fig. 11.1). Full usage of this test method shows the extreme situations, taking into account normal distribution.

Following the data approximation rules, data sets can be verified by estimating changes using the Grubbs test method.

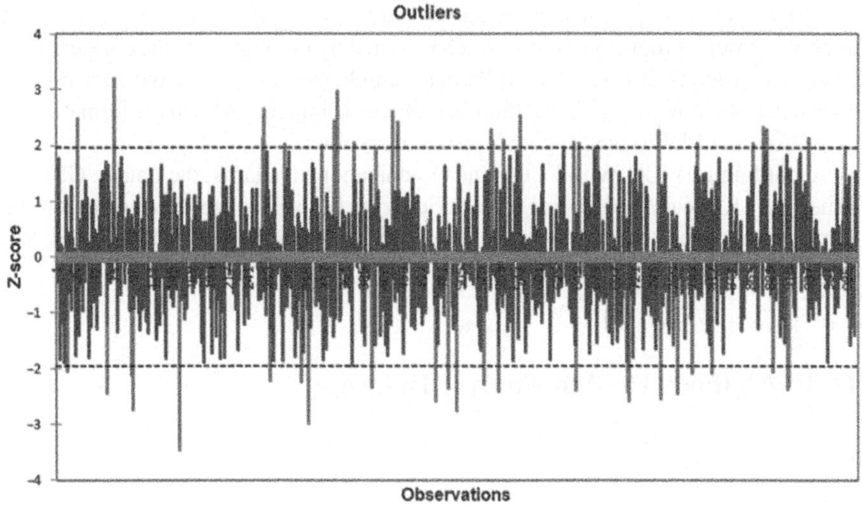

Fig. 11.1 Grubb test

The usage of the Grubb test method is of practical value and could be applied to define freight transport datasets.

Detecting emissions with the Grubb test and evaluating data points help optimize historical data. Once emissions are detected for points of sale outside distribution boundaries, data within healthy distribution boundaries are taken into account.

The test of the null hypothesis tells that data has no outlier, compared to the alternative hypothesis that the minimum or maximum value is an outlier. There are three steps to detect outliers:

Step 1. Sort data points from smallest to largest;

Step 2. Finding the mean (x¯) and standard deviation (S) of the dataset;

Step 3. Calculate test G statistics (minimum, maximum, and critical values) for a two-way test by applying any of Eq. (11.1):

$$G = \frac{\underset{i=1,..,N}{\max} |x_i - \overline{x}|}{S}. \tag{11.1}$$

The statistics of the *G* test for the maximum value in the dataset may be an exception, so it is better to use test statistics (right-handed test) (11.2):

$$G = \frac{|x_{\max} - \overline{x}|}{S}. \tag{11.2}$$

If the minimum value in the dataset is an outlier, it is better to use test statistics (left-handed test) (11.3):

$$G = \frac{|\overline{x} - x_{\min}|}{S}. \tag{11.3}$$

G_{crit} is the critical value of G which is calculated for the test as presented below (11.4):

$$G_{\text{crit}} = \frac{N-1}{\sqrt{N}} \sqrt{\frac{t^2_{\alpha/(2N),\ N-2}}{N-2+t^2_{\alpha/(2N),\ N-2}}}, \tag{11.4}$$

where α– significance level (i.e. 0.05); $N - 2$—the upper critical meaning of the distribution of t expressed with two degrees of freedom $N - 2$. https://www.statisticshowto.com/probability-and-statistics/t-distribution/

If the computed G is greater than Gcrit, the data point is not an outlier, and the zero hypothesis H0 is accepted. Otherwise, if G is lower than Gcrit, the alternative hypothesis is accepted.

The detection of outliers helps to avoid the upstream and downstream values that are represented by historical data points.

11.2.2 Analysing Big Data Data Sets

It is not easy to analyse and revise such an amount of data. The data could be accumulated, but sometimes is unstructured (Papadopoulos et al., 2017).

However, the review of Big Data requires analytical solutions. The data generation speed, the variety of data collected from different data sources, and formats matters a lot. Such data set could be generated by sensors and in different formats is being transferred frequently and treated as a large due to variety and update frequency. Revising and analysing this enormous size of data, which is interesting due to its content, can affect the decision-making process. The authors show how such analysis could improve the productivity and efficiency of operational decision-making (Chongwatpol and Chan, 2015).

A study of the same authors presents that data analysis brings alternative ways for reporting business problems (Chongwatpol and Chan, 2015). Analytics improves results by enlarging the visibility of situations, helps to find opportunities actual for process development, and supports the processes of freight senders and receivers (Papadopoulos et al., 2017). It synchronizes capacity and requests for services in small parcel delivery businesses (Croxton et al., 2002). Other researchers provided the investigations that revenues could be enlarged in the enterprise by 9.7%, after submitting forecasts using the insights from Big Data. The authors note Big Data linked with demand forecasts help to create added value for companies.

Table 11.7 Studies integrating Big Data in transport management research

Aspects	Early phase	1980	1990	2000	2010
Advanced statistical methods	Crisp (1968), Egger (1973)	Malmborg (1988)	Mineo and Plaia (1998)	Fang and Lin (2006)	
Data mining methods		Khan (1984), Malmborg et al. (1986)	Hendrickson (1999)	Han and Mei (2004), Jiang et al. (2005)	Chiang et al. (2014)
First-generation BI				Lawrence et al. (2005)	
Second-generation BI					Furtner et al. (2016), Singh and van Sinderen (2015), Jnr et al. (2020)
BigData analytics BDA					Bodden-Streubuhr (2016), Baker et al. (2017), Dinh et al. (2020)

While Big Data supports business, there are not many studies on transportation and operational management.

Although studies on Big Data have not yet become announced in the study area, its value delivered to companies has been extensively explored. The past decade has seen the rise in the adoption of information technologies used for the review of operations using BigData (Arunachalam et al., 2018).

For more than half a century, the evolution of BigData analysis has included techniques that included advanced data mining focused on performance metrics; Egner's work (1973) is one of the examples.

Techniques of data mining were introduced in 1980. Data mining means finding patterns and relationships between data variables and focusing on the identification of relationships (Waller and Fawcett, 2013). In the beginning, the modelling method was predominant as one, which used for data mining.

The companies started to develop individual tools which were called in term "business intelligence". Initial business intelligence tools focus on the description of the business and making solutions based on facts (Table 11.7). They were evident from 1990; however, in logistics field studies, they came 10 years later. Second-generation BI combined the company's internal and external data, which required more powerful tools (Davenport, 2013).

Big Data solutions have been used until they reached limitations by integrating data from multiple sources (Arunachalam et al., 2018).

Over the past 10 years, more and more studies are provided for transport cases. Many researchers investigated performance indicators, while others researched the functional field of transport management. BigData analytics can be used to revise the strategic, tactical, and operational levels in transport management. Strategic analytics techniques were used to design transport networks, while two other analytics methods were used for resource planning (Addo-Tenkorang and Helo, 2016). The author highlights below in the table studies relevant for the area of study.

Analytics for operations research has been used for many years. Three types of analytics are known: prescriptive, predictive, and descriptive.

Prescriptive analytics uses techniques supporting optimization and decisions. Due to complexity, enterprises use such analytics to optimize graphs, important for decisions and modelling results (Tiwari et al., 2018).

Predictive analytics has been used to predict the events in future. It is good at identifying trends or patterns and discovering causes (Gunasekaran et al., 2017). Predictive analytics predicts traffic volume, transaction costs, and customer order fulfilment based on historical data (Waller and Fawcett, 2013).

Descriptive analytics illustrates, provides investigations following old data and the one that is reported in real time. These techniques are applied to compare and model data. One of technique is modelling important for both types of analytics. Modelling allows companies to identify transportation and resource needs, reach higher utilization, and improve service (Chan et al., 2016).

Different authors cite various studies related to the topic (Choi et al., 2018). To revise the studies, the author chooses several topics. The author used the Publish and Perish tool to indicate articles corresponding to the "big data" word and headlined "operational management", "logistics management", and "transportation management".

Documents on "operational management" have been published since 2010; in logistics management since 2014, and transport management since 2016. Trends show a rapid increase in publications in the latter category.

Articles that write about logistics management account for 5.5%, and most of them cover the topics of urban logistics and 3PL services.

The authors analyse transportation, the use of systems developed for transport management (TMS), IoT application as sources that generate real-time data (Furtado, 2013). However, such studies are not covering all topics in terms of operations, the authors consider TMS as a system that generates BigData (Yao, 2016). Of course, transport management has many resources and takes a central role among other activities in business. The use of BDA in transportation management is highly important.

Big Data tools contribute to the creation of value and the improvement of the advantage of the company (Wamba et al., 2018). Investigations derived from the review of the huge size of information support decisions related to actual data for any transport function, and the output of these decisions increases the overall performance. In addition, transport managers should develop capabilities related to analytics and the technical part. These data analysis services should be used for operational freight management (Dremel et al., 2017).

Actions have been taken to obtain results, starting with the generation (1), acquisition (2), storage (3), and analysis of data (4). The data is generated and acquired from different sources. The structure of BigData can help identify delivery events that are not effective. BigData store data on the server with which services supports real-time updates. The quality of data depends on communication services which help transport chiefs organize data selection, grouping, storage, transmission, and distribution, resulting in the fastest possible visibility of output data (Wamba et al., 2018). The last step BigData can be used to revise strategic, tactical, and operational data. The author (Choi et al., 2018) proposed multifunctional methodological approaches

11.2.3 Revision of Operational Data of Cargo Delivery

There are three stages of supply planning. The first step is applied for the calculation of distances between the points of pick-up and drop-off and the distribution of demand. The second stage uses mathematical programming techniques to obtain a delivery plan. The last step involves adjusting the delivery plan. Some articles suggest taking 20% of excess capacity to periodically have good planning results.

The truck planning algorithm minimizes the number of trucks used, taking into account the limitations of trucks such as maximum truck availability. The truck planning tool uses the customer's time window to move the route forward or backward to minimize the number of transport units.

The key elements of the trucks planning technique are the minimum and maximum time between the assigned truck routes and the maximum time per day when the truck can be used (Fig. 11.2).

The decision hierarchy is presented in Fig. 11.3. During the planning stage, freight transport targets are established based on agreements. In the planning

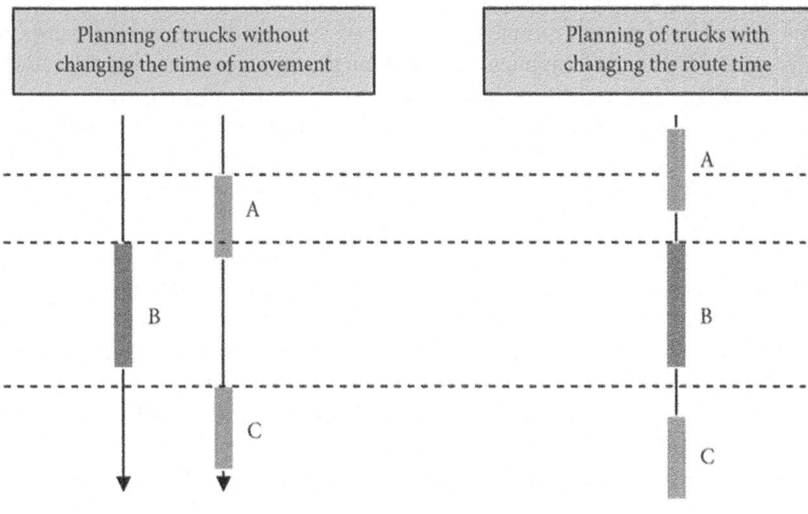

Fig. 11.2 Truck planning cases

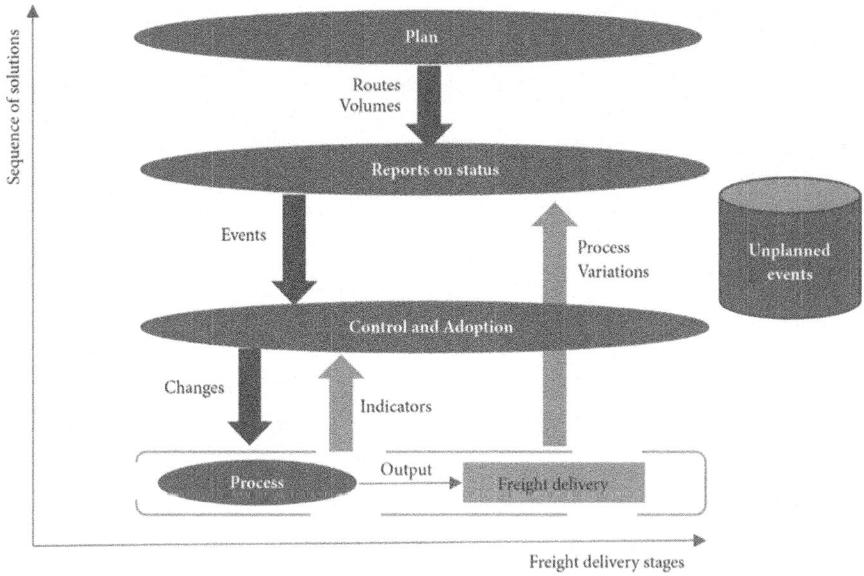

Fig. 11.3 The sequence of solutions in freight transport processes

stage, the freight transport sequence is defined and transferred to the management phase as target/defined points. During the data collection stage, variables are tracked related to the specified infrastructure values, and with other objects. In the controlling phase, decisions are made to resolve the conflict:

- Postponement of the primary plan ("Cancel plan");
- Changing the primary plan ("Correct plan");
- Eliminating incompatibility ("Resolve the issue").

The primary plan is for some reason incompatible—there is an equal probability of changing the primary plan or cancelling it. Most likely, there will be room for modifications, revision by setting time updates or transferring it without changes. If the primary plan cannot be cancelled or moved, its duration may be shortened.

The process of cargo transportation has a lot of complexity and is costly. Around half of logistics costs are associated with the transportation of goods. The authors constantly search how to minimize these costs. They apply algorithms to reduce the cost of transportation of goods including the distance of travel (Psychas et al., 2015). The problem itself is usually multi-purpose. The researchers use a couple of variables to reach the best results. In particular, the authors revise the problem aiming to reduce expenses and the number of trailers (Pierre and Zakaria, 2014). Other authors investigate the reduction of trip time and removal of trailer fleet restrictions (Zhang and Qinghe, 2014). Researchers use more than two variables: time window, the capacity of the trailer, and the maximal trip time indicated for the delivery of shipment (Kumar et al., 2014). Although the studies were only recent, several

Table 11.8 Multi-purpose cargo delivery tasks

Description of the problem of cargo delivery					
		A single goal			
Study	Multi-purpose	Multi-terminal	Multi-period	Types of multi-vehicle	Multi-time windows
Psychas et al. (2015)	X	X			
Pierre and Zakaria (2014)	X				X
Zhang and Qinghe (2014)				X	
Kumar et al. (2014)	X			X	X
Liu et al. (2014)		X			
Mohr (2014)					
Cai et al. (2015)	X	X		X	X
Karakatic and Podgorelec (2015)		X			
Ahmadizar et al. (2015)	X	X			
Wen et al. (2010)			X		

documents use the GA to find a solution in the field of cargo delivery. The studies are shown in Table 11.8.

Concerning freight delivery activities, the article presents a methodology for multi-purpose analysis covering the choice of a transport corridor in a road network.

This case is defined as a larger observation of the issue of vehicle routing (VRP), which includes such directions:

1. Earlier studies by Ahmadizar et al. (2015), Cai et al. (2015), Karakatic and Podgorelec (2015), Psychas et al. (2015), Pierre and Zakaria (2014), Zhang and Qinghe (2014), Kumar et al. (2014), Liu et al. (2014), Mohr (2014), Wen et al. (2010) focus on route improvement where the start and end points of the cargo delivery process are predetermined.
2. The authors Psychas et al. (2015), Kumar et al. (2014), Liu et al. (2014), Mohr (2014) apply Euclidean distance's approach to find the optimal solution in routing goods. Although, a path-based approach is used to provide indications about the length of the transport corridor.
3. Papers announced by Cai et al. (2015), Kumar et al. (2014), Zhang and Qinghe (2014) analyse the change of mode cases and possibilities to deliver the freight without the mode change.
4. Researchers Ahmadizar et al. (2015), Psychas et al. (2015) revise the nature of transportation in the time frame and suggest that the activity of cargo transport over the long-distance routes is inherently linked with several time intervals.
5. Authors Ahmadizar et al. (2015), Cai et al. (2015), Karakatic and Podgorelec (2015), Psychas et al. (2015), Liu et al. (2014) include a multi-terminal aspect that is applied for the consolidation of various cargos. The provided methodology includes terminals dedicated for temporary storage as elements of the transport

system that may be needed to transship cargo from one type of cargo transport to another, or temporally used for the customs tax declaration.

11.2.4 The Resolution of Conflicts During the Planning

Vehicle planning has been researched for the past four decades. Many authors explore aspects of route selection and planning. The first step is to identify the actual conflicts that arose during the creation of the planning. During operations, inconsistency arises when an unplanned situation happens. These cases are usually considered as not conflicting but could request multifunctional works.

By identifying the issues in the dataset, the responses supporting the choice of strategy and its application for resolving such situations are needed. A further review step is taken to continue the work to determine whether the conflicting activity can be modified. For each conflicting case, the type of resolution is determined by the other change actions. If no action has been taken, you should monitor changes in several activities and later come up with the solution. After the situation is resolved, the potential conflicts list is updated by identifying that conflict is solved. All conflicts that occur for reasons other than changing activities are removed from the conflicts list to correctly resolve the primary incompatibility due to overlapping situations and are rescheduled.

Theoretically, many possible activities contradict currently planned activities that will not be included in the schedule because they could be postponed or cancelled. Due to that, the rejection of conflicting actions is not a good choice, but more the postponement in taking solution.

The model for conflicts resolutions could be created, you first defined the main parameters that are considered during the process. Key features can be classified as follows (Auld et al. 2009):

- Characteristics related to the driver. This group covers the behaviour and personality traits of drivers. Characteristics related to the driver of the truck give the identification about the ability of the driver to replan the activity.
- Characteristics related to the activity. These characteristics include the length of the trip, the location of the heavy road transport, consumption of fuel, the goal of the route, and the detailed trip information that reflects the needs and the degrees of flexibility.
- Characteristics related to conflicts. The characteristics associated with conflict usually are identified in recurring situations. The model uses basic descriptive statistics about conflicts and suggests decisions.

Each conflict requires decisions that follow resolution strategies. Activities with greater flexibility (flexibility of leisure or only personal flexibility of the driver) have a higher probability of modification. When it comes to the planning of time horizon, activities could be somewhat postponed.

The timing of the operation is the main characteristic with regard to freight delivery activity. The reduction of the volume and modification of the activity could lead to an increase in duration. In case, short duration operations that require less than two hours are compared to long duration operations that last 6 h and are revised, after revision some number of operations could be removed. In the case of conflicting operations, some operations could be prolonged after the revisions and modifications.

Resolution should be submitted to fully reflect conflicts. The model has to be constructed which presents the influence of variables on each other, changes of variables over time, as well as their presence, which was previously ignored.

The revision of the studies on routes construction shows that there is not one study that follows the latest techniques proposed in solution taking area. Since many types of research are published in the latest years, a review of the latest approaches is presented. The author separated the models into three groups:

- Multi-objective oriented which combine to the search of solution several goals;
- Optimization oriented which suggest the best decision alternative;
- Output oriented which investigate the accuracy of results.

Typically, the models dedicated to planning focus on costs minimization and miss additional constraints. Such constraints can be divided into the following sets:

- productivity: quality of service, penalties, navigations, congestions;
- routing: time per operation, available resources, time balance, management of uncertainties;
- driving: places of visit, the quality of infrastructure, speed limitations, cargo type, lancet, and trip taxes;
- reliability of vehicle: the last breakdown, the readiness of the truck, the behaviour of the driver;
- sustainability and environment aspects: consumption of fuel, emissions, safety criteria.

These criteria can be weighted, hierarchically or multifunctionally formulated or used as separate criterion.

11.3 Validate Trust and Privacy Data in a Context-Sensitive Service Application

Service providers use a contextual service platform in their services, adapting it to the current situation of service users and applying contextual data received from contextual data providers. Such a service delivery platform reveals the importance of issues, trust, and confidentiality, as different organizations handling various tasks must cooperate in the provision of service delivery. By its very nature, contextual information is sensitive to privacy, and thus the transmission and processing of this data pose a potential threat to privacy. Acceptance by the users of contextual data depends on their perception of how contextual service delivery platforms follow privacy regulations. Users of contextual service delivery platforms should control the rights to access the contextual data and revise how the context is transmitted and updated for users having access. Users having control of their private data are dependent on particularly complex contextual service delivery platforms because users' privacy preferences depend on the contextual situation. To make consumers feel in control of their privacy, specs of their privacy preferences alone is not enough to reach a compromise between confidentiality and contextual adaptation of services. Users should also be sure that the entities of the contextual service delivery platform responsible for the transmission and update of their contextual data, such as contextual service providers, ensure compliance with these privacy provisions. Thus, trust is an integral part of the user concerns in privacy that arise in the context of the application of intelligent technologies. Contextual service providers are concerned about the ability to deliver reliable contextual service. Service providers searching for reliable solutions providing contextual information about users of these services. The aspects of privacy are also important for service providers, as reporting privacy incidents related to their service delivery infrastructure also affects their core business reputation. Trust and privacy solutions for contextual service platforms do not take into account various aspects of trust and dependencies between entities involved in the contextual service provider's activities. Exclusion decisions focus on at least one aspect of trust at the same time, such as confidentiality or identity certification, and do not include the interdependence between the various aspects of compromise. Existing solutions do not include the concerns of service providers and users, such as the reliability of the use of contextual services or the relationship between the quality aspects of contextual data and trust, in a comprehensive manner. In addition, existing solutions for managing trust and privacy offer context-sensitive service delivery platforms, poor support for the management of context-based privacy issues. The functionality of a contextual service platform consists of three main attributes focusing on trust and privacy from the perspective of consumers and service providers. A trust-based selection mechanism is required that helps users of contextual services select reliable service providers with whom they can communicate. This mechanism assists consumers by taking their targets, beliefs, and dependencies on trust between the users and entities that provide service for them. The objectives of users of the services used in the development of the mechanism are related to the trade-off between the protection of confidentiality and the adaptation of contextual services. In addition, there is a trust-based selection mechanism that supports selecting trusted contextual information providers. This mechanism helps select contextual information providers based on their reliability in providing contextual information about a particular user and quality level. These mechanisms help to increase the adaptability of context-sensitive information providers. Finally, the overall context-aware concept of policy management, known as contextual management (CAMD), is important. The CAMD concept supports a trust specification and privacy policy with context-sensitive permissions and obligations. The CAMD concept needs to be implemented by using policy-based management and contextual data as inputs for managing strategic tasks. The CAMD concept is designed to help users and administrators of the system manage policies to control the authorizations of users of contextual data, the quality of data, and the actions of these context-sensitive information providers should take after granting access. Confidentiality obligation is the deletion of all information about the user's location when the contextual condition is met. Technical feasibility assessments and user surveys help to better understand the reliability and privacy requirements of the contextual service platform, such as context-sensitive policies and users' targets which are reached by using contextual services.

11.4 Conclusions

The chapter presents the data collected from different sources, its revision by using outlier detection methods and integration into operational freight delivery management for resolving conflicts.

This data is used for several purposes:

1. collecting contextual datasets and validating data that may be taken with Grubbs bias analysis, in particular, to clean up data and determine the statistical significance of events reporting data.
2. analysis of conflict resolution and hierarchy of solutions for freight services. Various purposes could be used in resolving conflicts where the most complex are multi-purpose cases.

The study shows that trust and privacy are important for contextual data management.

References

Addo-Tenkorang, R., & Helo, P. T. (2016). Big data applications in operations/supply-chain management: A literature review. *Computers and Industrial Engineering, 101*, 528–543.

Ahmadizar, F., Zeynivand, M., & Arkat, J. (2015). Two-level vehicle routing with cross-docking in a three-echelon supply chain: A genetic algorithm approach. *Applied Mathematical Modelling, 39*(22), 7065–7081.

Arunachalam, D., Kumar, N., & Kawalek, J. P. (2018). Understanding big data analytics capabilities in supply chain management: Unravelling the issues, challenges and implications for practice. *Transportation Research Part E: Logistics and Transportation Review, 114*, 416–436.

Auld, J., Mohammadian, A. K., & Doherty, S. T. (2009). Modelling activity conflict resolution strategies using scheduling process data. *Transportation Research Part A: Policy and Practice, 43*(4), 386–400.

Baker, P., Croucher, P., & Rushton, A. (2017). *The handbook of logistics and distribution management: Understanding the supply chain*. Kogan Page Publishers.

Bodden-Streubuhr, M. (2016). Warehouse-Management-Systeme im Spannungsfeld von Industrie 4.0. In S. Sohrt et al. (Eds.), *Handbuch Industrie 4.0: Produktion, Automatisierung und Logistik* (pp. 1–13). Springer.

Broach, J., Dill, J., & Gliebe, J. (2012). Where do cyclists ride? A route choice model developed with revealed preference GPS data. *Transportation Research Part A: Policy and Practice, 46*(10), 1730–1740.

Cai, Y. G., Tang, Y. L., & Yang, Q. J. (2015). An improved genetic algorithm for multi-depot heterogeneous vehicle routing problem with simultaneous pickup & delivery time windows. *Applied Mechanical Material, 738*, 361–365.

Carrion, G., Nitzl, C., & Roldan, J. (2017). Mediation analyses in partial least squares structural equation modeling: Guidelines and empirical examples. In *Partial least squares path modeling* (pp. 173–195). Springer.

Chan, H. K., Choi, T. M., & Yue, X. (2016). Guest editorial big data analytics: Risk and operations management for industrial applications. *IEEE Transactions on Industrial Informatics, 12*(3), 1214–1218.

Chiang, M.-H. D., Lin, C. P., & Chen, M. C. (2014). Data mining based storage assignment heuristics for travel distance reduction. *Expert Systems, 31*(1), 81–90.

Choi, T. M., Wallace, S. W., & Wang, Y. (2018). Big data analytics in operations management. *Production and Operations Management, 27*(10), 1868–1883.

Chongwatpol, J., & Chan, H. K. (2015). Prognostic analysis of defects in manufacturing. *Industrial Management & Data Systems, 115*(1), 64–87.

Crisp, R. D. (1968). *How to reduce distribution costs – A practical, scientific approach to increased selling efficiency*. Funk & Wagnalls Company.

Croxton, K. L., Lambert, D. M., Garcia-Dastugue, S. J., & Rogers, D. S. (2002). The demand management process. *The International Journal of Logistics Management, 13*(2), 51–66.

Dadashova, B., Arenas-Ramires, B., Mira-McWillaims, J., Dixon, K., & Lord, D. (2020). Analysis of crash injury severity on two trans-European transport network corridors in Spain using discrete-choice models and random forests. *Traffic Injury Prevention, 21*, 228–233. https://doi.org/10.1080/15389588.2020.1733539

Davenport, T. H. (2013). Analytics 3.0. *Harvard Business Review, 91*(12), 64–72.

Dinh, L. T. N., Karmakar, G., & Kamruzzaman, J. (2020). A survey on context awareness in big data analytics for business applications. *Knowledge and Information Systems, 62*(9), 3387–3415.

Djenouri, Y., Zimek, A., & Chiarandini, M. (2018, November). Outlier detection in urban traffic flow distributions. In *2018 IEEE International Conference on Data Mining (ICDM)* (pp. 935–940). IEEE.

Dremel, C., Wulf, J., Herterich, M. M., Waizmann, J. C., & Brenner, W. (2017). How AUDI AG established big data analytics in its digital transformation. *MIS Quarterly Executive, 16*(2), 80–89.

Egger, P. (1973). *Studien zur Grundlegung der Logik und der logischen Interpretationsmittel: mit besonderer Berücksichtigung von Texten griechischer Denker.*

Egner, C. G. (1973). *Experiment in worker productivity*, Working paper, University of Montana, Montana, 25 May.

European Commission. *Annual accident report* 2021. Available https://ec.europa.eu/transport/road_safety/specialist/observatory/statistics/annual_accident_report_en

Evangelista, P. (2017). Information and communication technologies: A key factor in freight transport and logistics. In *Training in Logistics and the Freight Transport Industry* (pp. 15–36). Informa UK Limited.

Fang, M. Y., & Lin, F. (2006). Measuring the performance of ERP system–from the balanced scorecard perspectives. *Journal of American Academy of Business, 10*(1), 256–263.

Furtado, P. (2013). Simulation and analysis of a physical internet network through multi-agent systems. Paper presented at the *13th WCTR*, Rio de Janeiro.

Furtner, E. M., Wildhölzl, H., Schlager-Weidinger, N., & Promberger, K. (2016). Impacts of SAP HANA on business intelligence (BI) strategy formulation. In *Multidimensional Views on Enterprise Information Systems* (pp. 163–171). Springer.

Greaves, S. P., & Figliozzi, M. A. (2008). Collecting commercial vehicle tour data with passive global positioning system technology: Issues and potential applications. *Transportation Research Record, 2049*(1), 158–166.

Gunasekaran, A., Papadopoulos, T., Dubey, R., Wamba, S. F., Childe, S. J., Hazen, B., & Akter, S. (2017). Big data and predictive analytics for supply chain and organisational performance. *Journal of Business Research, 70*, 308–317.

Gyulyev, N., & Dolia, C. (2017). The issue of probability of traffic road accident on the elements of the transport network. *American Journal of Social Science Research, 3*, 17–24.

Han, Q. L., & Mei, Y. X. (2004). Distribution center: BP artificial neural network based site selection, China. *Soft Science, 24*, 140–143.

Hazen, B. T., Skipper, J. B., Boone, C. A., & Hill, R. R. (2018). Back in business: Operations research in support of big data analytics for operations and supply chain management. *Annals of Operations Research, 270*(1–2), 201–211.

Hendrickson, B. A. (1999). U.S. Patent No. 5,930,784. U.S. Patent and Trademark Office.

Hui, X., & Kan, E. C. (2018). Collaborative reader code division multiple access in the harmonic RFID system. *IEEE Journal of Radio Frequency Identification, 2*, 86–92. https://doi.org/10.1109/jrfid.2018.2852484

Jackson, A. Y., & Mazzei, L. A. (2012). *Thinking with theory in qualitative research: Viewing data across multiple perspectives*. Routledge.

Jiang, W., Vaidya, J., Balaporia, Z., Clifton, C., & Banich, B. (2005, April). Knowledge discovery from transportation network data. In *21st International Conference on Data Engineering (ICDE'05)* (pp. 1061–1072). IEEE.

Jnr, B. A., Petersen, S. A., Ahlers, D., & Krogstie, J. (2020). Big data driven multi-tier architecture for electric mobility as a service in smart cities: A design science approach. *International Journal of Energy Sector Management*.

Karakatic, S., & Podgorelec, V. (2015). A survey of genetic algorithms for solving multi depot vehicle routing problem. *Applied Soft Computer, 27*, 519–532.

Khan, M. R. (1984). An efficiency measurement model for a computerised warehousing system. *International Journal of Production Research, 22*(3), 443–452.

Kumar, V. S., Thansekhar, M. R., Saravanan, R., & Amali, S. M. J. (2014). Solving multi-objective vehicle routing problem with time windows by FAGA. *Proceedings Engineering, 97*, 2176–2185.

Kuppam, A., Lemp, J. Beagan, D. Livshits, V., Vallabhaneni, L. & Nippani, S. Development of a tour-based truck travel demand model using truck GPS data. In *93rd Transport Research Board Annual Meeting* (pp. 1–10), January 2014.

Lawrence, F. B., Jennings, D. F., & Reynolds, B. E. (2005). *ERP in distribution*. South-Western Pub.

Lin, C., Tsai, P., Lai, K., & Chen, J. (2013). Cloud removal from multitemporal satellite images using information cloning. *IEEE Transactions on Geoscience and Remote Sensing, 51*(1), 232–241.

Liu, R., Jiang, Z., & Geng, N. (2014). A hybrid genetic algorithm for the multi-depot open vehicle routing problem. *OR Spectrum, 36*(2), 401–421.

Malmborg, E. (1988, June). Design of the user-interface for an Object-Oriented statistical data-base. In *International Conference on Scientific and Statistical Database Management* (pp. 314–326). Springer, .

Malmborg, C. J., Balachandran, S., & Kyle, D. M. (1986). A model based evaluation of a commonly used rule of thumb for warehouse layout. *Applied Mathematical Modelling, 10*(2), 133–138.

Mineo, A. M., & Plaia, A. (1998). Statistical multivariate techniques for the stock location assignment problem. In A. Rizzi et al. (Eds.), *Advances in data science and classification* (pp. 573–578). Springer.

Mohr, C. M. (2014). *Optimisation of warehouse order-picking routes using vehicle routing model & genetic algorithm*. Doctoral dissertation, State University of New York at Binghamton.

Nyiendo, S. O. (2018). An evaluation of RFID security systems for tracking of parcels in Kenya. *International Journal of Digital Information and Wireless Communications, 8*, 223–231. https://doi.org/10.17781/p002466

Ordu, M., Demir, E., & Tofallis, C. (2019). A decision support system for demand and capacity modelling of an accident and emergency department. *Health System, 9*, 31–56. https://doi.org/10.1080/20476965.2018.1561161

Papadopoulos, T., Gunasekaran, A., Dubey, R., Altay, N., Childe, S. J., & Wamba, S. F. (2017). The role of Big Data in explaining disaster resilience in supply chains for sustainability. *Journal of Cleaner Production, 142*, 1108–1118.

Pérez, M. M., Dafonte, C., & Gómez, Á. (2017). System based on RFID technology for product transport tracking. In *Proceedings of 2017 International Conference on Circuit, Power and Computing Technologies, 2017, Kollam, the Proceedings*; MDPI, Volume 1, p. 752.

Pierre, D. M. & Zakaria, N. (2014). Partially optimised cyclic shift crossover for multi-objective genetic algorithms for the multi-objective vehicle routing problem with time windows. In: *2014 IEEE Symposium on Computational Intelligence in Multi-Criteria Decision-Making* (pp. 106–115).

Psychas, I. D., Marinaki, M. & Marinakis, Y. (2015). A parallel multi-start NSGA II algorithm for multiobjective energy reduction vehicle routing problem. In: *International Conference on Evolutionary Multi-Criterion Optimization* (pp. 336–350). Springer International Publishing. doi: https://doi.org/10.1007/978-3-319-15934-8_23.

Shen, M., Liu, D.-R., & Shann, S.-H. (2015). Outlier detection from vehicle trajectories to discover roaming events. *Information Science, 2*, 294.

Singh, P. M., & van Sinderen, M. (2015). Interoperability challenges for context aware logistics services– the case of synchromodal logistics. Paper presented at the *Workshops of the 6th International IFIP Working Conference on Enterprise Interoperability*, Nîmes.

Tengler, J., & Pekná, J. (2020). Alternative methods of using RFID technology to identify transport units. In *Smart Infrastructure and Applications* (pp. 277–292). Springer Science and Business Media LLC.

Tiwari, S., Wee, H. M., & Daryanto, Y. (2018). Big data analytics in supply chain management between 2010 and 2016: Insights to industries. *Computers and Industrial Engineering, 115*, 319–330.

Waller, M. A., & Fawcett, S. E. (2013). Data science, predictive analytics, and big data: a revolution that will transform supply chain design and management. *Journal of Business Logistics, 34*(2), 77–84.

Wamba, S. F., Gunasekaran, A., Papadopoulos, T., & Ngai, E. (2018). Big data analytics in logistics and supply chain management. *The International Journal of Logistics Management, 29*(2), 478–484.

Wen, M., Cordeau, J. F., Laporte, G., & Larsen, J. (2010). The dynamic multi-period vehicle routing problem. *Computers and Operations Research, 37*(9), 1615–1623.

Yao, J. (2016). Optimisation of one-stop delivery scheduling in online shopping based on the physical.

Zhang, S. & Qinghe, H. (2014). Application of GA optimising neural network to coal sales forecasts. In: *2nd International Conference on Information, Electronics & Computer*. Atlantis Press.

Chapter 12
The Concept Towards Transportation Infrastructure and Its Capacity

Aurelija Burinskiene

Abstract The concept of transport infrastructure capacity is complex. The infrastructure of transport is ubiquitous and essential for the modern society and its economic and social well-being, as it ensures freight mobility, access to delivery services, and the transfer of products, and facilitates economic activities and social interactions. The infrastructure of transport includes harbours, road and rail networks, storage facilities, and facilities related to the networks. The transport system belongs to a set of distributed infrastructure physical components which are interconnected and placed throughout the region. This system exists in a connected network that spans that region, allowing vehicles and goods to move throughout the region.

Keywords Transport infrastructure capacity · Capacity-usage estimation · Demand for mobility · The life cycle of transport infrastructure

12.1 Transportation Infrastructure

Physical system components are placed in the connected network that covers that region, allowing vehicles and goods to move throughout the region. Thus, physical elements depend on the dominating environmental and climatic factors in the region. Land transport road and rail systems require connected physical equipment to operate. Water transport needs certain fixed assets other than ports. Intermodal exchanges are needed when transportation units and/or products move between different transport's modes.

The multimodal and multifunctional transport systems are inseparable from other infrastructure systems (such as energy and communication).

Communication systems are installed to increase safety and minimize congestions. They can integrate all modes of freight transport, which could lead to the

---

A. Burinskiene (✉)
Vilnius Gediminas Technical University, Vilnius, Lithuania
e-mail: aurelija.burinskiene@vilniustech.lt

D. Dzemydienė et al., *Development of Smart Context-Aware Services for Cargo Transportation*, International Series in Operations Research & Management Science 330, https://doi.org/10.1007/978-3-031-07199-7_12

creation of new services. Cooperative communication is a system that enables efficient data exchange between vehicles and other users.

These interfaces have to be considered in conjunction with infrastructure systems. The entire transport infrastructure of the urban area, which includes a network of streets and roads, and other technical measures, requires fundamentally new approaches to organizing traffic, traffic flows, and integrated systems that simultaneously cover arrays of heterogeneous data and provide multi-level interactions, subordinate complex subsystems for the construction of urban highways, and traffic flow development management methods, including the application of intelligent transport management systems (ITMS) (Comi & Rosati, 2013). A multi-tiered, complexly organized ITMS is a hybrid system consisting of many heterogeneous systems which interact in a complex way, including management, classification, forecasting, expert, decision-making, or support processes combined with visualization and installation of transport object infrastructure in the thematic layers of interactive material electronic map for a single purpose. The integrated ITMS acts as a road transport infrastructure control system, aiming to improve the level of traffic management, including road network characteristics, deployment of traffic management equipment to optimize process traffic management, adapting flows at all stages of traffic, reducing transport delays, and increasing traffic safety.

Capacity related to infrastructure and technologies could be investigated. The proper revision approach can increase physical throughput and consequently the capacity without building extra infrastructure, guaranteeing capacity extension.

The capacity is separated into nominal, static, and dynamic. The first one is evident when the infrastructure is reliable and is working on normal loads. Static one is in the case the infrastructure has a higher capacity. It is built on purpose as the enlargement costs more due to the land and the capital price.

Transport infrastructure operating above 80% of rated capacity starts facing dynamic capacity problems. Dynamic capacity relates to infrastructure, work, and technologies. It could be improved by regulating traffic lights, the direction of transport, the type of trucks for entry, and the timing in which the infrastructure could be used (Fig. 12.1).

Although the dynamic capacity could be increased slightly, the nominal capacity could be only increased by using extra static capacity (either by reducing mobility or by changing the mode of transport).

Optimal rated capacity cannot be achieved effectively, because the particular vehicle or infrastructure is linked to others, so capacity increases must be synchronized. For example, a port terminal operating near optimal nominal capacity faces serious congestion problems—queues at terminal access points; ships stand in line at the port side to enter the terminal, and trucks wait at the gate to pick up or deliver containers (Van Riessen et al., 2016).

Transport problems can be addressed from two different perspectives at the least: capacity and use. The capacity perspective focuses on the impact of infrastructure improvements on global transport system activities. These improvements range from modifications to the network, which are mainly analysed using equilibrium

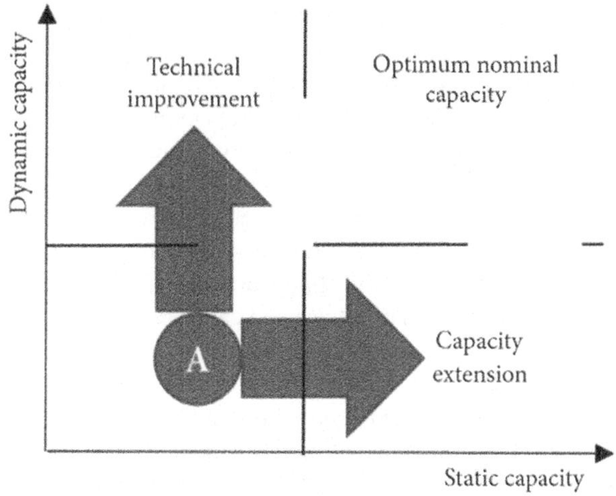

Fig. 12.1 The capacity revision of the infrastructure

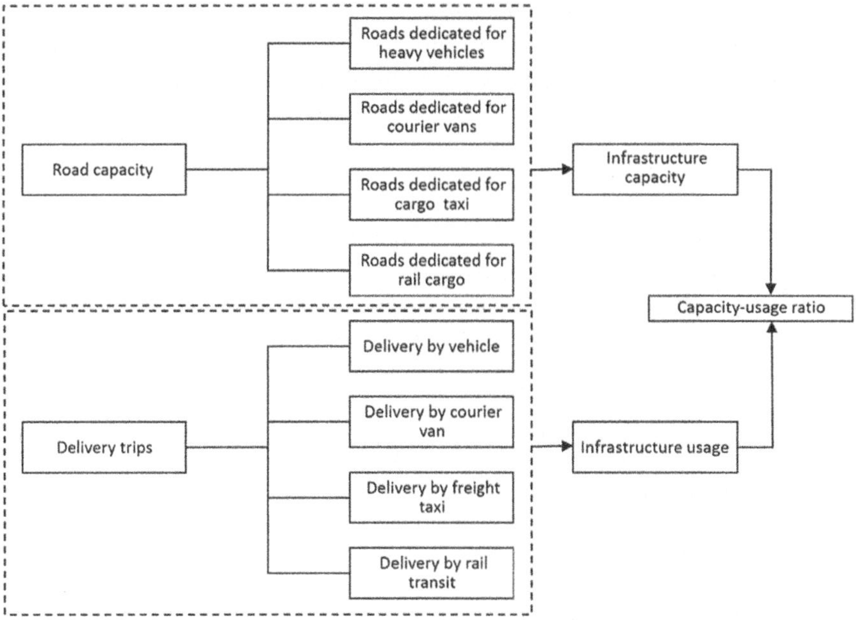

Fig. 12.2 Urban transport infrastructure capacity use ratio estimation model

allocation methods (Patriksson, 1994), to making effective use of advanced telematics (Argyrakos et al., 1991).

The urban transport infrastructure capacity and use ratio (TICUR) estimation model (Fig. 12.2) and quantified urban transport system conditions have been developed and support the theoretical basis for transport infrastructure development.

In short, TICUR represents the transport system capacity and transport flows and determines whether transportation conditions are balanced. In many documents, transport infrastructure capacity and its use are the means of traffic phenomenon analysis to elucidate high-impact transportation theories and methods.

Given that transport infrastructure capacity and its use cause travel time variations (Wang et al., 2009), a theoretical framework for building a reliable transport network was designed as a two-level setup, where the upper level indicates the goal concerning the optimized subject usage growth and economic constraints, and the lower one deals with the application of time-dependent equilibrium models.

After calibration of all use and capacity requirements, the dynamic traffic allocation model for heavily congested urban networks demanded a change in reception capacity management (Ben-Akiva et al., 2012). With stochastic use, the process for assigning the intended residual minimization model emphasizes the traffic perspective and the stochastic cell transfer model (Zhang et al., 2011), capturing the traffic density and spreading standard deviation in time and space.

Transport capacity comprises facilities (highways, roads), services (transit service delivery lines and timetables), rules (roads turnover and parking rules), and fees (transit fees, taxes) which allow travelling.

This means that measures, such as vehicle positions, improved quality, enhanced management, and a reasonable price, are aimed at offering additional travel services. Then it is necessary to plan transport following to the maximum freight loads capacity of transport system per unit of time.

Transport capacity is the load capacity of metropolitan transport systems; it has three characteristics: limited resources, intermodality, and multi-level service objectives. TICUR helps to define the dynamic equilibrium.

The dynamics of system revision is a way to understand the complexity of systems from a time perspective and revise time delays affecting the behaviour of the entire system (Borshchev & Filippov, 2004). This method has been suitable for strategic issues and could be a useful support tool for policy analysis and decision-making in the transport area (Shepherd, 2014). Its areas of application include assimilation of alternative fuel consuming vehicles that have an impact on capacity chain management transport, highway maintenance, strategic policy, and newly emerging areas of use.

Equilibrium of capacity and use was discussed in the context of the road hypernet (the network where the route is located), including the expected reliability of the road network: travelling time and capacity reliabilities (López & Monzón, 2010).

The mathematical approach has been used to forecast the scope of the origin–destination movement in a given network, model performance, and dynamic flow assignments (Lin et al., 2008) and the benefits of responsive pricing and the use of quantified travel information (Gardner et al., 2011). When increasing the capacity contributing to transport capacity, a dynamic congestion impact assessment system pricing policy (Liu et al., 2010) could show how capacity dynamics affect travelling needs, the choice of modes of transport, and the design method (Estrada et al., 2011). Efficient transit networks could be used to create a high performing network.

12.2 Demand Exceeding the Capacity of Transport Infrastructure

Transport infrastructure faces congestions and the goal is to increase the capacity ends without a solution.

Traffic spikes are the main challenge for freight delivery, as currently there are unprecedented transport needs that are increasingly dependent on mobility. The reasons why congestions form are clear, but solutions are not available. Vehicles and locations are involved in congestions which form for two reasons:

- Firstly, transport units overload the roads, which lacks capacity but are highly prioritized by drivers for mobility.
- Secondly, in case of accidental situations that lead to temporary disruptions in the provision of services, such an occurrence of traffic or environmental accident is of short duration.

Congestion cases are not limited to urban traffic (Nathanail et al., 2016). Over the world, twenty container ports reload more than 50% of containers delivered by maritime transport (Van Riessen et al., 2015). However, traffic concentrations in many of these networks already cause capacity problems, especially in reaching some parts of their more remote areas. International trade has grown faster in recent decades than GDP-measured economic growth, while the congestion related to trade flows (or long-distance freight transport) is expected to remain a problem in the future (De, 2011).

Traditionally, the decision for avoiding congestions is to give more capacity for transport infrastructure. This solution depends on engineering capabilities, i.e., the structure and construction of the infrastructure and its development. This requires the revision of a broader perspective, taking into account different objectives and possibilities, responding to various travelling needs and looking for the future.

- Increased demand and the increasing likelihood of congestion will strengthen new spatial responses, increasing the likelihood of the emergence of new spatial flows (Batty, 2012). They include:
- Demand management related to situations and places when mobility demand may be affected. Under the conditions that capacity is fixed but demand grows, prices are inevitably adjusted. In case the highways are free of charge, thus with increasing congestion levels, there is no road costs increase, and only time is wasted by users. This increase requires to re-assess priorities for the use of infrastructure, especially in urban areas (Taniguchi et al., 1999).
- Concentration before devolution. The improvements of the accessibility of infrastructure are called up. With the devolution of sea ports and terminals, choosing peripheral locations becoming more preferred.
- Demand management is a key issue in transport modelling. To forecast transport demand is difficult for several reasons. First, transport demand is called as

derived. Second, the complexity of human behaviour makes the selection process uncertain, unpredictable, and random.

Several types of demand forecasting models are considered. Both of them aim to recover observable data as much as possible, they differ in their basic assumptions and structure. The first model type is called origin–destination (OD) models that show transport demand through trip directions.

They indicate each pair of origin–destination and the capacity of drivers going on this journey. The second type of model transport demand is a sequence of choices: choice of activity, the purpose of destination, choice of mode of transport, etc. Here the focus is on one of the most widely used choice models following logistic model structure.

The existing methods for estimating vehicle traffic from one location to another are generally divided into several methods. Under the first model, the process is broken down into a sequence of four steps, and this is where the method takes its name from. The primary step of the multi-step model activities such as route generation, distribution, and mode selection is taken to construct origin and destination matrices including the counting of travels from origin to destination.

The transit time is estimated and using methods is calculated according to real-time data and information about transit infrastructure. The latest activity-based models are related to trip demand at the individual level. Assuming that transport demand is generated by the purpose to operate, the models apply similar survey data to evaluate travel performance curves and predict behaviour using probit or logit models.

Effective management of transport systems requires accurate travel demand data. Anonymous vehicles can sometimes be tracked on the way to and from a terminal through previous research of smart sensors, licence plates, and cell phones. However, often the main sources of transport data, such as entry and exit numbers, are aggregated, and the original demand needs to be recovered from these aggregates.

Choosing the right representation of transport demand depends on reciprocal transaction model complexity and provided data accuracy. On the one side, the list of travels in the region being surveyed, in, for example, on weekly basis, can be applied to forecast transport demand. Practical usage of such data is limited, or it could not be applied in general. The information about each travel, including origin, all stops, exact time, the destination of such trip, etc., provides the driver sufficient orientation.

The benefit of the collection of such data is questionable, especially when research covers large areas. Also, even if a large picture of reality would be available, an extensive amount of information could not be controlled, and the amplitude of measurement errors would probably be unacceptable. Hence, reasonable demand should be somewhere between the two extremes.

Algorithms and metrics are used for better travel behaviour understanding and the performance of transport infrastructure. BigData analysis methods have been used for processing the sets of data. Individual surveys, the presence of extraction (Asakura & Hato, 2004), origin–destination assessment, validation (Nie et al., 2005; Caceres et al., 2007; Wang et al., 2012; Iqbal et al., 2014), traffic speed estimation (Bar-Gera, 2007; Zhan et al., 2013), and performance modelling (Reades et al., 2009; Phithakkitnukoon et al., 2010) are investigated by applying the gathered data.

However, the studies typically present the alternatives with only several steps of traditional four steps or activity oriented models that are used to estimate travel demand rather than to compare results with estimates of travel demand from other sources.

Chow et al. (2010) note that the recording of logistics solutions at the overall regional level is a highly data-intensive model, which includes data about trade flows and various costs factors. In addition, very detailed transportation data is required for the evaluation. The authors argue that despite accurate modelling logistics structures, which are very important for freight flows and truck travelling, logistic models are still in their early stages of construction. Logistics models could exist if the basic data would be available.

The proposed methods could be transferred from a single network to many, requiring minimum extra data collection and calibration activities.

Forecasting Flows in Freight Transport Regional growth patterns are distinguished in the Tavasszy (1996) flow model. However, even with simplification, there are still some options for calculating the potential impact. As transport becomes cheaper, prices fall and demand rises. Another way is to study land prices; having reduced transportation times, the availability of new lands may increase and the demand for these lands will increase compared to the lands which are currently used for infrastructure. The third common method is to estimate the impact on the improved connection only by calculating the monetary value of how the utility of multiple users will be changed. This last approach covers all the consequences of infrastructure investment (Mohring, 1993) and could have a different effect than that in the case of the improved connection as identified by Hussain (1996).

To determine the function of a specific model and to link demand, user ratings of the transport solution must be considered, with a particular orientation to the benefits or value that a company in need of transportation brings about. When the value added to it changes transport attributes such as transport time, delay, damages, frequency of departures, incidents, the covered distance by mode, these analyses become focused on attributes rather than mode (Winston, 1983). In the attribute-oriented approach, each transportation service is treated as a set of parameters. The business selects attribute packages, which indirectly offers selection modes. In addition to those transport-related factors, the impact of other external attributes is evaluated at the company level.

Surveys of the said preferences are usually conducted as elective games in which an informant is asked to choose between transport alternatives, thus indicating the company's preferences. This could be estimated by using two transport alternatives: with one mode of transport and with several different modes of transport. As stated above, into characteristics-oriented approach can even be created without explicitly specifying the transport mode, as it focuses on attributes. However, in practice, it has so far been challenging to start an analysis of attributes.

Together with users' assessment of transportation services and infrastructure, forecasting demand flows is perhaps the most important aspect of the cost–benefit analysis. Changes in flows and changes in road safety are important tangible long-term improvement results. These flows are forecasted by applying various methods. Some methods are network-oriented and other ones are link-oriented. The STAN freight flow model which is used in Sweden's national freight planning implies a network-oriented approach. The model determines the optimal flow system, taking into account the minimum total transport costs. It has multimodal wording; Sweden's implications are gathered from road, sea, and rail networks.

The optimization of the system is the right way to model a railway system. However, it may not be suitable for carriage by road network involving many agents and may be unsuitable for railway transport in the face of competition. An alternative presents two attractive features: flow distribution and decentralized decision-making in models.

Models of links and user-oriented flows requiring transfer between two transport modes can be modelled in several ways, starting from macro aspects and ending with the micro ones. At the lowest micro level, there is a single approach that looks at the individual problems of efficiency gains [i.e. Small et al. (1989), Ben-Akiva and Lerman (1985)], where manufacturers select to sell their products to the countries that provide maximum expected benefits (or gave entities to achieve maximum profits).

The usage of this approach allowed to development of the macro-oriented gravity model. Using the macroeconomic approach, both demand and supply curves are modelled in the aggregate way. These models consist of parameters that describe the attractiveness and node-sized export flows. Their amounts are more linked with economic than geographic issues.

When assessing infrastructure, traffic forecasting is very important, as cost and benefits change depending on the traffic intensity. Herein, predictions are often made by using gravity models for problem-solving. In the gravity model, the appeal of nodes and the friction between them are applied for revising projected flows. The appeal is usually calculated according to the size of the node, e.g. population or gross domestic product. Nodes can represent towns, states, or other urban objects (Dablanc, 2007). The distance among nodes is analysed as a Euclidean distance or transportation time, but also as qualitative estimations, like the width of the road or the overall representation having a quality index, could be used (De Vany & Saving, 1983). Traffic forecasts must be made in light of the differences of methods that could be used depending on whether the planning horizons are short or long.

12.3 Policy Impact on the Life Cycle of the Infrastructure for Transport

Despite the particular solutions to minimize congestions being under consideration, increasing demand has led to the unprecedented need to invest in transport infrastructure. The main question faced by countries that are financing the development of the infrastructure for transport. The requirements for capital increase exponentially in its enlargement case. The private sector is usually funding transport infrastructure development. However, the sustaining demand growth makes it difficult to secure the scale of government funding needed to meet freight mobility expectations.

Requirements for funding are especially required on the infrastructure life cycle spectrum. In this regard, the highways of N. America and China show several main situations. Over the two last decades, China has developed the longest national highway network worldwide. However, the highway system of North America is approaching its usage end. Upgrading and maintaining systems require significant capital investment, including ageing bridges in highways. Many country highways are financed by the N. America government, most of China's instate roads are financed by companies. The role of various sectors is fundamental in the development of the infrastructure:

- Partnerships among the government sector and companies is one of the options. This is a key opportunity for developing countries, where modern transportation infrastructure is required and public funds are insufficient for large investments. Some models are developed: (1) private sector builds, operates, and returns to the government after a certain period (Build-Operate-Transfer); (2) private sector builds, leases to the government, and gets back (Build-Lease-Transfer); (3) private sector builds and retrofits an existing object to operate for a specified time before it is returned to the government (Remediation-Operate-Transfer).
- Pricing. Another way is the creation of fees covering the usage of highways. The market forces usually determine prices, especially in maritime (Mendelson, 1985). However, the road transport infrastructure is mainly financed by the public sector. Pricing is also important when the usage of transportation infrastructure, in general, is in high demand (Mendelson & Whang, 1990). Even there are different pricing models, road users must pay to use the highways. Under the rise of environmental issues, external vehicle taxation is becoming a reality in many jurisdictions. The effectiveness of these alternatives and their impact on travel behaviour is still to be seen.

Such funding models could be also applied for other types of infrastructure elements than highways.

12.4 Data-Driven Modelling of Transport Infrastructure

The transport infrastructure is planned usually for a 30 years period due to significant investments. Transport infrastructure has to be safe with implemented road and traffic signs, like traffic lights, speed limits, signs reporting intersections, separate lanes used for changing direction. They have to report road width, transport height limits, infrastructure maintenance works, and infrastructure's belongings (fuel refilling stations and the rest points for drivers). These signs have to be visible in different day time (morning, evening, night). The information about traffic conditions and congestions could help drivers to prefer other routing options. Among traffic conditions could be reported not only weather, accident cases but also the information about road quality that requires maintenance and forces the damages of vehicles.

The data collected about traffic intensity could also help infrastructure planners to make decisions on its improvement. By revising the information infrastructure operators could increase infrastructure capacity through the automated and optimized speed control. The application of similar solutions in terminals could reduce the movements of freights and minimize the waiting time for different transport modes transferring freight to and from terminal. Automation of movements can improve the quality of service.

The specialized software and other communication and computer systems could be used to plan transport infrastructure. After the planning the flow of traffic could be improved by reaching higher efficiency, sustainability, and safety. This could be proven by analysing and comparing flow at intersections and junctions of revised transport infrastructure.

The structure of infrastructure modelling could include the revision of such points:

- traffic intensity, congestions, and mode selection priorities,
- volumes of freight transportation in the area and connection requirements to other transport modes,
- expected network capacity and performance: travelling time and costs,
- future transport realization and sustainability needs (such as CO_2 emissions, accessibility options, and others).

The modelling and revision of transport infrastructure may include some steps:

- identification of criteria for decision-making (for example, the reduction of accidents, possibility to share transport modes, response to the growth of freight transportation demand, the total travel time in the network presented at the region or other criteria),
- collection of context data and information from survey,
- validation, calibration, and evaluation of collected data,
- sensitivity analysis of data inputs and the identification of relationships among the data,

- modelling and setting up the targets that are prioritized during the choice of proper solution.

Finally, the comparison of alternatives is performed including sustainable and safety elements, also advantages and disadvantages of new proposals are evaluated.

12.5 Conclusions

The chapter focuses on the capacity of infrastructure revision and presents the case study for urban infrastructure use aspects. The author also revises cases that demand exceeds capabilities of transport infrastructure, the life cycle of infrastructure, the importance of forecasting freight delivery flows, and policies to increase the infrastructure capacity.

To support this the paper presents the North America and China capacity increase models developed with business sectors such as partnership and pricing. The partnership model could have more versions in allocating private and public funds for renewing infrastructure separated into the build, operate, and funding activities.

Most of these activities present the evaluation and expansion of road network transport; however, the same approach could be used for other infrastructure which could be implemented as separate infrastructure objects.

References

Argyrakos, G., Carrara, M., Cartsen, O., Davies, P., Mohlenbrink, W., Papageorgiou, M., Rothengatter, T., & Toint, P. L. (Eds.). (1991). *Advanced telematics in road transport, commission of the european communities – DG XIII*. Elsevier.

Asakura, Y., & Hato, E. (2004). Tracking survey for individual travel behaviour using mobile communication instruments. *Transportation Research Part C: Emerging Technologies, 12*(3), 273–291.

Bar-Gera, H. (2007). Evaluation of a cellular phone-based system for measurements of traffic speeds and travel times: a case study from Israel. *Transportation Research Part C: Emerging Technologies, 15*(6), 380–391.

Batty, M. (2012). A generic framework for computational spatial modelling. *Agent-based models of geographical systems* (pp. 19–50). Springer

Ben-Akiva, M., & Lerman, S. R. (1985). *Discrete Choice Analysis; Theory and Application to Travel Demand*. MIT-Press.

Ben-Akiva, M. E., Gao, S., Wei, Z., & Wen, Y. (2012). A dynamic traffic assignment model for highly congested urban networks. *Transportation Research Part C: Emerging Technologies, 24*, 62–82.

Borshchev, A., & Filippov, A. (2004). From system dynamics and discrete event to practical agent based modeling: reasons, techniques, tools. Paper presented at the *Proceedings of the 22nd International Conference of the System Dynamics Society*.

Caceres, N., Wideberg, J., & Benitez, F. (2007). Deriving origin destination data from a mobile phone network. *IET Intelligent Transport Systems, 1*(1), 15–26.

Chow, J. Y. J., Yang, C. H., & Regan, A. C. (2010). State-of-the-art of freight forecast modeling: Lessons learned and the road ahead. *Transportation, 37*(6), 1011–1030.
Comi, A., & Rosati, L. (2013). CLASS: A city logistics analysis and simulation support system. *Procedia – Social and Behavioural Sciences, 87*, 321–337.
Dablanc, L. (2007). Goods transport in large European cities: difficult to organize, difficult to modernize. *Transportation Research Part A: Policy and Practice, 41*(3), 280–285.
De, P. (2011). *Trade facilitation in India: An analysis of trade processes and procedures (No. 95)*. ARTNeT Working Paper Series.
De Vany, A. S., & Saving, T. R. (1983). The economics of quality. *Journal of Political Economy, 91*, 979–1000.
Estrada, M., Roca-Riu, M., Badia, H., Robuste, F., & Daganzo, C. F. (2011). Design and implementation of efficient transit networks: Procedure, case study and validity test. *Transportation Research Part A: Policy and Practice, 45*(9), 935–950.
Gardner, L. M., Boyles, S. D., & Waller, S. T. (2011). Quantifying the benefit of responsive pricing and travel information in the stochastic congestion pricing problem. *Transportation Research Part A: Policy and Practice, 45*(3), 202–218.
Hussain, I. (1996). Benefits of transport infrastructure investments: A spatial computable general equilibrium approach. *Umeå Economic Studies, 409*.
Iqbal, M. S., Choudhury, C. F., Wang, P., & González, M. C. (2014). Development of origin–destination matrices using mobile phone call data. *Transportation Research Part C: Emerging Technologies, 40*, 63–74.
Lin, D.-Y., Eluru, N., Waller, S. T., & Bhat, C. R. (2008). Integration of activity-based modeling and dynamic traffic assignment. *Transportation Research Record, 2076*, 52–61.
Liu, S. Y., Triantis, K. P., & Sarangi, S. (2010). A framework for evaluating the dynamic impacts of a congestion pricing policy for a transportation socioeconomic system. *Transportation Research Part A: Policy and Practice, 44*(8), 596–608.
López, E., & Monzón, A. (2010). Integration of sustainability issues in strategic transportation planning: A multi-criteria model for the assessment of transport infrastructure plans. *Computer-Aided Civil and Infrastructure Engineering, 25*, 440–451. https://doi.org/10.1111/j.1467-8667.2010.00652.x
Mendelson, H. (1985). Pricing computer services: Queueing effects. *Communications of the ACM, 28*(3), 312–321.
Mendelson, H., & Whang, S. (1990). Optimal incentive-compatible priority pricing for the M/M/1 queue. *Operations Research, 38*, 870–883.
Mohring, H. (1993). Maximising, measuring, and not double counting transportation improvement benefits: A primer on closed and open economy cost-benefit analysis. *Transportation Research B, 27B*(6), 413–424.
Nathanail, E., Gogas, M., & Adamos, G. (2016). Smart interconnections of interurban and urban freight transport towards achieving sustainable city logistics. *Transportation Research Procedia, 14*, 983–992.
Nie, Y., Zhang, H., & Recker, W. (2005). Inferring origin–destination trip matrices with a decoupled GLS path flow estimator. *Transportation Research Part B: Methodological, 39*(6), 497–518.
Patriksson, M. (1994). *The traffic assignment problem, models and methods*. VSP.
Phithakkitnukoon, S., Horanont, T., Di Lorenzo, G., Shibasaki, R., & Ratti, C. (2010). Activity-aware map: Identifying human daily activity pattern using mobile phone data. *Human Behaviour Understanding* (pp. 14–25).
Reades, J., Calabrese, F., & Ratti, C. (2009). Eigenplaces: Analysing cities using the space-time structure of the mobile phone network. *Environment and Planning B: Planning and Design, 36*(5), 824–836.
Shepherd, S. (2014). A review of system dynamics models applied in transportation. *Transportmetrica B: Transport Dynamics, 2*(2), 83–105.

Small, K. A., Winston, C., Evans, C. A., & Work, R. (1989). *A new highway pricing and investment policy*. The Brookings Institution.

Taniguchi, E. T., Thompson, R. G., & Yamada, T. (1999). Modelling city logistics. In E. Taniguchi & R. G. Thompson (Eds.), *City logistics*. Institute of Systems Science Research.

Tavasszy, L. A. (1996). *Modelling European freight transport flows*.

Van Riessen, V., Negenborn, R. R., & Dekker, R. (2015). *Synchromodal container transportation: An overview of current topics and research opportunities computational logistics* (pp. 386–397). Springer.

Van Riessen, V., Negenborn, R., & Dekker, R. (2016). *Real-time container transport planning with decision trees based on offline obtained optimal solutions*. Econometric Institute Research Papers.

Wang, G. D., Shao, H., & Cao, D. X. (April 2009). A mixed equilibrium traffic assignment model for transportation networks with ATIS under demand and supply uncertainties. In *Proceedings of the International Joint Conference on Computational Sciences and Optimization (CSO '09)* (vol. 2, pp. 132–136). IEEE Computer Society, Sanya.

Wang, P., Hunter, T., Bayen, A. M., Schechtner, K., & González, M. C. (2012). Understanding road usage patterns in urban areas. *Scientific Reports, 2*(1001). https://doi.org/10.1038/srep01001

Winston, C. M. (1983). The demand for freight transportation models and applications. *Transportation Research Part A: General, 17*(6), 419–427.

Zhan, X., Hasan, S., Ukkusuri, S. V., & Kamga, C. (2013). Urban link travel time estimation using large-scale taxi data with partial information. *Transportation Research Part C: Emerging Technologies, 33*, 37–49.

Zhang, C., Chen, X., & Sumalee, A. (2011). Robust Wardrop's user equilibrium assignment under stochastic demand and supply: Expected residual minimization approach. *Transportation Research B: Methodological, 45*(3), 534–552.

Chapter 13
Information Systems in Freight Transportation

Kristina Čižiūnienė

Abstract Transport sector faces many problems, including traffic congestion, accidents, environmental pollution, dependence on fossil fuels and the like, and these problems are becoming more severe and relevant with the annual vehicles number growth. The use of traditional measures to solve transport problems is too expensive and inefficient usually. IT have become an inseparable part of every person's life and they have increasingly been used in manufacturing and provision of various services, including transportation services. The creation of information infrastructure is like the creation of an environment for the functioning of information systems rather than the creation of one very large computerized information system. Information systems can ensure a smooth and uninterrupted work and a sustainable transportation process through their interaction only. However, many systems have also been used in the organization and management of the freight carriage process. This section will justify the need for information systems in freight transportation.

Keywords Decision-making · Information system · Intelligent transport system · Vehicle · Database · Digitalization

13.1 Areas of Activities for Transport Sector by Implementing ITS

Information society distinguishes its main requirements and trends for the further development of information. The most important thing is the speed of information processing, globalization of information, and the abundance of information. Of course, services in the transport sector are also very closely linked to the provision of information. According to Palšaitis (2010): "... *acquisition of materials, production planning, delivery of finished products to consumers and all information and*

K. Čižiūnienė (✉)
Vilnius Gediminas Technical University, Vilnius, Lithuania
e-mail: kristina.ciziuniene@vilniustech.lt

D. Dzemydienė et al., *Development of Smart Context-Aware Services for Cargo Transportation*, International Series in Operations Research & Management Science 330, https://doi.org/10.1007/978-3-031-07199-7_13

financial flows necessary for the management of material resources". Thus, this allows concluding that information transfer in transport is one of the key factors which allows to ensure proper service assessment in the "supply chain" and its quality provision. Only by improving and expanding information systems in the transport sector, regardless of the sector (maritime, rail, road or air transport), we can achieve the best results. Therefore, the development of these systems requires attention for both public and private business sectors.

Intelligent transport systems (ITSs) have a direct impact on the development of the transport sector. Intelligent transport systems are understood as the totality of the application of communication and information technologies in the field of transport. In this case, information system (IS) instruments comprise a significant part of the information system, i.e. IT is just a tool that can be applied in a broad human and organizational context. The adopted strategic decisions (related to information technologies and other determinants of innovation) and the introduction of new IT completely change the functioning and development of all modes of transport (Baublys & Jarašūnienė, 2010; Batarlienė, 2011; Bazaras et al. 2010; Dzemydienė et al., 2021).

The types of ITSs are very different; therefore, the following key groups of ITS activities can be distinguished:

- Information for travellers (websites, billboards, navigation systems, information stands, etc.);
- Traffic management (e.document systems, changing road signs, traffic lights, traffic detection, etc.);
- Vehicle services—driver assistance and vehicle control;
- Operations of freight transport and commercial vehicles (fleet management, accounting of working and rest time);
- Public transport services (ticket booking systems, electronic tickets, route planning and optimization);
- Emergency management (automatic call to emergency services, "black boxes");
- Electronic tolls (parking systems, road user charges);
- Personal protection means;
- Monitoring of weather and natural conditions (road pavements, monitoring of weather conditions, information about natural conditions in real time);
- Coordination and control of response to natural disasters;
- Information of national security.

Transport companies, especially those involved in international transport, were among the first in the new economy to realize the need to integrate information technology into their processes of management. The competitiveness in the transport services sector relating to the emergence of many small businesses and the active development of foreign transport has led to the need for transport companies to mobilize domestic resources. It is obvious that efficient operation of transport companies is inconceivable without the widespread adaptation of IT and personal computers (Jankauskaitė, 2010).

13.2 Decision-Making and Importance of Databases

Many different equipment and technologies have been used in today's freight carriage, including information technologies and systems.

Information systems (IS) are based on a computer communication network, which uses hardware and software to meet the need for organizational information and to support communication needs. IS for transportation management needs are classified. Management Information Systems (MIS), which allow managers of large companies to make strategic decisions, are computer-based systems that ensure the necessary information and support for management decisions. For example, when production department managers need to create a production schedule, they require, for example, data on forecasted volume of orders. Further forecasting is made taking into account market development trends, inventory levels, and the need of employees in various specialities. MIS have to provide all these data. A common form of a similar system is information accounting systems that help managers and employees involved in management to prepare the reports necessary for making day-to-day decisions.

Operational Information Systems (OISs), which allow managers of small enterprises to process data of daily business functional operations. OIS is a software add-on that can efficiently perform the planned operations in the prescribed manner in order to obtain the desired result. These systems are used to make sales and purchases of products from suppliers, calculate salaries, and estimate changes in inventories.

The ISs consist of DBs and knowledge of organizational management rules that help to create information products. For example, consumers, settlement for the products supplied, payment of salaries to employees, financial reporting, and the issuing of invoices. "Transaction" processing is an elementary procedure with DBs and the support of such systems identifies all needful operations with DB, i.e. store and arrange key budget information blocks that perform operations and generate key reporting documentation. In addition to data registration and processing, an operational information system (IS) can be created, which is used for standard decision-making and efficient control of processes.

Production process management systems provide monitoring and control of transformation processes, e.g. special sensors that continuously monitor changes in pressure and/or temperature and generate data that is transmitted for processing to the central computer. Automated systems of institutions, such as processors for editing operations, e-mail, and telecommunications, are also a part of the operational IS group.

Their main purpose is to replace manual work with e. documents and programmes. When necessary, employees transfer information to a computer database, and managers immediately adjust technological procedures. This allows avoiding inaccuracies and errors.

The working places of specialists are arranged by personal computers and enable different activities of managers; for example, they help to make purchasing actions. This requires two types of software:

- System software as specified by the computer manufacturer;
- Applications that manipulate data for specific purposes: creating a file for an existing purchase order or accepting supplier statistical information, processing information to evaluate supplier activity, and preparing a report.

Software packages for the implementation of orders have appeared recently. Most of these packages have their own set of applications for solving tasks designated for this system only (Fig. 13.1).

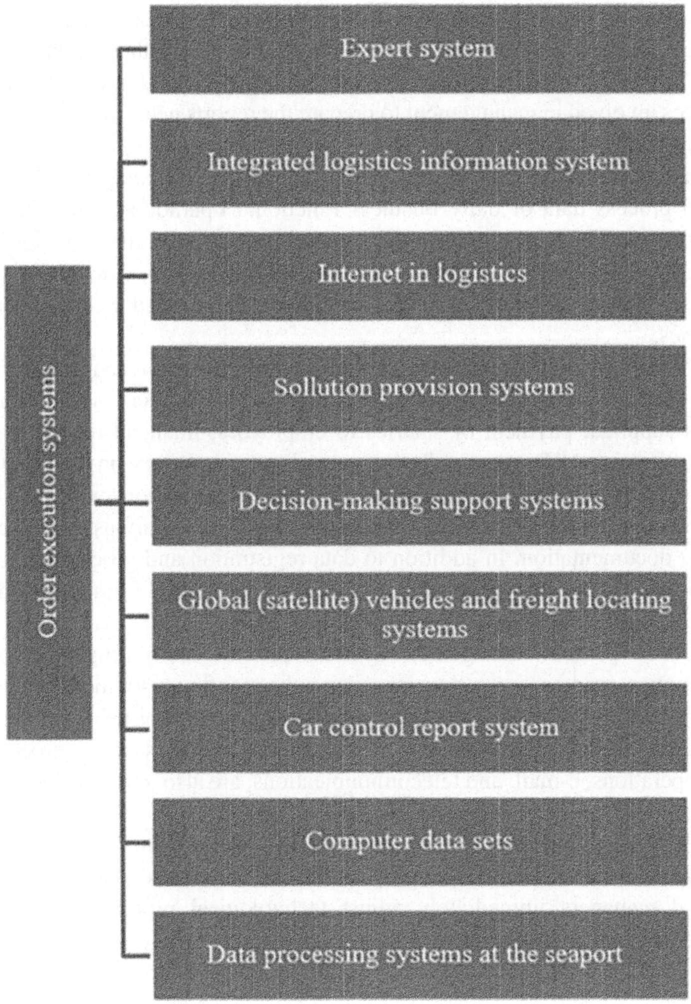

Fig. 13.1 Order execution systems

Most common systems include many types of MIS. The expert systems have the main features and abilities to process knowledge. Artificial intelligence research has increasingly been used in the development of expert system. The emergence of such systems is related to the formation of a new branch of research called knowledge engineering. A typical expert system covers a set of facts (database), a knowledge and interpretation base. For example, there are automated warehouse expert systems in Japan that are used to facilitate and streamline work in the warehouse, namely, for improving diagnostics, freight loading, freight transportation, crane management, and other logistics operations. By increasing the reliability of management and control systems and reducing time needed to find failures, the expert system indirectly affects the quality and quantity of products.

Integrated logistical information system is the information system which consists of a data bank and the connection of user terminals to other information systems when the Internet or e-mail is used to connect to the information systems of enterprises, ports, and transport. For example, if information about a vessel, train, or car carrying freight is entered periodically, users of the information system get data about the freight, including where it has been loaded, how it has been packaged, the final destination of the freight, specific freight storage, and loading conditions, etc. Such information on the time and place of arrival of the vehicle allows to prepare a place for the freight in the company, terminal port, or a vehicle in advance. Moreover, when having all the information about the arrival of a vehicle and a freight, the customs service can plan its work more accurately. Having combined the logistic information system with customs information, customs declarations (especially for transit freight) can be prepared automatically.

This logistics information systems allow the consignee to find out where exactly the freight is, they help in understanding what freight conditions are at the certain time moment, and help to plan its pickup. This allows to make much more efficient use of ordered vehicles by avoiding downtime. The integration of the logistics information system with the customs system allows more efficient tracking of transit freight movement and cuts down on their withholding at control points, at the same time increasing the attractiveness of a specific transport corridor. For example, data on vehicles and freight, contractual and fixed tariffs for the services provided, and state toll rates available in the port information system allow to issue invoices automatically, to make bank payments for these services, and to engage in cost control. Thus, a comprehensive port information system can significantly reduce the time needed for ship entry, exit, and freight clearance procedures.

The use of microcomputers in procurement procedures nationwide allows using barcodes for information exchange in any terminal equipped with the respective software, telephone connection, and modem. This connection now helps to connect to another computer thousands of miles away. Information is transmitted at extremely high speeds. Having such direct contact with suppliers, the buyer can find out the prices, determine the presence of the required goods in the supplier's warehouse, transfer orders for purchase, obtain control department information on order execution, provide the supplier with information on any changes in requirements which the raw materials purchased are subject to and which form having adjusted the supply schedule, also receive information regarding the handling of orders, and send letters and other messages.

The Internet in logistics is the key tool for information exchange in exploring the potential of e-commerce. Some companies send information which employees need online. The Procurement Department can retrieve the necessary information and to read computer files in electronic; to e-mail messages around the world; to search for the necessary commercial information; to provide data on companies, private persons, and public sources; to search for suppliers and distributors, and to purchase products using e-catalogue. The legal and commercial importance of the Internet has been constantly growing.

Solution Delivery Systems (SDS), which help managers to assess possible solutions and to justify their choice, using various modelling and analysis methods to this end. An important part of solution delivery systems is the processing of data from operational systems. SDS are used for the assessment of alternative investment schemes, credit risk analysis, selection of alternative suppliers, etc. However, with the introduction of the latest technology is a possibility that a company can turn into an "information swamp" when employees receive excessive volumes of information and are no longer able to sort it out. System deployment stages or the sequence of steps (algorithms) used by system developers is the most effective strategies for the development and implementation of IS (information systems). First of all, the needs of potential users and requirements for technology are determined, followed by design and actual implementation of the project (acquisition of hardware and software). The system is installed after all these steps have been taken.

Decision-making support systems are interactive computer systems that use a variety of decision-making models and specialized databases that help responsible persons to make decisions. The software of such systems allows users to ask the question "what if?" and to consider specific actions in the interactive mode stages. Information systems help managers to make critical strategic decisions. The software used in them allows users to use large arrays of complex data, provides the opportunity to analyse and present data to users. Such systems help managers of all, including the largest, companies to quickly obtain the necessary internal and external information, to identify problems and analyse situations related to them.

The global (satellite) vehicles and freight tracking systems are satellite-based communication systems, which allow to get in touch any time and to effectively manage the transportation process, especially when carrying expensive goods. If these systems operate in automatic mode, it allows to locate a vehicle or a container at any time, to contact a vehicle crew, and to clarify the current situation.

Modern global systems, such as NA VSTAR (USA) or GLONASS (Russia), which have artificial earth satellites that allow to locate the receiver of this system at the accuracy of 15–100 m at any time have been developed. The differential station of the global (satellite) vehicle and freight tracking system receives global location signals, calculates their errors and transmits them to the vehicle or container receiver. If the differential station is close to the terminal, the location of the receiver is determined at the accuracy of 15–30 cm. Moreover, all the information about freight movement and its storage location is stored in computer memory while operating in constant mode. Such accurate movement information reduces the likelihood of freight (containers) being mixed up or lost (stolen).

The Controlled Automobile Reporting (CAR) System has been used to optimize vehicles allocation until the product reaches the recipient from the manufacturer. It allows to control the following stages of the carriage and forwarding process: production, taking goods from the factory, carriage to the port, loading on a vessel, and notification of loading on a vessel. Participants in this system can always find out the information that is important to them.

Computer data sets are groups of various data compatible with computer data structures and data transmission interfaces developed by freight forwarders from the German Federal Association of Freight Forwarders to transmit data on the selected freight and stevedoring services. These data sets provide the following information: (1) on the carriage to the point of dispatch (date of dispatch or transfer, time of dispatch); (2) address data (addressee: e.g. consignor, customer, supplier, consignee of goods, number-coding these data; customer No., nature of activity); (3) consignment (a) data: consignor's (customer's) order number, order date, number of packages, gross weight, code of the forwarding order prepayment for freight carriage (franking): 01 = unpaid, 02 = franking destination, 03 = franking recipient company, 04 = franking German border, 05 = franking recipient freight forwarder, 00 franking under agreement. "00" marking requires indicating franking code, type of order, transport data processing system [DAVIS—German: Datenfernverarbeitungsorientierte Abwick-lung von Industrieanlagen-Geschäften im Seetransport (Jarašūnienė, 2011)] for comprehensive logistics of industrial equipment. This system is used as the basis to manage and control the entire transportation chain from production to delivery of materials to the construction site, preparing orders for manufacturers first of all, and finally informing about the dispatch of the equipment, providing information about the planning and installation works at the construction site. Freight forwarders, stevedoring, and other companies always receive timely information about freight shipments.

Systems for processing data at seaports (COMPOSS (German: Computer-Orientierte Methode für Planung und AblaufSteuerung im See-Hafen), DAKOSY (German: Daten-Kommunikations-System), and LOTSE (English: Logistic-Tele-Service) (Jarašūnienė, 2011)) are systems which allow to rapidly transmit information by way of electronic data processing when carrying goods by sea. When carrying goods, the consignee may receive data on its consignments in advance. Moreover, information is necessary for many cooperating port authorities to be able to exchange it with each other. Data is exchanged between a forwarding company or service, ship brokers, port administration, freight loading company, stevedoring company, freight volume control service, and customs office.

13.3 Integration of Order Management and Enterprise Logistics Management Information System

The order management system promotes the following directions of logistics activities:

- Determining the method of carriage, the carrier, the loading queue and procedure;
- Distribution of inventories, assembly and preparation of packaging;
- Assembly and packaging in warehouses;
- Updating data on inventories (crossing out assembled goods);
- Preparing lists of automatically replenished goods;
- Preparation loading and carriage documents (freight invoice, if services of the same carrier are used);
- Delivery of the order to the customer.

Computerized order processing can also be used to control inventory levels and to generate production, financial and special reports of management.

The flow of information required to process orders and connecting one department to another, also information from various files or databases, such as financial status of the customer, availability of inventories, and scheduling of transportation. IS can be fully mechanized or non-automated; many systems are automated average. The quality and speed of information flow will vary depending on the complexity of the order management system and the overall MIS (Management Information system). These things will affect the manufacturer's ability to combine batches and keep stocks to a minimum. Non-automated systems tend to run very slowly and tend to miss errors. The time taken to perform various operations is quite long and variable, so information often comes late. Such a system significantly hampers the company's ability to implement integrated logistics management, which would reduce overall costs, maintain or improve the quality of customer service. Ordinary problems are associated with failure to detect price identification errors or to obtain solvency information in a timely manner. These problems also have other adverse consequences, such as errors in bills of lading, late payments, rejection of orders due to incorrect inventory information. The producer's profitability is reduced by lost sales and, as a result, higher costs.

Thus, only information provided on a "just in time" basis is necessary for successful operation. Delayed data hinders timely performance of works. Automating and integrating the ordering process saves time and ensures that information is not delayed. Automation helps executives integrate their logistics system and reduces costs by reducing inventory and shipping costs. An integrated information system of the company is the key factor in the pursuit of the lowest total logistics costs.

13.4 The Main Needs of Information in Logistics

Information systems in logistics are needed for:

- providing managers with knowledge necessary to take advantage of new markets or to change packaging design;
- choosing hired or private carriers;
- making decision regarding increasing or decreasing inventory levels;
- determining customer usefulness;
- creating a high level of customer service;
- choosing shared or private warehousing;
- determining the number and size of the necessary warehouses;
- choosing the level of automation of the order management system.

For making these strategic decisions, managers need to know how various alternative operating methods will affect a company's profit and costs. Once a decision has been made, managers must evaluate its implementation to determine the following:

- whether the operation of the system is controlled and whether its level is in line with the initial estimates of likely profits;
- whether current costs justify a search for alternative systems.

The latest order management system can provide lots of information for different departments of the company. Logistics, manufacturing, and sales/marketing managers can use data terminals. The system can regularly generate various information in the form of reports and generate specific reports on demand. The system can submit queries about all current reports and various data, for example, on the status of orders, market, and inventories.

13.5 Classification of Information Systems for Logistical Management

Information system for logistical management planners starts with the establishment of the principal provisions of the logistics system, the analysis of customer needs, and the establishment of work standards that would allow those needs to be met. The customer's needs must then be aligned with the company's existing capabilities, while the conducted activities must be reviewed to identify areas for improvement. At this stage, it is important to engage in dialogues with managers of different levels. This allows the company to determine what strategic and operational decisions have already been made, what data is still needed to make decisions. Tables 13.1 and 13.2 present various strategic and operational solutions that managers must evaluate in each area of logistics activity.

Table 13.1 Logistics functions behind a strategic decision

	Logistics function				
Decision type	Customer service	Carriage	Warehousing	Processing orders	Inventory
Strategic	Determining the level of service	Determining the mode of transport	Determining the number and territorial layout of warehouses	Defining the level of automation	Replenishment systems
		Consignment combination programmes	Warehouse automation and mechanization	Centralized or decentralized	Backup inventory amounts
		Own or shared carriers	Own or shared warehouses		

Table 13.2 Logistics functions behind an operational solution

Decision type	Logistics function	Decision type	Logistics function	Decision type	Logistics function
Operational	Setting service criteria	Carriage tariffs	Pickup	Ordering paths	Forecasting
		Checking carriage invoices	Packaging	Checking the authenticity of orders	Monitoring inventory
		Processing orders	Assessing inventories	Checking invoices	Assessing current costs
		Routing	Adjusting inventory levels	Coordinating orders	Inventory turnover
		Price negotiations	Staff	Performing evaluation	
		Planning consignments	Warehouse layout and product layout		
		Planning consignment routes and deliveries	Selecting lifting transportation mechanisms		
		Selecting carriers	Performance evaluation		
		Performance evaluation			

Having selected the solution type (-s), processing of the available data is reviewed in the following stage to decide what changes are needed. Finally, shared data files must be created and management reports must be planned, taking into account costs and profits of each of them. A well-designed system must support these management functions and pass the information on to the relevant management services.

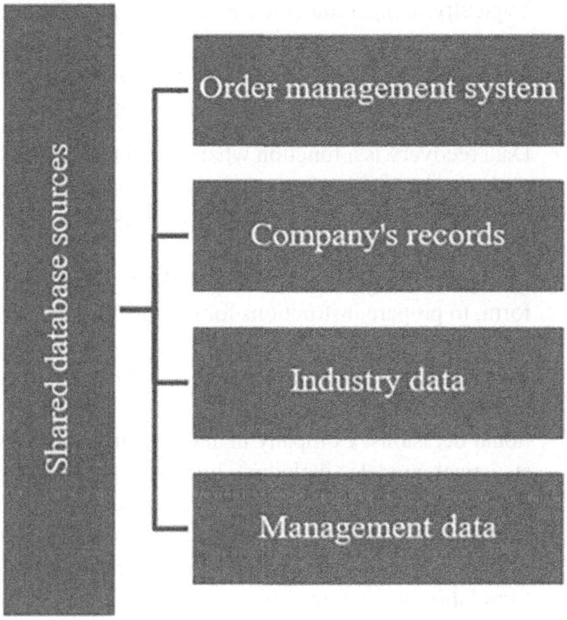

Fig. 13.2 Key database sources

Telephones, teletypes, personal conversations, and computer communications are just some of the ways to transmit information. A computerized information system must not only process information, but also store it to the extent necessary for decision-making.

Data necessary for a logistics management information system are obtained from a variety of sources. There are four key data sources for shared databases (Fig. 13.2).

Order management system can provide various data, including customer addresses, desired goods, customer revenue, and sale's forms (after the goods were already ordered), order size and the seller.

Company's records are administered to obtain information on production and logistics costs, data on capital costs, company resources, costs associated with the sale of various goods, such as insurance, taxes, obsolete goods, and losses.

General business data can be obtained from trade and professional organizations. Professional magazines and publications are a useful source of information. They provide reports on various studies and reviews of current working methods. Statistical publications of public institutions can also be a valuable source of information. They contain data on population migration, inventory levels, construction of residential houses, consumer credit costs, and so on.

Managers can also contribute to the creation of a computerized database. They could provide information on expected reaction of competitors, future sales trends, government policies, available suppliers, and the likelihood of success of alternative strategies.

Typically, a database consists of a variety of digital data files, such as a billing system for shipments, shipping records, stock status, pending and cancelled orders, and fixed costs for various logistics activities, marketing, and production.

Digitized IS must perform the following functions:

- Data recovery is a function when data on freight rates, typical warehousing costs, or the current stage of processing of a customer's order is obtained. Usually, these data are unprocessed; computer records allow to find information quickly and conveniently.
- Data processing is a function which allows to provide data in a more convenient form, to prepare instructions for picking goods in warehouses, generate shipment invoices, and print orders.
- Data analysis is a function which allows to obtain data about orders and to provide managers with the information they need to make strategic and operational decisions. Company managers can choose from several mathematical and statistical models, including linear programming and simulation models. The linear programming method is the most widely used tool for both strategic and operational planning in logistics management. It is an optimization technique that provides a variety of possible solutions to problems that managers identify. Simulation technique allows to model the situation so that managers can decide how the operation of the system would change having chosen various alternative strategies.
- Report generation is a function which allows to provide processed or systemized data both in an electronic and paper form.

13.6 Digitization of Freight Carriage and Its Relevance to Transportation Processes

A solution to every problem must be found. It is no surprise that once there is a major demand, logistics specialists, engineers, and developers start thinking about ways to accelerate and facilitate carriage, looking for ways to deliver goods of the right quality, do not mix them or lose them along the way. These problems have had a significant impact on the digitization of logistics companies and have led to rapid technological development.

Computers and the Internet initially emerged, facilitating efficiency and transparency in the collection and dissemination of information. Then the process of introduction of digital technologies into trailers started. Finally, digitization has had a major impact on the performance of logistical work. Warehouse, transportation process management, order processing systems help employees to make rational decisions, perform proper analysis, planning. By developing specialized programmes, they have significantly increased the efficiency of companies.

Therefore, in the field of logistics, innovations have been created to facilitate business processes, to reduce service costs, environmental pollution, and to contribute to more efficient work. Over the last few decades, the development of digital technology has influenced changes in the freight business.

Today, the international freight market is undergoing a major change. As virtually all freight documents have been on paper so far, the European Union passed a decision based whereon starting August 2024, public authorities will be required to accept information provided by carriers digitally in a common format valid in all European Union (hereinafter—EU) states. This is both a challenge and an opportunity that is very positive and promising, as this is an unprecedented focus on digitization and the Green Course in Europe. It is therefore important to design digitization processes in such a way that they operate not only within a country, but can also be shared as good practice with neighbouring countries, thus ensuring a smooth implementation of the logistics process. Therefore, ensuring a smooth movement of freight in the logistics chain requires a smooth flow of data. So far, some relatively good solutions have already been offered for individual modes of transport, but this data flow should be as smooth as possible with the involvement of several modes of transport and all participants in the logistics process.

The digitization of the transport and logistics sector is important for the economy also because of the ongoing rapid digitization of trade and the transformation in the industry. Customers in these sectors demand top transparency and real-time data on the entire supply chain. Therefore, better digitized, real-time data-based logistics would increase international competitiveness of each country's trade and industry. Currently, road transport has the lowest level of digitization, as, according to EU's data, 99% of the data used are paper data. It is therefore necessary to create legal and technical preconditions which would ensure secure exchange of data between the public and the private sector as well as different participants in the logistics process, not only in local but also in international markets. The European Commission estimates that the digitization of transport data would save up to 27 billion euros over 20 years, and, if estimated along with fuel savings and non-emitted pollution, it would save another 74 billion euros. Therefore, digitized freight documents are expected to solve two major problems in the transport and logistics sector:

1. Freight travel time will decrease by a few hours or even a day, as currently same data, including freight contents, trailer registration data, driver's full name, warehouse addresses, etc., are re-entered manually and reprinted many times, while drivers have to drive with freight in a trailer around various institutions in order to collect the necessary documents.
2. Digital data will allow to see accurate statistics on the movement of freight and, accordingly, to plan the work of border posts, to forecast congestion, to plan road repairs, to clearly forecast the tolls charged, etc.

In this context, it is crucial to find a solution that creates the greatest possible value for the sector as a whole.

EU digitization policy not only obliges B2G digital data exchange but also changes the legal framework for its authentication and validity.

The aim is for the freight transport process to use digital documents. The scanning of these documents would allow the relevant authorities to know the direction of the vehicle's destination, the freight which it carries, and to prevent stopping trucks for no reason. The use of electronic signatures would also come in handy, allowing to give up the collection and sending of paper documents.

In this case, digitization can only ensure a faster and cheaper freight carriage, but it will not reduce competition when comparing one person to large companies.

Therefore, now investments should be made in data (digital data) exchange platforms—EDI, which would be a bridge between paper and eCMR or e-invoice. The absence of a single standardized data server is a problem, and the eFTI regulation obliges EU states to develop such servers. 4.0 logistics will not be possible without digital data exchange. It is therefore important to ensure that digital documents reflect all the necessary information so that the documents of the transported freight are changed smoothly on the road, that the original shipper is not lost, etc.

On the other hand, the method of signing acceptable to all parties must also be installed together with digital documents. Just like a bank has many different options, freight carriage should also have several signing options.

If we were to look at processes in the transport sector, electronic identification, authentication, and blockchain technologies that protect against distortion will be important in future technologies. The development of a common data ecosystem rather than a technology itself is important in this field. Since transport and logistics cross borders, data must do the same, and the dissemination of this data among all participants in the logistics process must take place in a high-quality and continuous manner.

These data are also relevant for customers and their partners in the transport and logistics sector. Technologies for data collection and processing in real (or near real time) will have to prevail. They will provide customers in the sector with transparency and the ability to manage supply dynamically. This will allow transport and logistics companies themselves to use resources more efficiently.

Augmented realities (with the help of field workers) and virtual reality technologies have been used more extensively to represent these data.

At the same time, the abundance of digital data will enable the development of digital platforms and new data-driven business models (such as the adaptation and extension of sharing economy models).

However, to generate data, it needs to be either obtained from the existing systems or collected. This is where 5G communication technology will come in handy and create value, essentially and eventually activating the Internet of Things—from autonomous cars, drones to sensors on the roads, smart containers, and the like, which will allow to see the processes in real time.

Of course, artificial intelligence (machine, in-depth learning systems) will both control autonomous devices and vehicles and help people predict and manage traffic flows and traffic.

In light of the above, the need for the so-called edge computing has been growing accordingly. By automating and deploying more autonomous transport solutions, algorithms must work as close to the scene as possible to reduce decision-making and transmission delays. Generally speaking, vehicles will have to run extremely powerful computers capable of performing complex data processing algorithms.

Moreover, artificial intelligence is penetrating the world of logistics. Artificial intelligence-controlled trucks will soon be on the streets, and new systems are already being learned from people to analyse information, choose the right route, or anticipate potential problems in the future. Logistics companies will need to focus more on innovation and digitalization to stay competitive. In such a case, the main challenges which the transport sector will face are related to data protection, the improvement of electronic operations, the proper training of staff, and the need to keep pace with technological developments, also their mutual integration, which promotes smoother and faster logistics processes.

13.7 Conclusions

The transportation process uses a wide variety of technologies, software, and information technology to ensure successful process management both inside and outside the company. This allows for the development of good relations with all participants in the transportation process, as well as the development of online logistics services. Timely provision of information to customers is essential to ensure a high level of service. Given the importance of accurate and timely information to customers—global (satellite) vehicles and cargo tracking systems are inevitable in the transportation process of any company. However, it must be borne in mind that in order to apply different systems, managers need to know how the various alternative operating methods will affect the company's profits and costs, and what the benefits will be for the whole process.

References

Batarlienė, N. (2011). *Informacinės transporto sistemos*. VGTU leidykla Technika.

Baublys, A., & Jarašūnienė, A. (2010). Statistical Probability evaluation of operating ITS. *Transport, 25*(2), 163–170.

Bazaras, D., Ledauskaitė, K., & Palšaitis, R. (2010). Comparative analysis of information and communication technologies in Lithuania and Finland. *Transport.* Vilnius: Technika. *25*(1), 98–104.

Dzemydienė, D., Burinskienė, A., & Miliauskas, A. (2021). Integration of multi-criteria decision support with infrastructure of smart services for sustainable multi-modal transportation of freights. *Sustainability*. Basel: MDPI, *13*(9), 1–26.

Jankauskaitė, A. (2010). Jūrų transporto ir uostų valdymo informacinės technologijos. Magistro baigiamasis darbas. VGTU.

Jarašūnienė, A. (2011). *Specialybės įvadas. Transporto inžinerinė ekonomika ir vadyba Mokomoji knyga*. Technika.

Palšaitis, R. (2010). *Šiuolaikinė logistika*. Technika.

Chapter 14
Information Technology Used in the Road Transport Sector

Kristina Čižiūnienė

Abstract Many road transport infrastructure activities are unimaginable and impossible without modern information systems and technologies. These technologies allow every well-developed city to have video cameras, data transmission, data processing and automatic control systems. Residents can easily find out about traffic jams, accidents, or congestions in the city. Modern IT systems help to combat road defects, allowing to easily report newly formed pits, broken asphalt or fallen road signs. IT helps to regulate flows for both freight transport, passenger cars and private cars. This section will address issues related to road transport and the IT systems used in it.

Keywords Road transport · IT · Freight and transport platforms

14.1 IT for Road Transport

Scientific literature offers abundant information on the application of road transport ITS. According to Perego et al. (2011), Coronado et al. (2009), Marchet et al. (2009), Davies et al. (2007), DfT (2006) and Giannopoulos (2004), road transport is the most widely used as it is an efficient means of speed, direct delivery and flexibility. In addition, Dft (2006), in its work analyzed advanced of road transport information programmes that focus on supply chain planning and management, transport tracking, fuel registration, etc., to ensure efficient road transport operations. In addition, Davies et al. (2007) analyzed the impact of cyberspace on freight and transport platforms and ITS applications for general freight carriage.

Road transport is one of the most emphasized modes of transport, with a strong focus on its competitiveness and stringent requirements that carriers are subject to. Usually road transport companies face more problems when vehicles go abroad.

K. Čižiūnienė (✉)
Vilnius Gediminas Technical University, Vilnius, Lithuania
e-mail: kristina.ciziuniene@vilniustech.lt

D. Dzemydienė et al., *Development of Smart Context-Aware Services for Cargo Transportation*, International Series in Operations Research & Management Science 330, https://doi.org/10.1007/978-3-031-07199-7_14

In order to provide quality, efficient and attractive services, many carriers seek to invest in technologies that facilitate their work. Such technologies include:

- Hardware known as GSM modems
- Software solution management and information representation application
- Databases for processing data from the vehicle
- Digital maps with coordinates and route illustration
- Software modules for connecting to vehicles

Such systems offer many advantages to managers of transport companies:

- Allow to remotely monitor locations of vehicles.
- Analyse trip routes.
- Analyse driver's work regime.
- Improve drivers' work discipline: knowing that their working hours are fully controlled, drivers avoid violating work discipline.
- Improve work discipline of managers: knowing that their working hours are recorded and can be checked at any time, they take a more responsible view of the driver's work planning.
- Collect data about cars, which allows to monitor and assess the measures that would help to save.
- Increase vehicle and freight security in case of a theft and robbery.

Also, it is important to remember that advances in identifying radio frequencies (RF-m) eliminate significant shortcomings of former electronic identification technologies—a contact between the identifying person and the identifiable object. Radio frequency tools now allow to identify fast-moving objects (which in some cases even exceed 160 km/h), which opens up new opportunities:

- Collection of taxes at any mandatory stopping of vehicles. Many European countries, the United States and Japan now use credit card-sized radio frequency transponders ("toll labels").
- Identify trucks at the gates of a factory, port, or logistics centre and then use various information signs to indicate the vehicle's path to the designated loading place. This technology has widely been used as a part of the "Just in Time" (JIT) logistics concept at all car assembly plants.
- Identify truck trailers and containers (at ports, railway stations and truck terminals), thus avoiding "lost" trailers and containers and effectively combining the operation of loading equipment and cargo. For example during loading, a heavy container that requires a special semi-trailer for safe carriage can be identified and quickly adapted for a suitable semi-trailer so that the freight can be legally transported within the permitted axle loads.
- Pre-identify trucks at regular stops (border crossings, weighing stations, checkpoints, etc.). Regular stopping for checks is a major concern for road carriers in almost every country. Calculations were made in the US revealing that each minute of downtime (at the border, toll booths, or weighing stations) costs USD1 for a typical five-axle truck with a semi-trailer. RF-I technologies have also been

tested in the US (HELP project) to shorten or eliminate these stops by automatically collecting information from trucks in advance before they enter checkpoints. Identification can be used to check the compliance of vehicles with traffic requirements (payment of registration tax, permissible amount of fuel, permissible axle loads, total weight, etc.); trucks can be dynamically weighed while driving (to determine the correspondence between the gross weight and axial weight limits). If the vehicles passes the check requirements, it can go through the checkpoint without stopping.
- Identify drivers who violate work and rest regimes. The driving time of commercial vehicle drivers is limited in many countries. In some countries, drivers are required to record their driving and parking times themselves in control books, while many cars have built-in tachographs to record that data. Many countries require drivers to have driving licences for a particular type of vehicle. More stringent rules for working hours and driving require time-consuming, manual records, tachograph discs and licence checks. These manual checks can be time-consuming for drivers carrying goods. The use of R-ID technology, which also includes "smart" driver licence cards and electronic tachographs, does not require stopping the vehicle in order to check that the driver complies with the rules regulating work and rest times, and traffic rules.

Over the past decade, automatic vehicle locating (AVL) and long-distance two-way communication technologies have become widely used in carriage operations.

When properly used, such information technology-based equipment (EUR 2000–4000 per truck, plus monthly and message charges) can pay off in less than a year. The following is needed to achieve this:

- Costs and time spent on frequent "control calls" of truck drivers to their dispatchers must be eliminated. Many transport companies require drivers of non-scheduled and long-distance trucks to call their dispatcher at least three times a day (while drivers carrying very important goods may need to call every hour).
- Vehicles must be equipped with automatic vehicle location and two-way communication technology, so that calling a dispatcher is no longer necessary. The dispatcher knows the location of the vehicle, and the driver no longer needs to call and report the distance he has travelled. Vehicle locating systems usually "find" a vehicle by its location once per hour. If the dispatcher has information for the driver, for example about the place of a new shipment, change of delivery time or place, etc.—it can be transmitted to the driver when he is in the vehicle. If the driver needs route instructions, the vehicle is causing problems, or he has ended up in a traffic jam, he can inform his dispatcher about that without leaving the vehicle. Information is usually sent in coded or plain text.
- Reduced freight search time and idle mileage, allowing the dispatcher to better match freight with suitable vehicles. Dispatchers provide faster services, while drivers and their employers can make more money.

- More efficient provision of transport services. Since increasingly more companies require real delivery times to better meet customer requirements, transport companies that know exactly where and when to pick up the customer's freight come out as winners in the business. Vehicle location and two-way communication systems allow trucks to be routed directly to a new destination point for pick-up of express freight. Delivery times can change in order to adapt to production schedules or depending on the place of loading.

Automatic door-to-door delivery of freight, the so-called real-time route, is considered to be the most promising innovation in logistics stemming from advances in information technology. The information technologies discussed above allow:

- To locate freight, vehicles and drivers
- To locate vehicles for long-distance carriage and to ensure two-way communication
- To manage current traffic information, which is a prerequisite for real-time route planning

The ability to quickly and efficiently match shipments to vehicles routes comes as an additional requirement.

The real-time route planning process begins when the consignor warns the carrier about the type of freight (size, weight, special loading requirements, destination, pick-up time, planned delivery time, etc.). Electronic exchange of this information (from the consignor's computer to the carrier's computer) speeds up the execution of routing algorithms in order to properly choose the freight for proper and currently available vehicle. If special permits are required to carry goods, they are obtained electronically from certain regulatory authorities.

After everything has been approved, and a vehicle has been assigned for freight carriage, the driver will be directed to pick up the freight, taking traffic conditions into account. If traffic conditions become worse, the driver will be directed to take a different route, if possible. If he cannot be directed to another route or doing this is inexpedient, the dispatcher may assign the freight to another vehicle or warn the customer of a delay in delivery.

Since the dispatcher knows the distance travelled by the truck travelling throughout its journey, he can plan the following pick-up and inform the driver of the truck thereof. Having received information from the dispatcher in advance, the driver can go to pick up the next shipment.

The information technology innovations discussed above and their perceived benefits depend on the ability to collect, process, store freight and to exchange information (data) in real time to automate operations and/or decision-making.

The combined individual IT innovations become networks and grow into an electronic infrastructure used for production, distribution and transport management. As information flow and processing capabilities are the key feature of the new economic context, these networks first make each individual system accessible to a wide range of geographically dispersed users and, second, such services can be made available simultaneously so that they complement and reinforce each other.

This new opportunity to use IT-based equipment interactively allows to strengthen the market through new innovative solutions. These specific opportunities determine the potential of IT and its economic appeal for the transport sector, logistics and industry in general.

Specific systems that are specifically related to urgent transmission of emergency and other alarms, the most efficient route planning, and other options that help ensure efficient operations have also been offered. The most well-known tracking and control systems include:

- Volvo's *Dynafleet*. The use of this system allows to find out the exact location and status of drivers, trucks and property at any time, and to obtain quality data so that activities can be planned and analyzed in advance. Moreover, *Dynafleet* highlights areas that need to be addressed. It offers seven platform services: (1) the *Dynafleet* Fuel and Environment service allows to easily monitor truck performance parameters and efficiency of drivers. (2) The *Dynafleet Driver Times* service allows you to see the driver's working hours. (3) With *Dynafleet Positioning*, driver and truck data can be displayed directly on maps. (4) The *Dynafleet Messaging* service allows the company and drivers to communicate quickly, easily and economically. (5) The *Dynafleet Vehicle Status* service provides an overview and detailed information on the conditions of heavy transport, regardless of their location. (6) Severe braking and the intervention of the truck's Active Safety system can indicate a difficult traffic situation. Dynafleet Safety helps to identify such situations. (7) *Data Access* service. It allows sharing Volvo's heavy truck data with other fleet management systems. Using the open rFMS standard, regardless of the platform, the Data Access service allows for a flexible and efficient management of the existing fleet (https://www.volvotrucks.lt/lt-lt/services/optimized-business/fleet-management/data-access.html).
- *MobiSafe*. This system is mainly used to protect vehicles, but it can also be used in a wide range of other areas where high levels of protection and safety are required, for example when carrying valuables or in order to protect property or a person in higher risk conditions. The vehicle is equipped with certain equipment to determine the location of the vehicle by satellites in geostationary orbits. The information received from them is sent by means of communication to the central security panel. Twenty-four satellites are positioned so that every car with a MobiSafe unit can be reached by signals transmitted from three satellites.
- *Navisat*. The Navisat ECTS system gives companies an overview of how easily they can better manage their mobile workforce and freight in the supply chain. With this system, companies can locate their transport and cargo at any time, thus providing cost-effective feedback. Navisat ECTS programmes are personalized to meet the operational needs of the organization. This versatility will allow all business units to take advantage of the real-time monitoring and availability of information (https://play.google.com/store/apps/details?id=com.navisat_gps.ectsclient&hl=en_US&gl=US).
- *Tracker*. Tracking equipment is designed to track the location or route of vehicles and other objects in real time.

- *Akis-AVL*. Some AVL systems designed for use in urban areas provide highly accurate information and thus a dispatcher can advise the driver on how to bypass a traffic jam. AKIS-AVL is one of such programmes which can monitor vehicles on the road. AKIS-AVL is an application design to automatically locate and track many vehicles using the Global Positioning System (GPS) and radio or GSM communication. AKIS-AVL performs the following key functions: (1) receives messages from vehicles with GPS data (coordinates, speed, direction, time) and additional information (e.g. sensor readings and alarm signals); (2) shows locations of vehicles on a digital map; (3) saves the travelled route and other data that can later be displayed on a map or table and analyzed; (4) sends messages and commands to cars (e.g. sets GPS data transmission intervals, blocks the engine, etc.); (5) responds to status changes in the car, alarms and other events (http://www.akis.mii.lt/?fuseaction=spec_versijos.browse#avl)
- Other used systems include: Business Navigator, Track Master, Logistic, WebFleet, (https://www.webfleet.com/en_gb/webfleet/fleet-management/vehicle-tracking/); AutoRoute and many others, the main activities of which include installation of such systems and development of the latest innovation.

Safety system eSafety is one of intelligent transport systems used in road transport. These are electronic devices designed to help the driver of the vehicle to avoid the danger by activating brakes when a car is too close to another vehicle or by sending warning signals. Advanced speed adaptation systems, electronic stability control and seat belt reminder control are also frequently used. Another road ITS, the Galileo system, is a satellite navigation system that provides a highly accurate and guaranteed global positioning service. It interacts with GPS (USA) and Glonass (Russia) global navigation satellite systems. Its purpose is the management of European transport infrastructure, and it can also be used in the fields of telecommunications and civil protection.

EGNOS—European Satellite Signalling System. EGNOS provides free enhanced satellite navigation signals across Europe, ten times more accurate than GPS. Such better accuracy can be useful in all areas of applications that use location and speed data: all modes of transport for better infrastructure management and provision of information on ground location, logistics, precision farming, civil protection and crisis management, cartography and land registry, fisheries, energy, natural resource management, mining, natural sciences, meteorology, climate change simulation, environmental protection, justice and law enforcement, border control, etc. Integrity is another advantage offered by EGNOS that is important for civilians. It is a reliability standard that ensures the accuracy of the information provided by the system automatically alerting the user when a system error is outside certain confidence limits. Integrity is important in safety-of-life systems used in vehicles (for all modes of transport, airport management, automated vehicles), in sensitive commercial areas of applications (high-precision locating of oil platforms, logistics, carriage of hazardous goods) or in areas important from responsibility perspective which require legal assistance (recreation after accidents, applying tolls on road users, synchronization of electricity or telecommunications networks).

Europe has a share of the global GNSS consumer market. The current EGNOS has improved the infrastructure together with GPS and later also with GALILEO to create a global market for GNSS products and services known as consumer market.

First of all, this market includes products and services related to basic locating and time setting signals, but the authentication and coding of signals planned to be installed should also bring benefit. Experts estimate that products and services related to mobile telecommunications and personal mobile equipment will account for a total of 75% (52% of revenue) of this market, intelligent transport systems will account for 20% (44% of revenue), and other areas of application will make up the remaining 5% (4% of revenue).

Road transport is the second-largest GNSS market (accounting for around 20% of the global market for GNSS products and services). ITS for road transport can lead to greater use of GNSS services. For example the use of GNSS is already recommended in the European Directive on the interoperability of electronic road toll systems or in the Regulation on the conditions for the transport of live animals. The European Commission's ITS Action Plan and the accompanying Directive discuss ways to promote the provision of GNSS services through specific ITS (emergency call system eCall, tolls, truck parking lots, etc.). The Logistics Action Plan is another option for the development of GNSS services to track containers in real time or to engage in theft prevention. However, the ITS sector is very poorly informed about EGNOS.

A distinctive feature of the GALILEO architecture is its integration with the current EGNOS satellite radio navigation system, the first European satellite navigation project. EGNOS is a joint project of the European Space Agency (ESA), the European Commission (EC) and the European Organization for the Safety of Air Navigation. The EGNOS system consists of three satellites in geostationary orbits and a network of ground-based control stations. The purpose of the EGNOS system is to complement the GPS and GLONASS systems to meet the requirements for air and sea navigation.

Despite European investment in its own GNSS infrastructure and possibility to use EGNOS, European industry has only a small share of the global GNSS application market compared to other high-tech sectors (almost a third of the market), which is a problem because:

- The use of EGNOS and GALLILEO would have a decisive impact on the development of knowledge society and the creation of high value-added jobs in the European Union. Europe would miss a very good opportunity to reap some economic benefits from GNSS applications. Moreover, if GALILEO and EGNOS do not become the core GNSS standard in Europe, the development of many areas of applications will be limited by technologies that do not allow to reap added value from new advanced services.
- Limited use of EGNOS and GALILEO systems leads to a high degree of dependence, as GNSS is highly permeable, allowing to obtain information on position, navigation and timing which is very important for many day-to-day activities, European security and socio-economic development. By relying solely

on GPS, the European Union may face a problem where there is no GPS signal, and the European Union cannot control this, as the primary GPS function is to assist in military operations in a third country.

These new circumstances related to European Union's GNSS programmes require a comprehensive action plan to increase people's confidence in the programmes, to promote the development of areas of application of EGNOS and GALILEO and to pursue the fastest, deepest and widest possible development of these services in all areas of application to ensure the maximum benefit of the infrastructure in the European Union.

The comparison of the price of development of the GALILEO system with the estimates for the construction of motorways or railways allows for concluding that it is not an expensive project. The GALILEO system costs approximately the same as building a 150 km section of highway through a moderately built-up area or installing one railway line on the high-speed rail line between Lyon and Turin designed in the main tunnel.

The use of road transport ITS, including advanced driver-assistance systems (ADAS), will be based on the GALILEO integrity and authentication functions. This will require the establishment of a certification body and certification procedures, which will be developed in cooperation with vehicle manufacturers, electronics suppliers and road traffic service operators.

Intelligent Driver-Assistance Systems (ADAS) can provide personal assistance on the road, but it does not always take into account certain limitations and the abilities of an older driver. An analysis of the strengths and weaknesses of older drivers shows that the main need for assistance stems from the difficulties faced by older drivers:

- When deciding whether offer traffic, participants are approaching the same intersection and at what speed.
- Noticing other traffic participants when connecting and changing lanes.
- Paying attention to traffic signs and signals.
- Promptly responding in a complicated traffic situation (https://ec.europa.eu/transport/road_safety/specialist/knowledge/old/what_can_be_done_about_it/adas_en).

Elderly drivers experience such difficulties due to their functional limitations such as worse movement perception, peripheral vision, head and neck flexibility, selective attention, speed of information processing, and decision-making. ADAS, which can compensate for these limitations, can help to reduce accident rates of elderly drivers. Such ADAS should have one or more of the following features (Davidse, 2006; https://ec.europa.eu/transport/road_safety/specialist/knowledge/old/what_can_be_done_about_it/adas_en):

- Notice approaching vehicles
- Notify about other traffic participants when the driver is in a blind spot
- Help the driver to focus on important information and/or
- Provide prior information about another traffic situation (Davidse, 2006)

ADAS with these features could improve the safety of elderly drivers. Using ADAS to improve driver safety or mobility should not only ensure that the additional task supported is performed safely. It also means that the assistance provided does not have any negative impact on other elements of the driving task. Examples of adverse side effects include increased workload due to poor human–machine interface design and behavioural adaptation effects. ADAS could improve the safety of elderly drivers on the road.

Several studies referred to ADAS as being able to provide tailored assistance to elderly drivers (Mitchell & Suen, 1997; Shaheen & Niemeier, 2001; Färber, 2000).

There are many aspects of road transport activities that are related not only to transport work planning but also to procedures, i.e. customs document processing, accounting data processing and analysis of working time of drivers. The handling of customs documentation is a rather complex matter which requires very careful inspection and coordination. Customs declaration application MDG2 has been used most widely by customs and carriers. This programme allows to generate the following documents electronically: import–export declarations; CMR consignment notes; specification sheets; bills of lading. MUITINĖ (English: CUSTOMS) is one of the most developed customs applications, which allows to complete the main documents related to export, import, transit, CARNET TIR and other documents necessary for freight reports and border crossing procedures.

The TachoScan programme is designated for the analysis of working time data of drivers, the main task of which is to examine the tachograph discs. Thus, this programme allows to examine information about overtime, to analyze damaged floppy disks and also to record relevant data related to driver rest or to mark the cause of a violation of the driver's working time regime.

In road transport, the benefits of IT innovation stem from the improved use of electronic data and the exchange of information in real time. These data exchanges will require much more than the so-called Electronic Data Interchange (EDI), a term used for the electronic transmission of order and dispatch documents to manufacturers and dispatchers, transport companies and recipients. Information technology innovations also allow to:

- Identify freight, vehicles and drivers
- Locate vehicles
- Choose the optimal route, taking into account the availability of freight and proper equipment for its carriage, traffic and weather conditions, the type of the carried freight
- Anticipate the optimal route of freight carriage (a part or the entire consignment); the optimal route of unit shipments (semi-trailers, containers, swap bodies, etc.)
- Provide for the optimal use of labour

Implementation of the latest IT technologies in road transport reduces the number of accidents. The following technologies help to prevent this:

- Systems that do not allow the driver to fall asleep. One of the main causes of accidents, which is especially common among professional drivers, is falling asleep at the wheel. A system has been developed to prevent drivers from falling asleep. The essence of the system is very simple: to detect when the driver starts falling asleep and to wake him up by activating an annoying buzzing signal. Sensors and video cameras monitoring vibrations of the driver's eye lashes and his head position have been used to this end. At the same time, the driving position of the car and the force which the driver uses to press the accelerator pedal have been analyzed. The analysis of all these algorithms allows the system to understand that the driver is falling asleep.

- Adaptive cruise control system (ACC). The system operates by setting the desired speed and distance from cars driving in front. Then the radar sensor of the system adjusts the driving speed in each traffic situation and ensures that the car always maintains the required distance.
- Rear-end-collision warning system. A millimetre-wave radar mounted in the rear bumper continuously scans the area behind the back of the car and determines the likelihood of a collision. Having determined a high likelihood of collision, the system activates a rear emergency alarm. In case of an inevitable collision, the system automatically pushes the front seat head restraints forward upwards to reduce neck injuries, compensating for the impact.
- Emergency pedestrian protection system. A video stereo camera mounted behind the windshield captures people and animals on the road in front of the vehicle and applies brakes. Such short-term braking allows to draw the driver's attention, while the braking system prepares the brake booster and the variable gear ratio steering system if a sudden collision avoidance manoeuvre needs to be made.
- Lane-keeping assist system. This system helps drivers to stay in the same lane while on highway. Depending on the weather, climate and road conditions, the system monitors the white road line markings via a video stereo camera and does not allow the driver to get out of the lane he is in.

An automated emergency call system is another alternative to ensure safe travelling. Having recorded a strong impact, the eCall device automatically dials the single European emergency number 112 and informs emergency services of the approximate location of the vehicle. Number can also be dialled by pressing the button. In both cases, the data are transmitted not only automatically but also via a voice connection between the car and the emergency centre. This way, the driver and passengers can provide more information about the accident if they are able to answer questions. Having installed the system throughout Europe, it is expected to save 2500 lives annually and reduce serious injuries by 15% at the least. Automatic emergency call devices would only be installed in new cars, as the European Union does not require them to be installed in existing vehicles.

Much is currently being done to provide drivers with better information on current traffic conditions. Although almost all the focus on "road transport informatics" (Europe) or "intelligent vehicle/road systems" (US) is focused to decide on information that car drivers need, drivers of commercial vehicles are likely to benefit from these technologies first of all, and they will be the first ones to invest in them.

Accurate information on current traffic conditions can reduce costs which road carriers incur due to congestion. Dispatchers can use this information to avoid sending trucks to major traffic jam sites or to redirect them to another route.

Companies with automatic locating function and two-way communication systems will know where their trucks are (i.e., whether they are driving towards a traffic jam) and can warn their drivers to choose another route.

Thus, in summary, the advantages of all systems are clear—to facilitate, speed up and ensure safe work in road freight transport. Unfortunately, these information systems are not interoperable.

14.2 Freight and Transport Platforms

14.2.1 Principle of Operation of the Freight and Transport Platform

Virtual enterprises are one of the new forms of enterprises. This way, online freight and transport platforms/exchanges (hereinafter platforms) have emerged over the last 10 years, which are currently responsible for the majority of road freight flows.

The freight and transport platform is an online exchange that meets the needs of carriers and freight forwarders. Figure 14.1 illustrates the principle of its operation.

The freight and transport platform has been designed to resolve daily problems when, for example:

1. There is a need to urgently carry freight from Kaunas to Frankfurt, but there is no returning freight. Of course, going back empty and doing nothing is one option, but there also is a possibility to increase the company's profitability and find freight for return on the freight and transport platform.
2. There is an urgent need to carry freight from Vilnius to Moscow, but there is no available transport. The freight and transport platform allows to easily and quickly offer your freight and find transport for it.

Advantages offered by the freight and transport platform:

- Optimization of planning processes
- Maximum use of the company's capacities

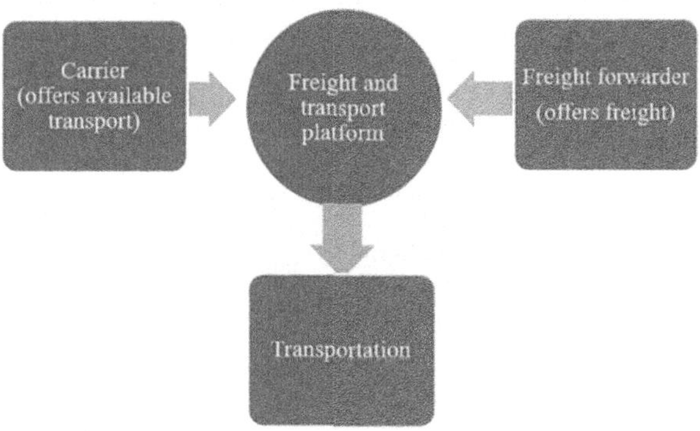

Fig. 14.1 Principle of operation of freight and transport platform

- Reduction of empty miles
- Increasing profitability
- Business advantages over competition
- New customers and business partners

Those working in the logistics sector are familiar with freight and transport platforms, which can be divided by regions, for example Europe, Asia, the USA, etc.

At first glance, the principle of operation of each of them may seem to be the same, with one party posting a transport or a freight offer and the other party accepting it.

However, most companies use at least two platforms, as each of them offers different value added to its users.

14.2.2 European Freight and Transport Platforms

Since there are many transport platforms in Europe, this section will examine the ten most popular transport platforms in Europe used by most companies on the continent. Tables 14.1, 14.2, 14.3, 14.4, 14.5, 14.6, 14.7 and 14.8 below present a detailed

Table 14.1 Romanian freight and transport platform

Platform	Platform description
123 Cargo	The platform was established in 2001 and is based in Romania. The system offers fast, simple and reliable solutions. The monthly platform fee posted on their website is EUR 50/month with a 6-month subscription or EUR 40/month with all-year subscription. Both of these packages include the possibility to submit and review offers of trucks and trailers, an unlimited number of users, a list of all exchange users and their contact details, a legal advice on non-payment for services or on other matters during transportation, and the possibility of correspondence with other users.

Table 14.2 Czech freight and transport platform

Platform	Platform description	Service packages
CargoCore	The platform was stablished in 2011 in the Czech Republic. This platform is small—it has a few employees only, who administer all the activities of the platform. The monthly platform fee may vary and depends on the plan chosen by the company.	Free 1-month trial period, when the company is allowed to test the platform, but access is restricted—users can only view and submit their truck (trailer) offers;
		The plan costs EUR 19/month. Having chosen the plan, companies can view and submit their offers. They can have three platform users, access the entire database on the platform, and can export its offers directly to the platform from an existing system, without having to enter any information manually;
		The best plan costs EUR 49. Having chosen the plan, companies can view and submit their offers, have an unlimited number of users, get access to the entire database on the platform, have the opportunity to export its offers directly to the platform from the system already in use, use the debt recovery service and have the company's logo displayed on the platform page as a partner, which is additional advertising.

Table 14.3 Freight and transport platforms of Great Britain

Platform	Platform description	Service packages
Euroloads.net	The platform is free, and any company can register on it. Offers can be placed on the platform for freight and transport across Europe, using an interactive map. The company is headquartered in the Great Britain. This platform does not offer different plans or accesses—they are the same for everyone.	
Haulage Exchange	Transport platform based in Great Britain where companies from all over Europe can offer their freight and transport.	The platform offers the following for carriers: • a package for EUR 114.73/month, which includes 5 users, up to 20 trailers, up to 20 registered drivers, up to 10 subcontractors, 2 mobile app logins and correspondence, messaging, vehicle tracking, support on the phone and other accessories; • a package for EUR 154.13/month, which includes 10 users, up to 50 trailers, up to 50 registered drivers, up to 20 subcontractors, and other services as those offered with a cheaper package; • a package for EUR 160.51/month, which includes 15 users and other services as those offered with the previous package.
		Freight forwarders are offered the following options: • a package for EUR 115.89/month, which includes three users, a map function, trip directions, messaging, on-the-phone support, but all this is available in Great Britain domestically only; • a package for EUR 154.13/month, which includes six users, while all other services are the same as those offered with the cheapest plan. The only difference is that offers are from all over Europe; • a package for EUR 192.39/month, which not only allows to see orders, but also offers a possibility to post own offers.
Return loads	A British transport platform which offers freight from all over Europe. This platform offers its users four access points, making an unlimited number of offers, real-time vehicle tracking, mobile app, interactive map, feedback on other users,	

(continued)

Table 14.3 (continued)

Platform	Platform description	Service packages
	possibility to provide documents of own company on the profile, integration of telematics systems with the platform and user support. There are two payment options—yearly or monthly. Having chosen a monthly payment, the price is EUR 46.30/month, while a yearly payment is EUR 502.32, which is equivalent to EUR 41.86/month.	

Table 14.4 German freight and transport platform

Platform	Platform description
TimoCom	TimoCom has added a new communication feature Messenger, which is not the hottest news for the German capital exchange and is nothing new for platforms themselves, but allows chatting on offer-related matters. This platform has also been actively integrating with telematics systems. According to information posted on their website, there are more than 200 telematics providers now integrated with their platform to make it easier for customers to track their car fleet. Moreover, TimoCom created a free platform to find out about bans on heavy vehicles on European roads. Also, the exchange known as the largest transport platform in Europe has expanded its functionality by adding a document upload and download function. It is the largest transport platform in Europe, headquartered in Düsseldorf, Germany. The company was founded in 1997. The price is the same for everyone a package costs EUR 249.90/month and—there is a one-time administration fee of EUR 250. For this price, users receive three accesses to the platform, access to the mobile app, unlimited number of offers, a possibility to take part in announced transport tenders, access to the entire database, an interactive map, vehicle tracking, integration of telematics systems, opportunity to store documents, collect debts, search for or submit warehousing offers, inspection of companies before granting to them access to the platform or their inspection at a customer's request, and help to customers.

Table 14.5 French freight and transport platform

Platform	Platform description	Service packages
Express-online	The platform was set up in France to reduce CO_2 emissions and increase the efficiency of trailers. The platform offers several plans grouped by the type of activity of registering companies rather than by the advantages offered by packages.	The price of the package offered to freight owners is EUR 25/month with a 24-month agreement. The programme includes the opportunity to submit a freight offer, review the existing trailer offers, and carrier evaluation.
		The price of the package offered to the carriers is EUR 20/month with a 24-month agreement. The package includes vehicle registration, the opportunity to submit offers, to review freight offers and access the list of platform users.
		There is a third package for freight forwarders for EUR 26/month, which includes all the options listed above. Also, there is an option to purchase an opportunity to hide information about your company in the directory for an extra charge of EUR 10/month.

Table 14.6 Spanish freight and transport platform

Platform	Platform description	Service packages
Wtransnet	The platform was established in Spain in 1996. Currently, it is one of the largest transport platforms in Europe offering as many as four service packages to its customers.	A free demo version where customers will be able to get acquainted with the services provided.
		A package for individual carriers for EUR 584/year (EUR 48.67/month), which includes the opportunity to offer a vehicle, search for freight offers, use the platform's customer database, write reviews about other users and receive recommendations, conclude long-term transactions, access to the mobile app, a convenient communication platform, and support on the phone.
		A package for transport companies for EUR 977/year (EUR 81.42/month), which includes two users, unlimited transport offers, freight reviews and other benefits, which also come with the cheaper package.
		A package for logistics companies for EUR 1462/year (EUR 121.83/month), which includes the opportunity to offer freight and search for vehicle offers, enter into long-term contracts, use the platform's customer database, write reviews about other users and get recommendations.

Table 14.7 Polish freight and transport platform

Platform	Platform description
Trans.eu	The transport platform headquartered in Poland was established in 2004. The platform offers one payment plan only—an annual subscription for EUR 1068, which is EUR 89/month. For this price users receive an unlimited number of offers and their reviews, an inspection of companies, a communication platform, a possibility to leave feedback about other users and their solvency indicators, debt recovery and other tools needed for everyday work that have not been identified. While the local exchange Cargo.lt is expanding abroad, the constantly evolving European platform Trans.eu offers a tool for the internal market, encouraging users of the exchange to work not only on international routes, but also in closed groups in the local market. Moreover, last year, the Trans.eu system invested in technology and offered many exciting innovations, including two free apps for smartphones TransExpress and TransParking. The first is designated for searching for goods throughout Europe, and the second one is for searching for safe parking spaces for truck drivers.

Table 14.8 Lithuanian freight and transport platform

Platform	Platform description
Cargo.lt	The transport platform was established in 2000 in Lithuania. The platform offers such services as truck and trailer offers and their review, a directory of companies, an interactive map, ads, feedback on other users of the platform, user assistance, publication of and participation in auctions, interactive field of correspondence with other users. As freight and transport platforms compete with each other, Cargo.lt took a big step forward. Companies that have accounts on this local platform have noticed many Ukrainians offering transport and freight. Also, Cargo.lt subsidiary OOO "Фиронет" ("Fironet") has developed in cooperation with the Ukrainian exchange Lardi-Trans a tool that allows posting offers on both platforms at once. All Ukrainian companies need to get access to the Cargo.lt platform (which is currently offered to Ukrainians for free), or Lithuanian companies need to register on Lardi-trans, and working on both platforms at once becomes possible.

analysis of freight and transport platforms which is based on publicly available information on freight and transport platforms (Kolosova, 2019; Čižiūnienė et al., 2021).

14.2.3 Asian and North American Freight and Transport Platforms

Thus, transportation platforms are used in both Europe and North America and Asia. Although Europeans have not heard and do not know much about them, it is very important to discuss what services are offered by the developers of this software and how they differ from European platforms. Transport platforms (Čižiūnienė et al., 2020) available in North America (Table 14.9) and Asia will be discussed below (Table 14.10):

The examples provided show that freight and transport platforms operate in Europe, Asia, North America, and that the services they provide are very similar to those offered by European platforms. However, we also see that two out of the seven identified platforms operate on several continents at once, thus covering a larger consumer market. The search for transport platforms in the North American and Asian regions revealed that there are very few of them, and it is quite difficult to find such platforms as the most common freight and transport platforms operate in Europe. It was also observed that the platforms of these markets do not provide prices—only one out of six platforms had a price list, with European market platforms openly share prices of their services and offer several different payment plans. Five out of the seven platforms offer an opportunity to write a review or rate partners, and only a half of the discussed European platforms have this feature.

Table 14.9 North American transport platforms

Platform	Platform description
Supply Chain Wizzard	The platform established by Dr. Evren Ozkaya believing that people now live in phenomenal times when the pace of deployment of technological innovation is matched by exponentially increasing data availability at its highest level of complexity and volatility. This speed offers significant opportunities for global business leaders to take action to change their end result for the better, thus improving lives of billions at lower prices of goods and/or services with greater availability. The platform is more aimed at manufacturers (sellers), as the main emphasis is on the safety of goods, the possibility to monitor their travelling and the conditions under which the product is carried, also monitoring if its storage conditions have been maintained. Customers can monitor their freight 24/7 both when carrying it by truck and when storing it in a warehouse. There is no information on the website of this platform whether the freight and transport exchange service is offered (www.supplychainwizzard.com).
CX North America	The transport platform was established in 2000 and is mainly available in the United States and Canada. This platform is intended for carriers, manufacturers or sellers, and freight forwarders. The platform offers real-time monitoring of carriers and freight routes, freight delays in real time and allows to get alerts on possible delays; the system can calculate the estimated time of arrival itself, automatically sending delivery confirmation documents after each shipment and generating invoices, which significantly reduces the work of the manager in issuing invoices and uploading the necessary documents to the system. Also, having selected a particular carrier each time, the manager can see the history of its shipments, contact other companies that have collaborated with the carrier, and get recommendations. Moreover, each user can have a circle of its own carriers, freight forwarders or manufacturers, thus creating a sense of security and reliability on the platform (www.cxnamerica.com).
Uship	The platform was established in 2004 in the USA. The company has 17 branches worldwide, mainly in North and South America. Carriers, manufacturers and freight forwarders can work on this platform, and choosing the right profile is the key. The platform emphasizes the importance of how companies present themselves in their system, as the first impression can determine whether the carrier will receive the order or vice versa, whether the manufacturer will look reliable enough and whether its freight will be transported quickly and safely to the right destination point. Therefore, the platform allows completing your data, providing a detailed description of the company, uploading a photo or a company logo, documents confirming insurance and company registration. The platform also offers a mobile app that helps to effectively manage orders. The system allows companies to pinpoint their location or share their vehicle locations for customers to be able to decide whether or not this partner is suitable for them. There is also a possibility to take part in transport tenders, to submit own price offers and rate partners. Moreover, the platform's website states that they ensure that the services provided are paid for, which creates a greater sense of security on the platform. In order to avoid overpayment for the service received, the customer can check how much similar shipments have cost recently and assess whether the amount is appropriate (www.uship.com).

Table 14.10 Asian transport platforms

Platform	Platform description
Freightos	The transport platform was established in 2012. It is headquartered in China, while its branches serve customers in the USA and Great Britain. The platform offers services to all actors in the supply chain worldwide. This platform also presents all possible carriage options, including carriage by trucks, ships or air transport—the customer needs to choose the most suitable option for him. This is easy to do, as the choices include both the price for the service and the estimated transportation time, which determines the customer's choice. The platform allows giving feedback about partners and rating service quality. The platform allows communicating with each other, tracking freight in real time, has a customer service centre open 24 h a day, 6 days per week which can answer user questions. The price per user indicated on the website of this platform is EUR 44.07/month (www.freightos.com).
Ezyhaul	The transport platform is headquartered in Singapore with branches in Malaysia, the Kingdom of Thailand and India. The platform mainly focuses on the East Asian region and offers trucks of various sizes and features as the main mode of transport. This platform is available to all participants in the supply chain. Their website states that the platform ensures competitive prices, all carriers are inspected by platform staff, each carrier has insurance, there are no paper documents—all information is submitted electronically, freight or trucks can be tracked in real time, and there is a 24-h customer service centre, which can be contacted for any questions of concern (www.ezyhaul.com).
AvtoTransInfo	The transport platform operates in Russia, and it is the largest platform in this country established in 1996. The platform offers its customers the opportunity to post and view offers for transport or freight, also assistance in disputes between partners, user rating, their so-called traffic light, which allows checking potential partner company on as many as 25 different websites, a forum which posts information about fraudulent users, and an opportunity to purchase insurance online on the platform itself. The platform's website states that the platform does not distinguish any customers, so both carriers and freight owners find out about new offers as soon as they are posted on the platform, also offering a possibility to publish all documents online, to receive and issue invoices on the platform, which saves managers time and cuts costs. This platform also has its own mobile app, which allows to use the platform anywhere. One of their advertised benefits is the ability to create a circle of well-known partners and be notified immediately when a truck or a vacant vehicle arrives. The platform has a customer service centre that can answer any questions (www.ati.ru).
20tonn.kz	The transport platform was established in 2016 in Kazakhstan. It is the first platform of its kind in the country, which should help the country to modernize its transport system and reduce shadow economy in this area according to information posted on the web portal www.kapital.kz. The platform offers a user rating system, a cargo tracking capability, assures that the customer can get an offer with the right price within 10 min, offers a possibility for each customer to view freight history and its transportation prices, also posting notifications about newly registered freight. This platform has developed its own mobile app that runs on all operating systems, also offering an interactive map that shows road conditions, truck movement and can calculate the necessary route. The platform connects logistics companies from Kazakhstan, Russia, Belarus and Kyrgyzstan (www.20tonn.kz).

The analysis of freight and transport platforms revealed that finding these platforms in Asia is much easier than in North America, and there are even more of them in Europe—almost every European country has its own transport platform, which reveals that transport systems are much better developed in this market, while the transport market itself has been growing and improving. Therefore interconnection between these markets is necessary.

Thus, an electronic freight and transport platform really facilitates the work of logistics market participants and promotes operational efficiency. The following common advantages of electronic freight and transport platforms can be distinguished: a rapid exchange of relevant information; internationality; ability to monitor the market and changes in it; opportunity to find new customers and partners; opportunity to negotiate and save on freight carriage (Batarlienė & Jarašūnienė, 2020).

14.3 Conclusions

Road transport is the most common mode of transport, with endless information systems (from transport safety to satellite tracking). The need for these systems is constantly growing. Therefore, in today's market, a company needs to use them to gain a competitive advantage, especially for carriers that are subject to high demands.

The use and integrity of IT in road transport allows for a faster execution of the entire transportation process: routing, toll collection, vehicle/container location identification, and ensuring the inviolability of drivers' work and rest, and so on.

Specific systems are also proposed that deal specifically with the rapid transmission of alarms, the planning of the most efficient route, and other options to ensure efficient operation. The best-known tracking and tracing systems include Volvo Dynafleet, MobiSafe, Navisat, Navisat, Akis-AVL, Business Navigator, Track Master, Logistic, WebFleet, AutoRoute, and more.

There are also systems to ensure the safe running of the transport process, as they prevent the driver from falling asleep, adaptively adjust speed and distance, protect against rear-end collisions, provide pedestrian and animal protection, and so on.

There are a number of systems for managing freight flows as well. One of these is freight and transport platforms, which are widespread around the world. However, the analysis of such platforms has shown that they are segmented by markets such as Europe, Asia, North America and so on.

References

Batarlienė, N., & Jarašūnienė, A. (2020). *Intelektinės technologijos transporte*. Technika.

Čižiūnienė, K., Kolosova, M., Gomienė, Ž., & Burinskienė, L. (2020). Azijos ir Šiaurės Amerikos transporto platformų analizė. *Darnios aplinkos vystymas, 17*(1), 25–31. Klaipėdos valstybinė kolegija. ISSN 2538-9017.

Čižiūnienė, K., Kolosova, M., & Liebuvienė, J. (2021). *Europos transporto platformų vertinimas. Studijos kintančioje verslo aplinkoje: straipsnių rinkinys* (pp. 104–109). Lietuvos ekonomikos dėstytojų asociacija. ISSN 2029-2805.

Coronado, A. E., Lalwani, C. S., Coronado, E. S., & Coronado, C. E. (2009). Facilitating multimodal logistics and enabling information systems connectivity through wireless vehicular networks. *International Journal of Production Economics, 122*(1), 229–240.

Davidse, R. J. (2006). Older drivers and ADAS – Which systems improve road safety? *IATSS Research, 30*(1), 6–20.

Davies, I., Mason, R., & Lalwani, C. (2007). Assessing the impact of ICT on UK general haulage companies. *International Journal of Production Economics, 106*(1), 12–27.

DfT. (2006). *Freight best practice guide: Information technology for efficient road freight operations*. Accessed July 23, 2021, from http://www.freightbestpractice.org.uk/information-technology-guide-for-efficient-road-freight-operations

Färber, B. (2000). Neue Fahrzeugtechnologien zur Unterstützung der Mobilität Älterer. *Zeitschrift für Gerontologie und Geriatrie, 33*, 178–185.

Giannopoulos, G. A. (2004). The application of information and communication technologies in transport. *European Journal of Operational Research, 152*(2), 302–320.

Kolosova, M. (2019). *Transporto platformos plėtros galimybės Azijos rinkoje*. Bakalauro baigiamasis darbas. Vilnius, VGTU.

Marchet, G., Perego, A., & Perotti, S. (2009). An exploratory study of ICT adoption in the Italian freight transportation industry. *International Journal of Physical Distribution and Logistics Management, 39*(9), 785–812.

Mitchell, C. G. B., & Suen, S. L. (1997, June 16–20). *ITS impact on elderly drivers*. Proceedings of the 13th International Road Federation IRF World Meeting, Toronto, ON, Canada.

Official transport platform 20tonn.kz website. [interactive] (2018). Accessed December 29, 2018, from www.20tonn.kz

Official transport platform AvtoTransInfo website. [interactive] (2018). Accessed December 29, 2018, from www.ati.ru

Official transport platform CZ North America website. [interactive] (2018). Accessed December 30, 2018, from www.cxnamerica.com

Official transport platform Ezyhaul website. [interactive] (2018). Accessed December 30, 2018, from www.ezyhaul.com

Official transport platform Freightos website. [interactive] (2018). Accessed December 30, 2018, from www.freightos.com

Official transport platform Supply Chain Wizzard website [interactive] (2018). Accessed December 30, 2018, from www.supplychainwizzard.com

Official transport platform Uship website. [interactive] (2018). Accessed December 29, 2018, from www.uship.com

Perego, A., Perotti, S., & Mangiaracina, R. (2011). ICT for logistics and freight transportation: A literature review and research agenda. *International Journal of Physical Distribution and Logistics Management, 41*(5), 457–483.

Shaheen, S. A., & Niemeier, D. A. (2001). Integrating vehicle design and human factors: Minimizing elderly driving constraints. *Transportation Research Part C: Emerging Technologies, 9*(3), 155–174.

Chapter 15
Information Technology Used in the Railway Transport Sector

Kristina Čižiūnienė

Abstract Rail transport is used to carry mass freight over long distances and regular shipments. In order to ensure good service quality and avoid obstacles, rail transport, just like other modes of transport, uses information technology. Given the breadth of the spectrum for rail transport, it contains a wide range of information systems, from management to technical safety, optimization of the process to be organized with various systems and technologies, and so on. This section will address issues related to rail transport and the IT and telematics systems used in it.

Keywords Railway · IT · e-Freight · ERTMS

15.1 IT for Rail Transport

Rail transport is a mode of transport that offers many opportunities and has a high potential for development. The RailBaltica project in the Baltics only shows that connection between the west of the European Union, the Baltic States and Scandinavia is a really important step towards efficient transport. The comparison of rail transport with road transport allows concluding that rail transport is safer, much less harmful to the environment, more efficient than road transport, but it requires huge investments in transport infrastructure.

The use of GIS (Geographic Information System) in rail transport is useful in helping to evaluate the environment: new designs, growing number of high-speed trains, the flow of carried freight and goods. The Document Management System (DMS) allows railway directorates, departments, and services to exchange electronic documents instead of paper ones. All this allows grouping users into groups, which usually overlap with structural divisions of the organization. Thus, depending on the rights granted to different participants and groups, employees can:

K. Čižiūnienė (✉)
Vilnius Gediminas Technical University, Vilnius, Lithuania
e-mail: kristina.ciziuniene@vilniustech.lt

D. Dzemydienė et al., *Development of Smart Context-Aware Services for Cargo Transportation*, International Series in Operations Research & Management Science 330, https://doi.org/10.1007/978-3-031-07199-7_15

- Create new documents.
- Read an existing document.
- Modify an existing document.
- And delete an existing document. It is worth noting that all documents are stored working on the intranet.

There are many information systems used in rail transport. They are all linked to alignment and efficiency of passenger and freight transport:

- Waggon transfer station computer information system "VPS KIS" is used to manage waggon fleet. Wagons are accepted and sent using the system, also transferring information when crossing borders of the European Union. The right to use wagons and continue their transportation is handed over to the administration of another country according to the approved procedure. Information system of the distribution station is used to arrange work in stations and to manage the technological process, to increase its effectiveness and efficiency and to reduce labour costs. This can be achieved because of the system connecting the entire technological process of specialized workplaces.
- The information system "FREIGHT" is an information system for the management and accounting control of documents accompanying the transportation of goods by railways of the Republic of Lithuania. The system is designed to collect data on freight transported by rail into a single central database and to ensure prompt retrieval and printing of the necessary information from it at all workplaces. The system also connects the entire freight carriage cycle—from concluding a contract with the customer to controlling the issue and payment of VAT invoices for provided services. The development of the "Freight" system was started in 2004, and today it is the core system of the Freight Carriage Directorate of AB Lietuvos Geležinkeliai together with the portal for customers "e-Freight". It includes more than 15 interconnected modules that automate a variety of processes: from ordering a freight carriage to customers or locating a freight. Another important step was taken in 2014—this is when an electronic services portal was developed. It was designed with customer needs in mind and focused on faster and more convenient work in managing freight orders. The one-stop-shop principle has been implemented in the customer portal, which allows to access and manage all the necessary information.
- "e-Freight" is an information system used to ensure control over the management and accounting of documents accompanying freight transported by rail in the Republic of Lithuania. According to Jarašūnienė et al. (2019), the e-Freight information system to ensure that users can collect and collate the collected data on the transported goods on a one-stop-shop basis, obtain the necessary information and, if necessary, print it out from all workstations. This IS allows users: (1) to coordinate process plans; (2) to simplify the ordering of wagons; (3) complete in and submit CIM (Rail Consignment Note) at the railway station; (4) fill in customs declarations; (5) to adjust documents with the railway station; and so on. The system "e-Freight" has five electronic services installed: (1) Backup of freight services: possibility to coordinate transport plan, order

services, order rolling stock, submit an application for the organization of transport, and track the long-term operative balance of programmes. (2) Tracking the progress of rail freight carriage process: a possibility to obtain information about the movement of goods by rail, also about the opening/driving of wagons to access roads in real time, and to register electronic documents. (3) Automated preparation and submission of rail freight documents to the service recipient: possibility to prepare and submit declarations, bills of lading, review additional service documents (KR-34, KR-12, KR-53, KR-142), provide other document necessary to initiate and execute rail freight carriage services and documents required for enforcement (KR-28, KR-29, KR-30). (4) Automated preparation of financial documents for rail freight carriage and their submission to the service recipient: a possibility to review normative, contract, and invoice information. (5) Automated declaration preparation and submission to customs: a possibility to generate NCTS declarations by automated means from the company's information systems, and to submit and view transit declarations.

Advantages offered by E-freight: all information in one place; electronic signature; one-sheet concept; suitable for most web browsers.

- OPKIS (Operational Computer Information System for Carriage) is a system that allows exchanging information between the main stations of the Baltic States and the railway information centres of the CIS countries, and collecting information on freight train composition, traffic, assembling, disassembling, arrival and departure of rolling stock, crossing of border stations and all data on the waggon fleet. The OPKIS system is an invisible guarantor of the freight carriage process. Without this system, train running on Lithuanian railways is unimaginable. The system accounts time of own and leased as well as foreign railway wagons used in Lithuania. These data are used to generate financial settlements between railways for the use of wagons. The OPKIS system is integrated with national systems of administrations of other countries that perform a similar function. The exchange of data between the railway administrations of other countries is essential for the international carriage of goods. An OPKIS system failure would halt the process of carriage of freight by rail and cause significant financial losses to companies. OPKIS ensures mutual information communication with the following systems: the waggon transfer station computer information system (VPS KIS); the station computer information system (STOKIS); Radviliškis and Vaidotai station computer information system (KIS–Rd, KIS-Vd); Draugystė station system; Bugenių station system; database management system of freight wagons of the CIS, Lithuania, Latvia, and Estonia SOSNAG; database management system of containers of the CIS, Lithuania, Latvia and Estonia KDVKONT; ASOUP system of Kaliningrad, Belarusian, and Latvian railways; the system used by the Operational Statistics Division of the Freight Transport Board to receive train sheet information by regulation from OPKIS and to transmit it to OPKIS; the system "Loading—unloading"; IS "Cargo". In summary, the exchange of OPKIS information can be said to take place with the help of information messages, inquiries, and certificates (Zabielienė, 2020).

- X-ray control system designed to check goods carried by rail without unloading them and to identify possible cases of smuggling is another system used in rail transport. Ionizing radiation on a computer monitor forms an image of the object being inspected, which can be stored in computer memory. By using this programme specifically, for example at the Kena customs checkpoint, customs officials believe that this system will help to detect illegally carried goods more effectively and increase the security of the international trade chain.

15.2 System of Traffic Management of European Rail (ERTMS)

The European Commission has also contributed to the development of all these information systems, while the Single European Transport Area Roadmap provides for funding for research that would help to develop intelligent mobility systems. Implementation of such systems as ERTMS, railway IS, ITS, as well as for newly deployed multimodal flow management and information systems (including charging), have been planned (Zabielienė, 2020).

Railways use the ERTMS (Fig. 15.1), which provides benefits in the following areas: increased capacity of existing lines and a faster and more efficient ability to respond to growing transport needs, higher speeds, higher reliability, lower production costs and lower maintenance costs.

However, in rail transport, the control and management, alarm and traffic management systems of the Member States are often still not aligned and are one of the reasons why crossing borders remains a major obstacle to the development of

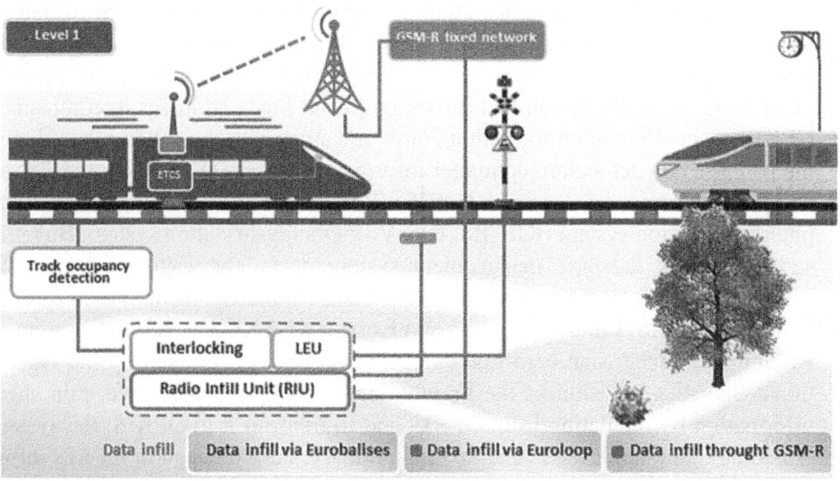

Fig. 15.1 ERTMS system (Source: Compiled by the author)

railways in Europe. The European Rail Traffic Management System (ERTMS) was developed to service this purpose.

Currently, ERTMS has two components:

- GSM-R—radio system used to exchange voice and data information between tracks and the train.
- ETCS—European Train Control System, which is used to align speed control systems. The existing differences between these systems now are one of the biggest technical problems for trains running on international routes. It consists of the modules installed in the train and next to the tracks.

Signalling systems offer significant economic benefits by allowing more trains to operate safely on a certain specific track section. Having introduced ERTMS, many incompatible systems on the European network will be replaced by a single more modern, more sophisticated and secure system compatible at EU level.

Moreover, next generation rail traffic management systems which would help to transform current technologies into an integrated rail traffic management system will have to be developed.

15.3 Telematics Freight Carriage Applications

Telematics Applications for Freight (TAF-TSI) have been developed in rail transport. TAF TSI is a European regulation needed by the railway sector to develop and implement a common standard for the mutual dissemination of information required for the carriage of goods by rail. The TAF TSI is designed to facilitate the sharing of information between businesses on freight services, especially across borders, in order to improve the quality and efficiency of rail freight in Europe (Table 15.1).

Table 15.1 Functional and technical standards laid down in TAF TSI, which infrastructure managers, railway companies and other stakeholders use to exchange the approved information

TAF TSI functions to be defined	TAF TSI defined	TAF TSI reduces the complexity of IT for individual players
• when (in a particular process) • what (what data and content) is to be sent • who (participant or participants) • how (in what form) partners must exchange data • where (reporting point) according to a contractual agreement the partners are to exchange information.	• what we want to convey (defined TAF TSI notifications) • how we want to convey (TAF TSI notification structure) • when we want to convey (TAF TSI process) • who we want to convey with (TAF TSI process) • where is the place we are talking about (TAF TSI place reference F).	• who my partners are (TAF TSI company reference F) • where (IT world) my partners are (TAF TSI metadata) • how to connect to my partners (TAF TSI metadata) • how to interpret certain notifications (TAF TSI metadata and a common interface).

Telematics applications are functional subsystems of the railway system. This subsystem consists of two elements: passenger and freight carriage services (https://ec.europa.eu/transport/modes/rail/interoperability/interoperability/telematic_applications_en).

When it comes to freight carriage service applications, it should be mentioned that it is a system that includes the following systems: (1) information for real-time monitoring of freight and trains; (2) sorting and distribution; (3) reservation; (4) payment and invoicing; (5) communications management; (6) generation of electronic accompanying documents, etc.

15.4 Train Protection System

The train protection system ALSN (continuous automatic locomotive signalling) has been used in the Baltic and CIS countries. All neighbouring countries, except for Poland, have an installed ALSN system that ensures technical interoperability of railways, i.e. locomotives with ALSN system can run freely through the territory of Lithuania.

The ALSN has been installed on sections with an automatic road blocking system and on some sections with semi-automatic road blocking ahead of a station and on the main station roads.

Locomotives have been equipped with automatic braking (hitchhiking) devices to improve train traffic conditions and ensure greater security. Locomotive signalling devices transmit views of level crossings and station signals to the locomotive driver.

Hitchhiking devices automatically stop trains when a stop signal is on, if the driver does not take measures to stop the train in a timely manner.

The Automatic Locomotive Signalling (ALSN) is a continuously operating system. The technological system algorithm depends on the transmitted information and its complexity. The application of ALSN allows to perform additional signalling functions, i.e. allows temporary traffic in irregular directions on two-way sections without following traffic light signals. Current ALSN systems are reliable, but abundance of outdoor equipment, its decentralization and the information provided via this system, mainly about the occupancy rate of the section, reduce the efficiency compared to modern signalling systems.

For example continuous locomotive signalling (ALSN) used by Lithuanian railways continuously transmits signals from traffic lights which the train is approaching to the driver's cab along the entire length of the blocked section. Track circuits act as a communication channel between the track and the locomotive in the ALSN system. The system uses digital code signals of the automatic road blockage. The system uses four-light signals, where a yellow-red signal is also used in addition to the red, green, and yellow lights.

15.5 Conclusions

Rail transport is used to transport mass freight over long distances and regular shipments. Like other modes of transport, rail transport uses information technology to ensure good quality of service and avoid barriers. These information systems are related to ensuring traffic safety, as well as filling in documentation and process management.

The most widely used European Rail Management System (ERTMS) consists of two parts, i.e. radio system and the European train control system. In the Baltic and CIS countries, the ALSN train protection system is the most widely used. It should also be noted that telematics applications for freight are being developed in rail to ensure the interoperability of the information needed for this mode of transport.

References

Jarašūnienė, A., Čižiūnienė, K., & Petraška, A. (2019). Research on rail and maritime transport interoperability in the area of information systems: The case of Lithuania. *Transport*. ISSN 1648-4142/eISSN 1648-3480; https://journals.vgtu.lt/index.php/Transport/article/view/11236/9386

Zabielienė, A. (2020). *The research on the problematic fields concerning the interaction between rail and sea transpprto carring acrgo abroad*. Master thesis.

Chapter 16
Information Technology Used in the Water Transport Sector

Kristina Čižiūnienė

Abstract Water transport is divided into inland waterway and maritime transport. However, not every country has access to the sea and not every country has a well-developed inland waterway network to carry goods. In any case, the choice of each of these modes of water transport to carry freight is subject to proper management of information and material flows, with information technologies playing an important role in its successful development. Application of information technology tools in water transport can also help to optimize routes and to better plan the fleet and freight carriage. This is actually one of the objectives of electronic marine and river information service system. Reduction of administrative maritime costs is one of the advantages of the e-maritime initiative. This section will address issues related to waterborne (inland and maritime) transport and the IT systems used in it.

Keywords Inland water transport · Maritime transport · Vessels · Ships · Route · Scheduling

16.1 Inland Water Transport

16.1.1 River Information Services

The River Information Services (RIS) system (Fig. 16.1) (www.ris.eu, online, 2021) has been used in the water transport sector, which aims to increase safety in inland water ports and rivers, improve the efficiency of inland navigation, optimize resource management in the waterborne transport chain by exchanging information between ships, terminals and ports, make better and more efficient use of inland waterway infrastructure, and improve environmental protection by providing more efficient traffic and transport information in the event of emergencies, better integrate

K. Čižiūnienė (✉)
Vilnius Gediminas Technical University, Vilnius, Lithuania
e-mail: kristina.ciziuniene@vilniustech.lt

D. Dzemydienė et al., *Development of Smart Context-Aware Services for Cargo Transportation*, International Series in Operations Research & Management Science 330, https://doi.org/10.1007/978-3-031-07199-7_16

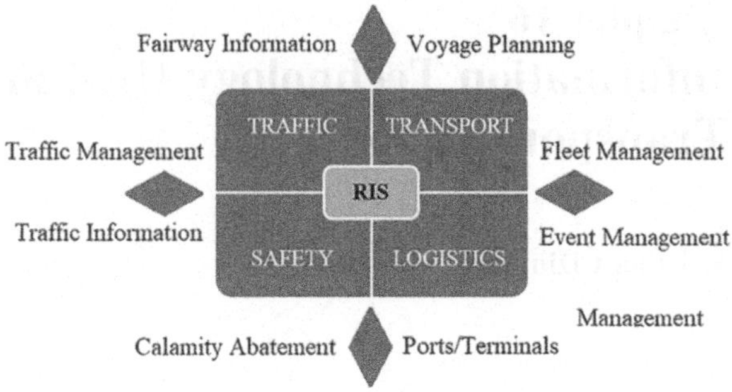

Fig. 16.1 RIS (Source: Compiled by the author)

inland waterway transport into a multimodal chain and provide timely information on transport management.

The RIS concept consists of advanced information services and functions that are supported by: different technologies, the Internet, positioning system, digital mapping, display information, and automatic identification systems (European Parliament and the Council of the European Union, 2005; Schilk & Seemann, 2012).

River information services are harmonized information services that help to manage inland navigation traffic and transport; the provision of these services provides geographical, hydrological, and administrative information on waterways allowing to provide electronic data on freight and voyages, locate and track ships. The aim of RIS is to facilitate the exchange of information between waterway operators and users. More broadly, the following RIS objectives can be distinguished:

- Improvement of inland navigation security in rivers and ports.
- Provision of traffic information to ensure (at tactical and strategic levels) safety monitoring.
- Increased efficiency of inland waterway transport and best use of available resources, management of chain of waterborne transport, allowing the exchange of information between ships, locks and bridges, terminals and ports.
- Better use of inland waterways—provision of information on the condition of fairways.
- Environmental protection—provision of information on traffic and transport in pursuit of efficient disaster reduction process (www.ris.eu, online, 2021).

Since 2005, the EU Framework Directive sets out minimum requirements for cross-border harmonization of national systems (www.ris.eu, online, 2021). The provided information helps, inter alia, to engage in maritime navigation, traffic management, accident prevention, fleet management, transport planning, and take execution and monitoring actions. Some inland waterway transport studies indicate that such measures can cut costs by 40%.

In order to ensure maximum RIS benefits, RIS services and systems must be deployed continuously, also developing services and specifications, integrating new technologies, using RIS data for logistics purposes, ensuring interoperability and interconnection with other traffic and transport management systems, in order to continuously monitor and support the development of water (both river and maritime) transport.

It can also be noted that RIS related to traffic and safety aspects cover (www.ris.eu, online, 2021):

- Information on the planning, execution, and monitoring of voyages of ship captains and flotilla managers (e.g. level of water, road signs and lock opening hours). IS contain administrative data, geographical, and hydrological data.
- Services of traffic information also include traffic information related to the movement of existing vessels on a restricted section of waterway, as well as the characteristics of strategic information about traffic such as ships and their display over a wider geographical scope, including analysis and forecasting of future situations for traffic.
- The aim of traffic managements is to optimize the exploitation of infrastructure, as well as to ensure safe navigation. The Vessel Traffic Service (VTS) centres are designated for improving vessel traffic efficiency and safety and protecting the environment.
- Disaster risk reduction services monitor the recording and updating of the ship and its transport data at the beginning of the voyage, applying reporting system of the ship. Responsible authorities (in case of an accident) can immediately provide data to rescue team, emergency team and to provide resource for terminal and port processes.
- Cargo and fleet management information essentially covers two information types, i.e. information about ships and flotilla, as well as details of the freight carried.
- Customs services and statistics: River Information Services simplifies makes it easier the production of inland water statistics in the Member States.
- Port and waterway charges: ship voyage information is used to calculate the fee and start invoicing procedures.

The growing need for a smooth flow of information and exchange between stakeholders in the inland waterway industry, its own (i.e. the inland waterway) sector, and other modes of transport, the river information services (RIS) have become increasingly more important in the transport logistics industry. Even though RIS traffic information is mainly intended for the safety of public stakeholders, RIS-related information mainly focuses on ensuring the effectiveness of the results for commercial consumer, e.g. transport and/or logistics companies using inland waterway transport.

16.1.2 Transport Logistic Service of River Information Services

Some countries have developed inland navigation information systems (this was done in the late 1990s). The lack of continuity, given that the work of this system was not very coordinated, may have affected the implementation of different technologies in each country. European research has focused on the development of RIS in all countries. European research and policy-making went hand in hand. Dissemination and implementation of research results have contributed to the organization and removal of RIS policies and barriers to effective implementation (www.ris.eu, online, 2021). Now RIS is once again the subject of research, this time exploring various requirements, logistics, and transportation services. The European Commission-funded project "RIS for the Integration of Inland Waterway Transport into Intermodal Chains (RISING)" aims to identify, integrate, and further develop information services, e.g. RIS to effectively support IWT (inland waterway transport) and logistics operations (plan-do-finish). Among other projects, the RISING project focuses on the planned and current needs of the European logistics and transport sector. Thus, inland waterway transport (in multimodal transport) has become an integral part of the door-to-door chain. Therefore, existing RIS needs to be used and new ones introduced at almost all stages of the inland waterway transport process, in order to increase the efficiency of intermodal freight transport and this mode of transport (Schilk & Seemann, 2012).

In the field of transport logistics, a typical process life cycle begins with transport planning, before and after the physical transport. The EU co-funded RISING project has examined processes of this business in defining, testing, and developing new services of RIS that primarily support the sector of transport logistics in Europe (Fig. 16.2).

16.1.3 Transport Logistic Service of River Information Services Users and Stakeholders

RIS help to plan and manage transport and traffic. Optimizing logistics and electronic data exchange operations ensures safer and more efficient use of waterways, bridges, locks, terminals, etc. This is an invaluable assistance for authorities of waterway in carrying out traffic management tasks the management task of traffic, also monitoring hazardous goods, which will also be very useful to commercial entities. RIS will increase competitiveness and security, and therefore it needs to be coordinated for implementation and further development on trans-European networks (European Commission, 2006; Schilk & Seemann, 2012). The transport logistics sector in Europe has recipients of RIS TLS, which are divided into the following groups:

Fig. 16.2 Services of RIS that support the transport logistics sector in Europe

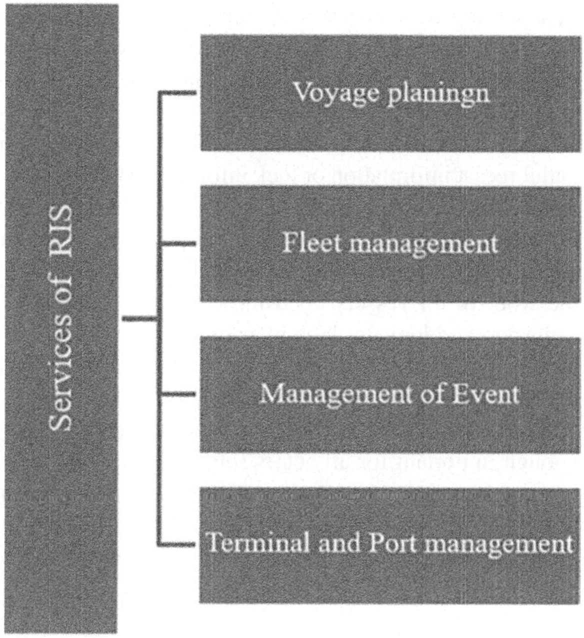

- Inland navigation companies (fleet operators, inland waterway operators)
- Inland port and terminal
- Providers of logistics service

Inland waterway companies are interested in information on inland waterway infrastructure to increase efficiency and optimize their transport route. In order to optimize transhipment procedures, the inland port requests as accurate information as possible on transport and operating status (this is similar to logistics companies that have to manage the entire logistics chain). Given the diversity and different interests of various stakeholders, an appropriate RIS TLS (RIS Transport Logistics Services) must also be offered to meet the specific requirements and needs of transport logistics operators (in Europe) using inland waterway transport (RISING, 2011; Schilk & Seemann, 2012).

Cargo owners or logistics providers, as well as port and terminal operators, need continuous tracking of cargo using the RIS transport and operation platform. As this information on the relevant inland waterway vessels and motorized inland waterway vessels (barges) is accurate and factual, it is necessary for the detection and identification of a special cargo unit. The RIS fleet management programme allows logistics providers and fleet managers to track the movements of their ships. By providing the so-called one-stop-shop RIS programme and further integrating existing systems and services into the fleet management programme (such as existing RIS traffic services and information), the programme meets the needs of fleet operators. RIS offers a scheduled arrival time service because the existing traffic information is essential for the quality management of traffic and can be used for transport logistics purposes. For example, in estimating the estimated time of arrival (ETA), a maritime navigation company uses relevant information sources and generated ETA data, which are usually based on its own algorithms. With this new RIS transport logistics service (RIS TLS), fleet operators can obtain more accurate information based on available traffic information (such as water level and forecast information or lock information). RIS information can also be improved for port and quay operations. This system integrates port administrations into the RIS information chain as a data user and provider. The addition of RIS information is therefore useful when the data available in inland ports are used for the planning of the work of all vessels, focusing on specific inland vessels and/or consignments. Transport and logistics lock management services. Factual and clear information on traffic conditions is essential for all modes of transport. With regard to inland waterway transport, it should be noted that locks are also part of the infrastructure that has a significant impact on inland waterway transport. Actual data of traffic is enough important for all actors (operators of lock, as well as inland waterway and users of logistics) who pass through these locks in their process of transportation. In this way, the information provision (on the condition and availability of locks) creates added value for actors in the transport and logistics sectors.

16.2 Maritime Transport

Maritime transport is a mode of transport carrying goods (or people) by sea. In some cases, maritime transport may include pre- and post-navigation activities. Mankind has used waterways to carry goods and people for centuries. This allows distinguishing the following advantages of maritime transport:

- Large capacity
- Trade volume (large vessels)
- Low unit transportation costs
- Good adaptability for loads of various sizes, weights, and shapes (https://www.coursehero.com/file/14085306/CC-Unit-6-Transportation/).

Maritime transport system comprises the following key physical components: (1) navigable waters, (2) ships (public and private), (3) ports (port and land facilities), (4) intermodal transport connections (highways and railways), (5) shipyards and (6) repair facilities (https://www.researchgate.net/figure/Components-of-Marine-Transportation-System_fig1_273694536).

16.2.1 SafeSeaNet system

SafeSeaNet system has been used in maritime transport. SafeSeaNet ships traffic surveillance and IS has been developed to increase:

- Fleet protection
- Security of ports and maritime navigation
- Protection of maritime environment
- Efficiency of maritime transport and traffic

SafeSeaNet was developed as a centralized European data exchange platform for maritime data, connecting maritime authorities from all over Europe. This allows EU Member States, Iceland, Norway, to enter and get information about ships and their movements, and transport of dangerous goods. AIS (Automatic Identification System, where the location report and messages are sent to the designated authorities, is the main source of information. Automatic Identification System (AIS) is an autonomous and continuously operating ship identification and tracking system used to ensure maritime safety and security. It allows ships to electronically exchange vessel identification, position, course, and speed data with other nearby vessels and on-shore authorities.

There also is VTMIS in place, while the Automatic Identification System (AIS) and the Long-Range Identification and Tracking (LRIT) System are being installed.

All these systems are being developed to improve the processes of freight and passengers carried by water transport, and to provide a greater protection and timely reporting of problems that can be eliminated.

16.2.2 Port Information Systems in Transport

The structural scheme of the port information system can be described as a data bank and a link with other ports. Paulauskas (1998) states that the data bank consists of the following elements between which information is exchanged:

- Stevedoring companies
- Port administrators
- Customs
- Ship agents
- Freight forwarders
- Rail transport
- Road transport (road carriers)
- Other (Paulauskas, 1998).

Ship agent or port administration (having received a ship's declaration) enters ship information into the database, if a particular ship agency company has not been included in the port information system. Freight forwarders or the port administration enters information on freight having received a freight from forwarders (or another information system or port). If information about a ship carrying freight is entered periodically, users of the information system know what the freight is, where it has been loaded, packed, what the final destination of the freight is, what are the specific storage and loading conditions (dangerous, refrigerated goods, etc.).

Information about vehicle location and its time of arrival allows to prepare for goods a place in the port in advance. With all the information about the arrival of a ship and freight, the customs service can plan its work more accurately. In cases where the freight is not inspected, customs documents can be arranged, while a customs declaration number can be assigned to a specific cargo while the ship is headed to port. Customs declarations are generated automatically, having combined information stored in the port information system and customs information (especially for transit cargo). If the consignee knows exactly where his cargo is and in what condition, he can plan its departure from the port more accurately, thus using the ordered vehicles much more efficiently, i. e. avoiding downtime. The integration of the port information system with the customs system allows for efficient tracking of the movement of transit freight and reduces their detention time at control points, at the same time increasing the attractiveness of a specific transport corridor (Jankauskaitė, 2010).

Port information systems and their operation are quite expensive, as satellite communication and hardware are mostly used, and it requires high qualification of service staff. Thus, costs of installation and operation of the port information system include:

- System development (acquisition of equipment and software).
- Operation: employee salaries, communications, repairs, taxes, etc., (Saulis & Vasilecas, 2008).

Revenue or "benefit" of the port information system (if the system belongs to the administration and the "benefit" is calculated due to the increase in cargo traffic, the attractiveness of the port, etc. only) consists of:

- Reduction of freight carriage time costs.
- Reduction of the freight insurance premium to the minimum amount (which is directly related to the buyer of the goods).
- Reduction of actual costs of freight forwarding companies, which can be estimated by monthly costs of salaries of employees.
- Attracting additional freight flows to the port, especially expensive ones.

Port information systems facilitate the work of all companies working in the chain, have a direct impact on the security of cargo, especially expensive cargo, thus increasing the attractiveness of the port, and cargo flows. Moreover, the installation of the port information system brings economic benefits to all participants in the transport chain and freight owners, which directly determines the choice of transport corridor, and the increase of cargo flow (Jankauskaitė, 2010).

Information systems in water transport have been used widely as in other modes of transport:

- The use of a Geographic Information System (GIS), for example in Klaipeda State Seaport, allows managing navigation data and combines them with connecting waterways. The marine GIS system includes oceanography, coastal zone management, navigation maps, the ocean industry and security. Advances in

GIS now allow the integration of many vessel sensors that show vessel direction, speed, navigational information, road lines and routes.

- The ship information system "LIS-GIS" (Jankauskaitė, 2010) allows to register and control ships entering and leaving the port and anchored in it, manage ship mooring and re-mooring operations, register pilots' work and calculate ship fees. The system also allows real-time monitoring of ships in the port water area and its accesses, visually present this information on the map of the port water area. The use of spatial information allows to better control of the ships in port and their movement. This system is most often used by the Vessel Traffic Service, Port Supervision Division, Shipping Channel Service, Port Surveillance and Rescue Service, and the port dispatcher office.
- The freight and goods information system KIPIS (Klaipėda Seaport. KIPIS; Klaipėda State Seaport Authority) was created in an effort to implement a freight and goods information system that would enable companies and institutions operating in the port to exchange electronic data while carrying freight through the port territory. Thus, the system was changed, which abolished the use of paper documents that accompany or relate to the formation of freight. The main functions of this system include the provision of information required by customs and other authorities; data exchange between the system and users during the temporary storage of goods; electronic exchange of works with stevedoring companies when placing and executing stevedoring orders.
- Electronic charts called "Integrated Information System" (ECDIS) have been used in modern navigation. It is a next generation electronic navigation device consisting of a special display and a specialized control device. All hydrographic information stored in the permanent system memory is used for navigation, which is displayed on the screen. Variable information is related to navigational parameters of the ship, such as the ship's coordinates, speed and distance to the set point.
- The information system Baltic Way has been used to disseminate information to stakeholders.
- The most innovative vessel traffic monitoring and control system OPEN EYES, which has already been installed in the port of Barcelona (Jankauskaitė, 2010).
- The Container Control Logistics System (CCL). The Container Controlled Leogistics (CCL) system was developed in conjunction with the international freight carriage company Transfracht and Deutsche Bahn. It is designed to monitor the path of each container from the port to the European Economic Area and to regularly provide information on the condition of the container (e.g. whether it has been loaded, empty, damaged, etc.). The system can be used for container control of shipping companies worldwide (Jarašūnienė, 2011).

16.3 Overview of Ship Routes and Scheduling

Maritime transport is one of the parts of transport in the majority of supply chains. However, the scheduling of operation of ships involves a great deal of uncertainty.

It is therefore important to remember that a risk- and cost-based scheduling system is designated for one supplier of freight, who has several customers. The necessary costs and risks need to be taken into account when looking for and deciding on alternative routes (Siddiqui & Verma, 2015). Such risks and costs continue to increase in intermodal and multimodal transport. It is therefore important to ensure the timely delivery of cargo to and from the terminal. In such a case, participants in the process must know the exact times of arrival of freight.

Therefore, problems relating to scheduling and routing are closely related to problems of work planning of the fleet and other process participants. Taking this into account and assessing the future development prospects and the use of optimization-based solutions, the use of support systems was, is and will be one of the key components in creating and scheduling ship routes, as trends show that the needs and benefits of such systems have constantly been increasing.

In scheduling, the following are the most important elements in maritime transport:

- Fleet size and combination that are associated with the strategic planning problem.
- Fleet deployment, i.e. assigning ships to routes. The deployment of an existing fleet is treated as a problem of tactical planning.
- Ordering freight, i.e. the choise of which freight to accept or reject for a particular voyage. This factor is attributed to planning (operational) problems. Given the growing number of ships in the world, the capacity of the container fleet has almost doubled over the same period. Despite such rapid growth, little research has been done on the selection, planning, and implementation of liner shipping routes.

Many shipping companies have merged in the last decade. This is because many shipping companies have wanted to focus and work together to gain flexibility in service delivery and thus increase their market power. In such cooperation, many shipping companies combine their fleets into a basin and manage them together. Not so long ago, the ocean fleet was planned manually, i.e., based on the experience and knowledge of planners, and using a pencil and a paper sheet. The above mergers and cooperation allow forming larger controlled fleets, which means that creating an optimal fleet schedule in application of manual scheduling methods alone has become increasingly more difficult.

Normally, fleet schedulers are experienced and have a maritime education. Therefore, Christiansen et al. (2004) say that such schedulers usually do a very good work, but as the fleet grows, manual planning becomes much more complex. Nevertheless, planners are often very sceptical about computers and systems designed to make and maintain optimization-based decisions (especially when planning a fleet). Schedulers usually have a good reputation and work in shipping companies. Therefore, some schedulers may be concerned that managers may need to reduce the number of schedulers or to stop using their help altogether, having introduced such systems. This can also lead to shipping companies becoming less dependent on schedulers and therefore less vulnerable to employee turnover.

Scientific literature mainly offers information related to ship scheduling and routing problems arising in commercial freight industry. Therefore, decisions on fleet size and composition define the ships that can be routed, scheduled, and deployed.

For ages, significant attention has been paid to analyzing problems relating to oil transportation. However, with the expansion of markets, increasing freight flows and organizing intermodal/multimodal carriage more often, a need for substantial analysis of scheduling and routing processes has emerged. Opinions of researchers about the factors under consideration that are important for scheduling differed, as some of them assessed the types of shipping, others—the types of freight, intermodality/multimodality, etc. For example (Christiansen et al., 2004):

- Dantzig and Fulkerson (1954) aimed to reduce the number of tankers required for a certain timetable. This can be considered a problem due to the size of the fleet when there is only one type of vessel. Therefore, this problem can be described as a transportation problem.
- Cho and Perakis (2001) analyzed the problems of planning the size of a container shipping company fleet and optimal liner route scheduling. This problem has been solved by pre-scheduling various shipping routes.
- Xinlian et al. (2000) analyzed a problem somewhat similar to that raised by Cho and Perakis (2001). They proposed a model for fleet planning that aims to identify both the types of vessels to be included in the existing fleet and the optimal fleet layout plan.
- Bendall and Stent (2001) also provide a model on ways to determine the optimal ships number and the plan of fleet deployment in a container hub and spoke application.
- Fagerholt (1999) presented a study analyzing the issues involved in design of optimal fleet and respective weekly routes for each vessel for a system of liner shipping.
- Pesenti (1995) analyzed container fleet resources management. This problem is related to decisions on the purchase and use of ships to meet customer needs. A hierarchical model of the problem was developed, describing heuristic methods of solving problems at different levels of solution.
- Richetta and Larson (1997) analyzed the issue of planning a waste transportation system in New York. Garbage trucks are unloaded at land stations, and the waste is then loaded into barges. They are then transported by truck to the Fresh Kills landfill on Staten Island. They developed a stand-alone case modelling model that includes a sophisticated dispatch module to help make decisions about fleet size and business planning.

- Mehrez et al. (1995) performed simulation of industrial ocean freight carriage problem, offering a solution that examines the carriage of dry bulk minerals from the primary source of raw materials to customers. The decisions taken include the number and size of chartered vessels (for each scheduling period), the location and number of transhipment ports used and the transport routes from the ports of departure to the ports of destination.
- Imai and Rivera (2001) used modelling to solve the problem. This study involved planning the size of a refrigeration container fleet, i.e., deciding on the containers number needed to meet foreseeable traffic future needs. The tactical and operational planning of the industrial shipping problem mainly focuses on scheduling and routing.
- According to the definitions of Ronen (1993), a route can be understood as the assignment of a sequence of ports (which ships are to visit). The term "scheduling" is used when a time aspect is included in a route. Therefore, the schedule includes the time of various activities on the vessel's route. Most of the reported vessel scheduling and routing problems comes due to industrial operations. Industrial shipping faces various ship routes and scheduling problems. Here, a timetable is understood as a visit to a sequence of hubs, including the time of arrival at each hub. Using the SP (set splitting) formula, the sequence of visits and the start time of the service at each hub are determined according to a schedule so as to cut costs for a particular set of hubs.

However, due to some ship planning problems, the fleet schedules number may be excessive to list the variables in detail. In the face of such problems, heuristic rules can only create viable timetables. However, when more than one mode of transport is involved, scheduling and planning become even more complex and responsible.

16.3.1 Scheduling and Routing of Commercial Ships

The majority of the studies relating to commercial freight routes and scheduling in industrial shipping describe scheduling problems that aim to cut costs of a fixed ship fleet. In this case, ships are fully loaded, and ports and terminals for loading and unloading are coordinated. The problem can be solved by creating all the possible timetables. Ships can have up to seven fixed compartments, so freight consisting of several products can be carried on the same ship. In addition to the specified set of transported cargo, several additional ones can be selected. These return voyages generate revenue and can be picked up if they are profitable. All dispatchers communicate through a spreadsheet regardless of the language and review the recommended timetables presented in Gantt charts. In this case, there may be a trade-off between channel charges and time of travel. For example Bausch et al. (1998) state that there are several tankers types with fixed section that allow to carry different products on the same voyage. In this case, the Lagrangian relaxation may be used to solve the problem in order to create a potentially good schedule with an optimal cargo schedule. Fagerholt and Christiansen (2000a) addressed a similar multi-product planning problem to that presented by Bausch et al. (1998). In contrast, each fleet ship has a flexible freight compartment which can be divided into several smaller holds in a certain way. Ship planning is a multi-level pick-up and delivery problem with time intervals. Division of flexible ship freight compartments and distribution of freight into smaller holds allows for multiple layouts. In such a case, SP model is provided for problem solving. Timetables are drawn up a priori, including therein optimal distribution of freight to possible compartments. Fagerholt and Christiansen (2000b) describe the algorithm for searching for individual optimal ship timetables in detail. In the face of this ship scheduling problem, ports do not provide services on non-working hours (i.e. at night) and on non-working days. Therefore, wide time windows can be considered to be multiple time windows. Moreover, freight loading and/or unloading may take a few days, which means that vessel will remain unused most of the time at the port, and all the time at the port will depend on the time of arrival of the ship. The aim is to find reliable timetables that make ships less likely to remain unused in ports on weekends and to impose penalties for arriving at risky times (i.e. closer to non-working days). They formulated the SP model as a problem. All the possible ship timetables are found a priori and are created based on two aspects: uncertainty and multiple time intervals. Results of the calculation show that the reliability of timetables increases after the increase of the price of transportation. Flexibility and reliability are the two components important in scheduling. Fagerholt (2001) considers the aspect of flexibility, introducing soft time windows into the model for ship scheduling. The reasoning behind the implementation of soft rather than heavy time windows is that allowing control over time window violations of some freight can make obtaining better schedules and significantly cutting transportation costs possible. In order to control violations of the time window, costs of inconveniences associated with handling of freight outside of their time intervals apply. SP model is also used here. The proposed solution method also allows determining the optimal speed at various stages of scheduling according to the ship's timetable (Christiansen et al., 2004).

16.3.2 Ship Scheduling and Routing in Freight Delivery

Although maritime transport is often a part of supply chain, there is very little focus on ship planning and routing issues, the focus is on the whole supply chain. Unlike the majority of ship scheduling problems, the number of calls to a particular port during the planning period that needs to be loaded and unloaded for each call to port is not predetermined. The production plan shall include (in addition to ship's capacities) the amount of entries at each port, the time of commencement of service and quantities of possible loads at each entry. This is an example where a shipper uses its fleet simultaneously in both modes: tramp and industrial. This problem can be justified through these additional problems:

- The each ship planning problem, and
- The problem of management of inventory for each port

Solutions to both types of tasks need to be synchronized. In addition to the main problem, additional problems introduce constraints on linking the time and the load for each call for port, variables depicting vessel timetables and sequences of port calls. Flatberg et al. (2000) also developed solution to the same vessel problem for scheduling (Christiansen, 1999). Their heuristics have been used to solve the complex ship routing problem, while the LP model has been used to find the service start time at each arrival and loading or unloading quantities. Liu and Sherali (2000) extended the problem described in the software, which is associated with high uncertainty. However, little analysis and research have been conducted on tramp ship scheduling. A large number of small operators in their market is the main reason for minimum attention on tramp ship work scheduling in literature. As previously mentioned, tramp transactions are similar to taxi operations. Ships are sent to places where there is freight. Moreover, tramp ship companies often conclude contractor agreements. These are agreements for the carriage of certain quantities of freight between the specified ports over a certain period of time for an agreed fee per tonne. Appelgren (1969, 1971) also analyzed a typical problem of scheduling for tramp ship. This task was the first to use the DW splitting method for ship scheduling and routing; this method was later used in many other studies. The majority of freights have been carried under contracts and must be carried by a fleet. However, some optional freight may be available on the market and may be carried by a fleet when this is profitable. Fleet vessels can carry one freight only at a time (Christiansen et al., 2004).

However, there are issues that do not have a direct impact on planning and routing, but are related to them. This includes, inter alia, the storage of containers on a container ship, the handling of containers in ports and the selection of environmental routes. During longer voyages by sea, environmental considerations (ocean currents and harsh weather) can affect cruising speed and fuel consumption, and thus the choice of route between the two ports.

Christiansen et al. (2004) say that many of the ship scheduling studies described in literature lack a supply chain perspective. Fleet is often scheduled under strict constraints when the sender describes freight and volume dedicated for loading and such takes some hours. Following the presented information, the operator strives to arrange schedule that matches different interests while increasing profits by reducing expenses. Being aware of the possibility to mitigate requirements, the author investigated the fluctuation of shipping volume and timing (Fagerholt, 2003). Extra charges associated with fixed timing and freight volumes can ultimately be shared across parties involved in shipment. Shipping companies need to see themselves as a logistics provider (or at least a part of the whole logistics supply chain) rather than a maritime transport service provider only. With this in mind, companies need to develop a system for decision support for scheduling of fleet, which can be a great success. Such shows it should be cooperation and integration in companies or other process participants and carriers. Inventory managed by vendor (VMI) model has advantage in setting timing and freight volumes, and its application popularity in carrying land freights. The model delegates stock ordering, management, ordering responsibilities to the seller or third party service providing company (3PL) in full. Service company evaluates volume and the time required for freight delivery. Even more shipping companies are expected to act as sellers in the future. The companies have barriers for switching from tradition to VMI ordering model. The model includes inventory monitoring in order to predict usage and identify shipping time and volumes. Although such tracking technology is in place the customer might not want passing such data for the seller. Such may be due to several issues other than mistrust alone. However, fleet planning and shipping charges could be reduced even when the booking system cannot be eliminated. In order to avoid unnecessarily short intervals, flexibility, and coordination of working time between different modes of transport, the parties involved in shipping could collaborate in sharing the minimum amount of above-mentioned data. It is now possible (at least technologically) online. Therefore, some research has been conducted over the last few years to explore how ocean shipping can be integrated into the supply chain of multimodal transport, which can guarantee the principle like "door-to-door".

16.4 Conclusions

Water transport is divided into inland waterways and maritime transport. Information systems are used in both modes of transport.

The RIS is the most widely used inland waterway transport system, with the aim of ensuring and enhancing safety in inland rivers and ports, thus optimizing the management of available resources in this mode of transport chain in order to increase the efficiency of shipping.

In the field of maritime transport, the following systems are used: the "SafeSeaNet" vessel traffic monitoring and information system, the port information system, the geographical information systems, the ship information system, the cargo and goods information system, electronic charts, etc.

References

Appelgren, L. H. (1969). A column generation algorithm for a ship scheduling problem. *Transportation Science, 3*, 53–68.

Appelgren, L. H. (1971). Integer programming methods for a vessel scheduling problem. *Transportation Science, 5*, 64–78.

Bausch, D. O., Brown, G. G., & Ronen, D. (1998). Scheduling short-term marine transport of bulk products. *Maritime Policy & Management, 25*(4), 335–348.

Bendall, H. B., & Stent, A. F. (2001). A scheduling model for a high speed containership service: A hub and spoke short-sea application. *International Journal of Maritime Economics, 3*(3), 262–277.

Cho, S.-C., & Perakis, A. N. (2001). An improved formulation for bulk cargo ship scheduling with a single loading port. *Maritime Policy & Management, 28*(4), 339–345.

Christiansen, M. (1999). Decomposition of a combined inventory and time constrained ship routing problem. *Transportation Science, 33*(1), 3–16.

Christiansen, M., Fagerholt, K., & Ronen, D. (2004). Ship routing and scheduling: Status and perspectives. *Transportation Science, 38*, 1–18. https://doi.org/10.1287/trsc.1030.0036

Dantzig, G. B., & Fulkerson, D. R. (1954). Minimizing the number of tankers to meet a fixed schedule. *Naval Research Logistics Quarterly, 1*, 217–222.

European Commission. (2006). *Communication from the Commission on the promotion of inland waterway transport "NAIADES", an integrated European action programme for inland waterway transport.* COM 6 final, p. 2 ff.

European Parliament and the Council of the European Union. (2005). Directive 2005/44/EC of the European Parliament and of the Council of 7 September 2005 on harmonised river information services (RIS) on inland waterways in the Community. *Official Journal of the European Union, L 255*, 152–159.

Fagerholt, K. (1999). Optimal fleet design in a ship routing problem. *International Transactions in Operational Research, 6*(5), 453–464.

Fagerholt, K. (2001). Ship scheduling with soft time windows—An optimisation based approach. *European Journal of Operational Research, 131*, 559–571.

Fagerholt, K. (2003). A computer-based decision support system for vessel fleet scheduling—Experience and future research. *Decision Support Systems.*

Fagerholt, K., & Christiansen, M. (2000a). A combined ship scheduling and allocation problem. *The Journal of the Operational Research Society, 51*(7), 834–842.

Fagerholt, K., & Christiansen, M. (2000b). A travelling salesman problem with allocation, time window and precedence constraints—An application to ship scheduling. *International Transactions in Operational Research, 7*(3), 231–244.

Flatberg, T., Haavardtun, H., Kloster, O., & Løkketangen, A. (2000). Combining exact and heuristic methods for solving a vessel routing problem with inventory constraints and time windows. *Ricerca Operativa, 29*(91), 55–68.

Imai, A., & Rivera, F. (2001). Strategic fleet size planning for maritime refrigerated containers. *Maritime Policy & Management, 28*(4), 361–374.

Jankauskaitė, A. (2010). *Jūrų transporto ir uostų valdymo informacinės technologijos*. Magistro baigiamasis darbas. VGTU.

Jarašūnienė, A. (2011). *Specialybės įvadas. transporto inžinerinė ekonomika ir vadyba Mokomoji knyga*. Vilnius "Technika".

Liu, C.-M., & Sherali, H. D. (2000). A coal shipping and blending problem for an electric utility company. *Omega, 28*, 433–444.

Mehrez, A., Hung, M. S., & Ahn, B. H. (1995). An industrial ocean-cargo shipping problem. *Decision Sciences, 26*(3), 395–423.

Paulauskas, V. (1998). *Uostų vystymas ir logistika*. Klaipėdos universiteto leidykla. 162 p.

Pesenti, R. (1995). Hierarchical resource planning for shipping companies. *European Journal of Operational Research, 86*, 91–102.

Richetta, O., & Larson, R. C. (1997). Modeling the increased complexity of New York City's refuse marine transport system. *Transportation Science, 31*(3), 272–293.

RISING. (2011). http://www.rising.eu/web/guest/learn-more

Ronen, D. (1993). Ship scheduling: The last decade. *European Journal of Operational Research, 71*(3), 325–333.

Saulis, A., & Vasilecas, O. (2008). *Informacinių sistemų projektavimo metodai*. Technika. 247 p.
Schilk, G., & Seemann, L. (2012). Use of ITS technologies for multimodal transport operations – River information services (RIS) transport logistics services. *Procedia – Social and Behavioral Sciences, 48*, 622–631. https://doi.org/10.1016/j.sbspro.2012.06.1040
Siddiqui, A. W., & Verma, M. (2015). A bi-objective approach to routing and scheduling maritime transportation of crude oil. *Transportation Research Part D: Transport and Environment, 37*, 65–78. ISSN 1361-9209, https://doi.org/10.1016/j.trd.2015.04.010. https://www.sciencedirect.com/science/article/pii/S1361920915000425
Xinlian, X., Tangfei, W., & Daisong, C. (2000). A dynamic model and algorithm for fleet planning. *Maritime Policy & Management, 27*(1), 53–63.

Chapter 17
Information Systems in Transport and Forwarding Activities

Kristina Čižiūnienė

Abstract Information technology has become an integral part of the provision of transport services. This has ambiguous consequences on the management of transport services—they change the service provision environment, mutual relationship between consumers and service providers, pricing, etc. Customers request actual information from companies at a specific time and prior knowledge of what is happening to the goods during their carriage. Therefore, the quality of this information is very important. Activities of transport and logistics companies are very receptive to information technology—these companies work with large amounts of information, the management of which requires a quick response. Companies are convinced that new technologies have a positive impact on activities of their company, so they are rapidly implementing specialized programmes, although until a few years ago, technology was valued by the number of available computers only. This section will address issues related to the use of IT in transport and forwarding activities.

Keywords Transportation process · Forwarding · Information systems · Transport chain

17.1 Good Foreign Practice of IT Application in the Process of Transportation

The analysis of IT transportation process revealed that not only each mode of transport has different systems, but also each country has a variety of applied IT. Here are examples of some countries:

K. Čižiūnienė (✉)
Vilnius Gediminas Technical University, Vilnius, Lithuania
e-mail: kristina.ciziuniene@vilniustech.lt

D. Dzemydienė et al., *Development of Smart Context-Aware Services for Cargo Transportation*, International Series in Operations Research & Management Science 330, https://doi.org/10.1007/978-3-031-07199-7_17

- Belgium. Given the small size of the country but it is very densely populated area, annual traffic growth in Belgium cities has been about 3%, leading to traffic congestion during peak hours in case of accidents or road repairs. This is due to a large number of people who commute to work in cities by car. Belgium has been using various intelligent transport systems to combat these problems. For example the following ITS systems have been used in Brussels:
  - MORBIRIS is a mobility management centre, which sends all the collected information on traffic regulation and management of districts to intermediate users (administrations, operational services or radio stations).
  - SERBRU, which measures the load on the main sections of the road network and indicates travel time on information boards.
  - CORBU regulates regional traffic lights and controls regional traffic lighting.
- Croatia. There is a lack of modern national and regional road corridors in Croatia, and there is only one road link between Western and South-Western Europe, and South-Eastern Europe and Asia. Highway routes have not been completed in full, and the implementation of information and telecommunications technologies needs to be improved. Croatia has worked to address these issues. For example there are five video surveilled tunnels and ten viaducts in the 9-km section between Split and Zagreb also using traffic management systems in this section. The other section between the towns of Kupjak and Rijecka is equipped with surveillance systems, information boards and meteorological sensors. Telematics in the local road system has been introduced on roads of local significance to improve the interaction between roads and maritime transport in the city of Split. Also, Croatia has used semi-automated tax collection systems.
- Czech Republic. Road transport takes up a significant share in the Czech Republic, while cars are seen as a symbol of good life, therefore the use of public transport and rail transport has declined. The telematics project ATRACTIC targeted at the city of Prague consists of several interconnected systems. These include traffic adaptation regulation system MOTION, congestion and accident management, information boards for car navigation to parking garages, and informal information displays. Road tax collection has also been developed in the Czech Republic. Czech highways have used warning systems, speed control equipment, special signs and illuminated signs, information boards and video tracking and surveillance cameras.
- Denmark. About 90% of passengers travel by road in Denmark. Freight carriage by road has also been growing compared to water and rail transport. As a result, road traffic loads have also been growing rapidly. The TRIM traffic information system used in Copenhagen allows providing travellers with information on radio and during major repairs on highways. Meteorological stations have been used on the major road network to help with proper road maintenance or to provide 3-h weather forecasts to drivers.
- Finland. Metropolises of Finland face traffic congestion problems. Road and water transport are the main forms of transport in the country. There are also 25 airports in Finland, with Helsinki's Vantaa being the most important. Finland

also uses a traffic and road meteorological monitoring system that provides real-time information on meteorological conditions. Various signs and signals (VMS) provide information on dangerous driving conditions (slippery or snowy roads), road congestions and animals on the road. Water traffic uses the PortNet vessel traffic management system, which aims to generate one report on a vessel's departure and arrival only. This system is linked to other systems and is also used by Finnish customs, ports, ship operators and others. VTMIS (Vessel Traffic Management and Information System) is particularly common and useful in the Gulf Stream section of Finland.

- Sweden. Sweden is a country that uses many different intelligent transport systems. High road traffic flows and difficult traffic conditions in winter cause problems for the country's population and lead to a large number of accidents. The RwiS system is a multi-meteorological information system that uses cameras and automatic meteorological stations and provides real-time information. The traffic information support system allows to manage traffic data that are received from SOS alarm and the police. Another service is provision of information on road surface and temperature, which provides real-time information. The following are traffic management systems widely used in Sweden:
  - MCS system that warns of congestion on highways using detectors installed on road surface.
  - "Go Green" system allows cyclists and pedestrians to take advantage of green traffic lights that stay on longer.
  - The Red Check system "sees" cars that are overspeeding when approaching intersections keeping red light on longer to reduce the speed of such cars. Sweden also has an integrated and automatic billing system that allows to pay the required road tolls without stopping. There also is an important system related to drunk-driving, the alcohol interlock system (HMI), the operation of which is linked to electronic equipment in the car, which is connected to the ignition system. Thus, if the alcohol content of the metre exceeds the permissible limit, the car does not start, and the driver cannot use his car.
- Germany. German road network covers over 600,000 km, interconnecting densely populated areas of the country. Passenger carriage and freight carriage account for the major transport load of the country (80% and 70%, respectively). Some vehicles in Germany have been equipped with navigation systems. The cities have intersections and traffic lights that regulate traffic, while pedestrian crossings have special signals. Dynamic car parking systems are used in major German cities. There are information signs with information boards providing information on meteorological conditions, traffic jams, speed or certain directions. Also, radio stations provide information on congestion or accidents and the necessary detours.

17.2 Information Technology Used in Forwarding

The application of information technology in forwarding activities is directly related to the obligations which the freight forwarder assumes by signing contracts with customers. Forwarding activities are associated with freight.

Coyle et al. (2011) say that forwarding activities can be divided into import and export forwarding. Authors state that the following actions are performed in the course of import forwarding: completing customs documents, paying taxes, participation in the inspection of goods and their accounting together with customs officials. By providing additional services, freight forwarders ensure efficient organization of cargo transportation in the local and foreign markets, arranging freight carriage and cooperating with other companies providing forwarding services. The following is the processing procedure of freight for export when arranging export forwarding activities:

- Preparation of documents for customs
- Completion of freight export declarations

Freight forwarders can also provide other services related to the payment for export transactions under separate freight carriage authorizations.

For example many Lithuanian forwarding companies are small, and the assets they manage are usually defined by office-type premises, computers and other information technology (Turbienė & Jurevičius Jurkevičius, 2018).

Areas of use of information technology (Briuchoveckaja et al., 2013):

- Acceptance of orders
- Document management
- Identification of goods, transport and drivers
- Locating transport and freight
- Selection of the optimal route
- Search for freight and available vehicles

Many freight forwarders work with electronic transport exchange. Currently, there are many electronic transport platforms on the market, and their developers receive income for their use. Such electronic exchanges as Trans.eu, Cargo.lt, Ati.su, Lardi-trans.com, TimoCom can be distinguished (Turbienė & Jurevičius Jurkevičius, 2018).

Information technology in forwarding activities includes not only computer software, but also communication equipment, smartphones, headphones. Such programmes as Skype or Viber, also e-mail and social networks (like Facebook) have also been used in these activities. When transport logistics companies provide additional services provided for in the freight forwarding contract, computer programmes such as Klevas or Speda are used. Software Muitinė (English: Customs) can be used to complete documents and customs registers, while Nano Warehouse and other specially developed programmes adapted for the company's needs. Transport service companies also use international intelligent technologies:

- CRM is a set of business model, business process methodologies and duly prepared interactive technologies. CRM is also a concept used in the information technology industry defining methods, software, Internet capabilities, and solutions that help companies manage customer relationships in an organized way.
- TRIS means Transportation Research Information Services. TRIS provides access to more than 450,000 bibliographic records, including transport research published in books, journal articles, reports, and the media.

According to Batarlienė (2011), this allows "freight forwarders to join the information space of stevedoring companies". In this information system, the freight forwarder places orders with a stevedoring company or warehouse, monitors order processing, tracks freight movement, and can make requests.

DAKOSY is another system. As one of the leading suppliers of logistics platforms and software, it has been offering promising solutions since 1982. Communicating between companies, these digital platforms allow all participating authorities and companies to plan and effectively manage processes of transport. DAKOSY also offers digital customs and freight forwarding formalities. This includes, in particular, CargoSoft for international forwarding (by air, land, sea) and ZODIAK for customs formalities in Belgium, Germany (ATLAS), Austria, Switzerland, and the Netherlands. More than 2800 companies across Europe use state-of-the-art DAKOSY data centres. These companies include world-renowned trading houses, freight forwarders, industrial companies, shipping companies, airlines, carriers, liner agents and various institutions (https://ipcsa.international/about/members/members-europe-and-north-america/dakosy/).

17.3 IT System Compatibility in the Transport/Logistics Area

Each transport company chooses IT systems depending on its needs. Smaller companies choose simpler ones, while larger companies need more functionalities so they choose more advanced applications. However, every modern system provides the ability to exchange data with other systems. Methods such as EDI and API are used to accomplish this task:

1. EDI is a method of exchange of electronic data. When creating EDI between company A and company B, information to be transferred and its compatibility between two systems must be assessed to see if all the data sent will fill up the recipient's programme boxes. Having analysed the amount of information to be transferred, the development of a supplement to the system is started for processing the information received and assigning each variable to the required field. Usually, such supplements are called engines in English. Electronic data exchange has many standards, including AS1, AS2, HTTP, FTP, etc. (E-Procurement–Electronic Data Integration Comes of Age–Finance Director Europe (the-financedirector.com)). The following are the data exchange options:

   a) Consignor—consignee (when all information is sent directly)
   b) Value-added network (when there is one consignee but many consignors)
   c) Internet (document with information is sent and integrated into the system)

   Advantages of Electronic Data Exchange:

- Lower employee time costs as less data has to be entered manually.
- Environmental friendliness, as there is no need to print information on a sheet of paper.
- Quality, as it reduces the human factor when employees can make mistakes.
- Speed, as important data is quickly exchanged over long distances.
- Flexibility, as an EDI interface, can be created for almost any application.

2. API is a programming interface of applications. API allows to integrated of additional functions and tools in the company's IT programme, thus ensuring faster and higher quality work. Most popular integrations for logistics companies include: Google Maps, interface with e-mail, SMS sending and more.

Despite these systems, the issue of how to ensure the dissemination of all the necessary information in order to make the movement of material flows as efficient as possible remains sensitive in the transport chain.

17.4 Needs for Information Systems in the Transport Chain

Today, transport information systems connect several independent systems: data banks, information data exchange system, system for locating transport and freight vehicles, and satellite communication systems. Therefore, interoperability of the dissemination of information and information systems also plays an important role in ensuring the smooth running of the process.

Information on the location of freight and its condition in the transport chain is very important in any case. In the transport chain, especially if mixed (multimodal) transport is used, the work of separate sections must be coordinated, inter alia: ordering road transport, loading onto a vessel and carrying in it, unloading at another port, customs clearance procedures, ordering transport to carry from port to the consignee, cargo insurance, payment for freight and separate transport and procedure services (Paulauskas et al., 2001). In order to guarantee the safety of freight during transportation, it is very important to have all the information about the location of freight and its condition any time in the carriage process. When accurate information is not available, it takes a lot of time and money to deal with problems that arise, such as customs procedures (Jankauskaitė, 2010).

Information integration is essential to create a single information space in the transportation and logistics chain that ensures the required speed under the current conditions and a comprehensive and accurate information in transport services. The complexity of information integration into transport logistics is determined by many information channels and information flows of mutual cooperation. Information integration into transport logistics at the global level has been implemented according to international programmes. The formed excess information or system incompatibilities and duplication of work hinder business processes due to the need to process a lot of unnecessary data and to re-enter, additionally enter or check them. This task is just as important as the lack of information. Optimizing the flow of information to transport and logistics systems is possible through a new modelling of the information system, which is still being worked on in the field of research.

Information technology is very important in optimizing the operation of transport systems. Efficient use of information technology can improve the quality, safety, and efficiency of transport systems and increase their permeability, and reduce travel times without significant investment in new infrastructure. It can also help to use energy more rationally and reduce environmental pollution. Information technologies are used to make multimodal transport more efficient.

The development of multimodal transport requires to expand and integrate logistics centres in terms of various organizational forms and content. Using modern information technologies, logistics centres traditionally perform the following functions: reloading freight from one mode of transport to another, freight warehousing, customs operations, distribution of cargo in small consignments and delivery to their end recipients (Paulauskas, 1998, 2015). Information technology can improve the efficiency of all modes of transport and promote the sustainable use of services.

The problem of information in transport logistics is defined in the following directions:

- Analysis of information flows, which is dynamically exchanged with information on changes in the form of ownership in business, providing opportunities to compete in open transport services markets.
- Development of information and software equipment and systems to make decisions on business planning, activities of transport, freight forwarding and agent companies.
- Improvement of mobile communication systems in transport.
- Implementation of Internet technologies in organizations improving transport management processes.

17.5 Conclusions

Information technology is also becoming an integral part of the provision of transportation and its services (especially forwarding). Despite the fact that different modes of transport, companies and participants in the transport process use their own information technology, each country is characterized by the uniqueness of IT in this sector (e.g. Belgium: MORBIRIS, SERBRU, etc., Czech Republic–MOTION, etc.).

IT used in forwarding companies is more focused on process management solutions. Transport exchanges or freight and transport platforms such as Trans.eu, Cargo.lt, Ati.su, Lardi-trans.com, TimoCom, and so on are also applicable.

Despite the wide choice of IT, both the transportation process and the forwarding activity face a major problem due to the interoperability of these systems.

References

Batarlienė, N. (2011). *Informacinės transporto sistemos*. VGTU leidykla Technika.

Briuchoveckaja, I., Kaikaris, P., Krasauskienė, S., Ledauskaitė, K., Melaika, M., Šustickienė, B., Turbienė, J., & Zarankienė B. (2013). *Transport logistics*.

Coyle, J. J., Novak, R. A., Gibson, B., & Bardi, E. J. (2011). *Transportation: A supply chain perspective*. South-Western Cengage Learning.

E-Procurement – Electronic Data Integration Comes of Age – Finance Director Europe (the-financedirector.com). https://www.the-financedirector.com/features/feature1420

Jankauskaitė, A. (2010). *Jūrų transporto ir uostų valdymo informacinės technologijos*. Magistro baigiamasis darbas. VGTU.

Paulauskas, V. (1998). *Uostų vystymas ir logistika*. Klaipėdos universiteto leidykla. 162 p.

Paulauskas, V. (2015). *Jūrų transporto plėtra*.

Paulauskas, V., Barzdžiukas, R., & Plačienė, B. (2001). *Uosto technologija* (p. 255). Klaipėdos universitetas.

Turbienė, J., & Jurevičius Jurkevičius, P. (2018). Informacinių technologijų taikymo tendencijos Lietuvos įmonių transporto ekspedicinėje veikloje. Aukštųjų mokyklų vaidmuo visuomenėje: iššūkiai, tendencijos ir perspektyvos. *Alytus, 1*(7), 254–259.

Chapter 18
Information Technology Used in Intermodal and Multimodal Transport

Kristina Čižiūnienė

Abstract Hardware (special transport units, specific unit freight and intermodal terminals with specialized equipment) is needed first of all for intermodal transport to operate. It is also important to have sophisticated information and management software to monitor the various stages of intermodal transport and to process additional information efficiently. Finally, intermodal transport requires human resources, highly qualified experts who organize, control, and manage all stages of the intermodal transport process.

Keywords Intermodal transport · Multimodal transport · e-Freight · Big data · Decision support

18.1 IT Services That Promote e-Freight

Information technology must be widely used to simplify procedures of administrative that can ensure the tracking and tracing of goods, improve the management of schedules and traffic flows. This concept is also reflected in European policy framework strategies. The development of an appropriate intermodal transport system would allow real-time tracking of goods, ensure liability for intermodal and promote transparent transport freight. The deployment of such systems like RIS, TAF, ERTMS, and VTMIS shows the progress, which was made in other transport modes, while the deployment of advanced ITS in road transport is helping to improve infrastructure management. A coherent ITS implementation strategy could make a significant contribution to fundamental changes in the logistics chain (European Commission: White Paper, Roadmap to a Single European Transport Area, COM, 2011; European Commission: Commission Communication COM, 2007; Schilk & Seemann, 2012). The concept "freight Internet" is also known as "eFreight" and means refers to a vision of paperless freight carriage processes. The

K. Čižiūnienė (✉)
Vilnius Gediminas Technical University, Vilnius, Lithuania
e-mail: kristina.ciziuniene@vilniustech.lt

D. Dzemydienė et al., *Development of Smart Context-Aware Services for Cargo Transportation*, International Series in Operations Research & Management Science 330, https://doi.org/10.1007/978-3-031-07199-7_18

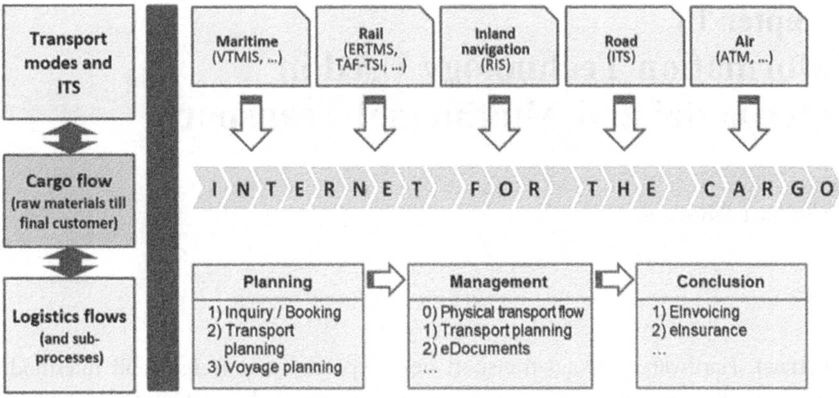

Fig. 18.1 Use of ITS in optimizing transport logistics flows (Source: Compiled by the author based on: Schilk & Seemann, 2012)

term "freight internet" is also known as eFreight and refers to a vision of computerized freight transport processes. In this case, information flow in cyberspace is combined with the physical goods flow. Freight transport will also be associated with changes that are expected to lead to (smart) freight transport in the future. These changes mean that material flows will become self-contextualized and perceived, inseparable from various information services, thus ensuring further automation of transport process management (see Fig. 18.1) (www.efreightproject.eu, online, 2011).

Today, all traffic infrastructures create the right ITS for dedicated networks of traffic. Although inland navigation implements RIS on European waterways, other transport sectors are doing the same. Given that these processes of implementation are not yet complete, the interoperability of traffic ITS will help to ensure an efficient and smooth flow of information for all transport modes:

- Essential and future (= forecast) information on traffic conditions is necessary for logistics planning activities to determine the route and volume of transport. For example, a multimodal transport logistics operator should have information on road traffic on certain sections, existing obstacles, real accidents, working hours and capacity of planned reloading hubs, level of inland water, information on lock condition and other information.
- For logistics management it is important integrated multimodal processes visibility. For example, operator of multimodal logistics should have an information on the current status of the vehicles he has ordered and the estimated time of arrival (ETA) (i.e. trucks, inland waterway vessels, seagoing vessels, railway waggons) in order to track its carried freight.

All available communication and information technologies (ICT), including advanced public transport services such as intelligent transport systems (ITS), need to be used and integrated to make freight transport smarter and to create a so-called freight internet and applications of transport logistics (i.e. software for planning route, tracing, and tracking) from sectors (especially commerce). To ensure the successful management of a large multimodal transport chain, all actors in the process must offer accessible information and ITS services.

18.2 Interoperability of Different Modes of Transport and the Related Problems

In order to ensure efficient and optimal delivery of goods to customers, freight carriage companies use two or more consecutive modes of transport in one transport chain to ensure efficient and optimal delivery of goods to customers.

With the emergence of each new mode of transport, its management and organization have been adapted accordingly. The work of each mode of transport has been organized according to certain carriage rules, traffic organization, infrastructure opportunities and in accordance with own regulations.

The author Baublys (2007) states that the theory of the transport system must examine the interaction of individual transport modes and their work in various transport terminals (Fig. 18.2).

Transport interoperability can be described as multimodal, intermodal, or combined carriage. Multimodal carriage is the transport of goods at least by two (different) transport modes. According to the author Vasilis Vasiliauskas (2013), the main features of multimodal transport: freight carriage by at least two transport modes, under a single carriage contract, one transport document, when there is one party responsible for the entire carriage process, which can conclude subcontractor agreements with the necessary carriers of all transport modes or freight forwarders. Gurnak et al. (2018) found in their research that currently, multimodal transport covers not only technological aspects of coordinated work of different modes of transport, but also economic, organizational, and legal assurance of the carriage process, efficient infrastructure operation, optimal use of vehicles and freight units during traffic flow.

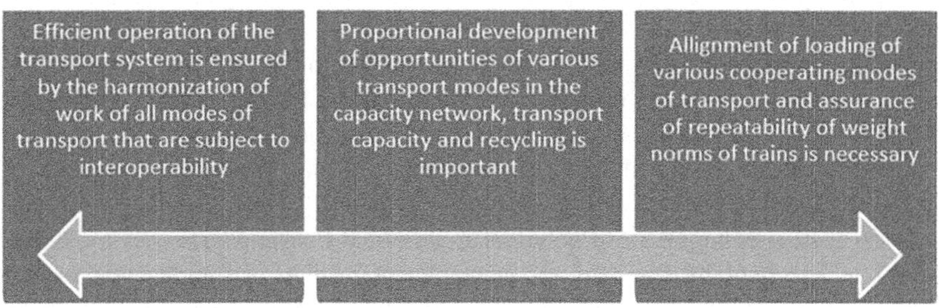

Fig. 18.2 Key principles of the operation of the transport system in the interoperability of all transport modes

Author Boris and Bureika (2017) define intermodal carriage as good carriage in container in several transport modes (road, rail, air, and water). The European Conference of Ministers of Transport (ECMT) defines intermodal transport as the carriage of cargo at least by two different transport modes in the same container or in the same vehicle, where the cargo itself is not reloaded, and only transport mode is changed. Intermodality is understood as a transport system that allows at least two modes of transport to be used in order to achieve the goal of delivering freight without reloading them in the same door-to-door freight unit. Author Vasilis Vasiliauskas (2013) says that intermodality will be implemented under the following conditions:

- The required freight concentration
- High-quality services provided throughout the entire chain
- Effective reloading points and their adequate network
- Standardized equipment, procedures, and documentation
- Efficient use of information systems
- Efficient promotion of carrier activities
- Favourable legal environment

"Standard" intermodal freight carriage usually comprises:

- Pre-shipment
- Distribution at the terminal
- Main carriage by rail
- Loading at the terminal
- Carriage to the destination point

Intermodal rail freight terminals are important for intermodal freight transport by rail due to cost, operational, and quality interaction (Wiegmans & Behdani, 2018).

ECMT defines combined transport as intermodal carriage in Europe, most of which is carried out by sea or rail, while goods are delivered by road.

In order to optimally and efficiently adapt the parts of multimodal transport in interoperability, two multimodal freight transport systems are used, which cover long-distance carriage by at least two modes of transport. The first system uses one transport network only, and the second uses several networks, with maritime or rail transport network acting as the basis. In the second transport system, road transport networks are used only to collect freight from different locations and then to deliver goods to terminals or from the multimodal terminals to the required point. Both of these systems have widely been used around the world, but the second system is the most sustainable because it uses three main parts of the multimodal network, namely, road transport, multimodal terminals and rail transport (Matamala & Salas, 2012).

However, it is important to note that multimodal transport networks are understood as a technology that requires management in light of freight units such as containers, demountable body, pallets, or semi-trailers (Jarašūnienė & Čižiūnienė, 2019).

The provision of information is still one of the most important elements of efficient transport use to facilitate and accelerate the movement of materials and information flows in the transport system by computerizing and automating the work of the elements of the transport system regulating this movement is one of the main goals of the creation and development of the information infrastructure of the transport system (Jankauskaitė, 2010):

- To implement technological means (information and methodological) that ensure reliable, fast, and high-quality transmission of information both in its interaction with the outside world and in the transport system.
- By systematizing and computerizing customer service procedures to improve service of customers of the transport system.
- To create databases of transport system branches and a national database of the transport system of common use, which will allow to obtain prompt, reliable, and comprehensive information on economic and commercial activities of interest to all their subscribers.
- To implement measures that would ensure the interaction of the elements of the transport system with other elements of the domestic and foreign information infrastructure (tax, statistical, customs systems, banks, etc.).

The strictly regulated and defined work process of a transport terminal is constantly affected by various random factors, which must be assessed, because a failure to assess them can severely affect the efficiency of the terminal. Random factors can form for both subjective and objective reasons. Objective reasons can be divided into three main groups:

- Technical (condition of vehicles, containers, lifting equipment, etc. and their technological level)
- Organizational (specialization, scientific and technical information, instruments, work of customs, moral responsibility for the quality of work, etc.)
- Economic (tariffs, salary forms and level, labour intensity, etc.)

Transport interoperability can optimize freight carriage, offer greater opportunities, especially when intercontinental carriage is necessary, and has many advantages. According to Palšaitis (2011), the main advantages of transport interoperability are the following:

- Higher speed of carriage
- Possibility to reduce overall carriage costs
- More efficient use of infrastructure
- Possibility to carry larger loads
- Standardization of container and semi-trailer carriage
- More sustainable carriage, reduction of environmental pollution
- Possibility to simplify the carriage process by using information technologies and systems.

However, it must be stated that there is a problem in this context that the integrity of information systems is not ensured.

Nevertheless, each mode of transport has several characteristics, so it can be said that the entire integrated multimodal process is a complex process because it involves many different participants such as third-party logistics service providers, freight forwarders, couriers and T.T. (Harris et al., 2015). For the best result, the interaction of these participants must be effective and rely on the principle of Just in Time. However, it becomes a challenge because each of the participants uses different technologies. Various activities in the management of multimodal transport, where each stage must be optimized and integrated with each other in order to ensure effective and efficient business operations, including:

- Processing transport orders (forecasting, delivery schedule)
- Preparation of the transport chain (selection and conclusion of service agreements)
- Preparation for carriage (loading, customs procedures)
- Carriage (loading/unloading reports, damages)
- Monitoring carriage (tracking vehicles and monitoring driver behaviour)
- Terminal transactions (managing loading/unloading, terminal of inventories) (INFOLOG, 1999; Harris et al., 2015)

In summary, the range of activities is very wide. Appropriate ITS solutions must therefore be in place to ensure a smooth process.

18.3 Information Technology Systems in Multimodal Transport

ITS was introduced in transport and logistics simultaneously, around the 1960s. And the application of these systems was most visible: in inventory management, transportation routing, planning, and billing. It should be noted that these systems are usually function-based and therefore independent of each other. There was a need for materials (MRP) and production resource planning (MRP II) planning in the 1970s. During that period, an enterprise resource planning (ERP) system was developed. And the development of interorganizational systems only flourished after the commercialization of the Internet in 1995. Until then, electronic data interchange (EDI) dominated only interorganizational communications (Grieger, 2003; Harris et al., 2015).

When it comes to multimodal transport, scientists show that ITS has a great impact on efficient and reliable management of intermodal freight transport in real-time (Dotoli et al., 2010; Boschian et al., 2011). Most scientists focus on certain types of technologies and their application in multimodal transport (Kengpol et al., 2012; Dullaert et al., 2009; Coronado et al., 2009; Bock, 2010). For example, in order to increase the cost-effectiveness, service, and security of various transport entities, Dullaert et al. (2009) provides an advanced mediator based on an expert communication platform.

Bock (2010) proposes a new real-time approach to the freight forwarder network management, which includes multimodal and multiple transhipments in the development of freight consolidation, in order to minimize the empty vehicle mileage and manage procedural disorders. Using a secure access architecture to improve visibility and connectivity in a multimodal logistics environment, it analyzes (Coronado et al., 2009) the possibilities of using special short-range communication (DSRC) and vehicle technologies (Harris et al., 2015).

However, several studies have analyzed the current application of ITS in terms of multimodal transport, with the exception of Giannopoulos (2004) and Perego et al. (2011). Both documents did not specifically address the development of ITS to facilitate intermodal freight transport and enforcement. The analysis of scientific literature and European Union projects (to support multimodal activities) allows to classify according to the TAP (2000) classification into the following main types: cargo resource management programmes and systems, port and terminal communication and information programmes and systems, cargo and fleet maintenance, management systems and applications and an integrated activity/information exchange platform/portal/marketplace. Freight resource management systems and applications use solutions to efficiently use the resources that support the organization, focusing on infrastructure, equipment and manufacturing, financial operations, human resources, transportation planning optimization, transportation routing and planning optimization and execution. etc. The aim of these programmes is to match demand and supply with minimum cost, by consolidating information in the dispatch centre and optimally e matching orders with vehicles (TAP, 2000; Harris et al., 2015). Examples of projects:

- The F-MAN (2005) project has produced a telematics system that gives information about waggon status and position, which allows vehicle fleet manager (railway) to make the economic choice of "its" waggons and to update this decision if a waggon is late.
- Project MarNIS (2009) represents the concepts of or the management of maritime transport information services for the management of the port (TAP, 2000). As many intermediaries are involved in this process such as cargo carriers, railway operators, port administration, cargo management companies and customs, so it is important to ensure one-stop accession system. It is often initiated by government agencies that allow traders to provide the entire import-export and transit information that Ryras are posed and not to provide and recycle the same information to different institutions (Choi, 2011).
- The project "Container Management Intermodal Centres (CHINOS)" of individual terminals or ports, dealt with tasks faced by terminals of container and operators of transport on security issues and volumes of cargo over the innovative information technology, for example RFID (Chinos, 2009).
- MIT (2011) project aimed to expand to a fully industrial and automated system intermodal transport distributed throughout the country and in the full processing of trains from a port to a dry port. Vessel and freight management and monitoring systems and tools aim to reduce uncertainty at each link in the multimodal

transport chain and to improve operational efficiency between connections. ITS management systems allow tracking, tracing and controlling goods and vehicles: they are based on appropriate reporting tools and real-time information, integrating various technologies such as on-board computers, web tools and various identification technologies.

- Project D2D (2005) focused on the integrated and comprehensive multimodal door-to-door management system, the development of a transport chain management system, a monitoring system of freight and the application of "advanced technologies" for improving the intermodal operations performance.
- The M-TRADE project (M-TRADE, 2007) main idea was to create an integrated comprehensive system to provide tracking services, tracking of goods, cargo identification, efficient handling at hubs and terminals, and tracking the transport of perishable and dangerous goods.

The door-to-door container transport chain has been implemented using advanced technologies in the SMART-CM (2011) project. The integrated operational/information exchange platform/portal/marketplace was intended to improve the overall performance of intermodal transport in order to create a seamless and secure information system combining the development of mobile and wireless, tracking, fleet and freight management and Internet technologies. Bringing together all actors to enable cooperation and exchange of information from place of departure to place of destination is a key objective of integrated platforms. Focusing on small and medium-sized enterprises to develop and integrate existing and new intermodal freight transport technologies into a single online platform (GIFTS Integrated Operational Platform–GIP) is a key objective of the Global Intermodal Freight System (GIFTS) (GIFTS, 2004). The activities of this programme relate to freight transport, administrative services, operational control and monitoring functions and e-commerce services. On the basis of the KOMODA project, the architecture of a visual pan-European e-Logistics system optimizing the logistics chain through ITS and intermodal transport was proposed (Fig. 18.3). The e-FREIGHT project aimed at the sustainable and optimal use of freight transport resources in Europe through the e-Freight platform. This platform includes a repository of e-Freight solutions and services and a "runtime" environment for interaction with solutions (e-FREIGHT, 2011).

The ITS varies across different transport modes or between various stakeholders in the chain of intermodal transport (PROMIT, 2009; KOMODA, 2009). In addition, operator may have different and distinct ITS applications provided by different providers of technology services (such as IBM and ORACLE), focusing on specific individual needs. According to PROMIT (2009), there may be little compatibility between these ITS programmes, and this is the main obstacle to the interconnection and integration of different programmes with future programmes. Such issues have a negative impact on communication and cooperation between all actors in the process of intermodal transport. According to Evangelista and Sweeney (2006), PROMIT (2009), KOMODA (2009), it can be argued that the problem of the lack of

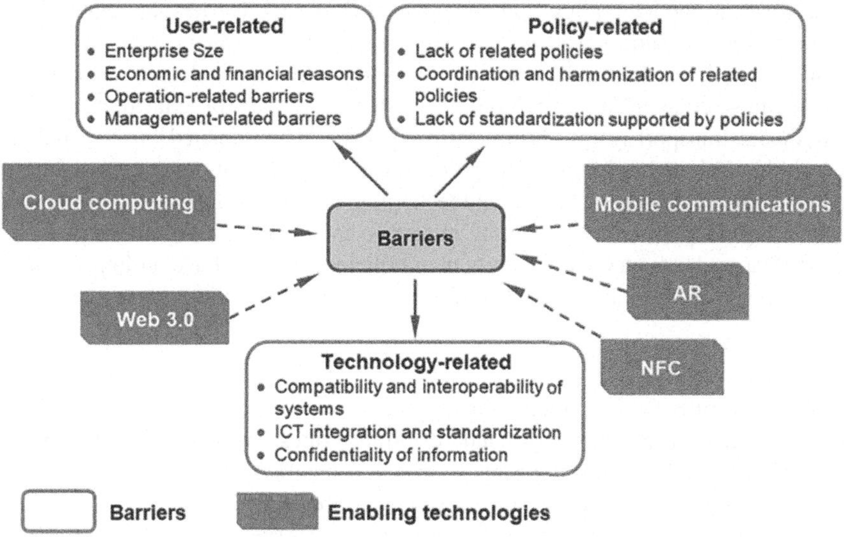

Fig. 18.3 Barriers to ITS adoption in multimodal transport (Source: Compiled by the author based on KOMODA, 2009; Harris et al., 2015)

homogeneous ITS standards affects the development of the entire multimodal transport chain and single-modal operating systems. Key challenges:

- Relate to the integration of all relevant transport modes into a single programme without standardization (KOMODA, 2009)
- According to Perego et al. (2011), Zeimpekis and Giaglis (2006) Pokharel (2005), GIFTS (2004), Hollenstein (2004) integration of ITS applications with old systems
- According to Jakobs et al. (2001), Pokharel (2005) cost of deployment and integration of new technologies

The interconnection of the various programmes used by intermodal operators is vital for the efficient and reliable movement of goods (PROMIT, 2009). Therefore, according to Piplani et al. (2004), Pokharel (2005), the integration of client applications is also considered an important barrier to ITS deployment. Other obstacles deal with the impact of other multimodal transport participants, lengthy implementation of projects of ITS and lack of interoperability of data transmission. In some studies of the freight market, researchers Perego et al. (2011), Pokharel (2005), Piplani et al. (2004) also point out that the apparent time required for full ITS deployment and the rapid ageing of technology are barriers to ITS deployment. It is argued that the mutual simplification of data transmission is mainly due to the reluctance of process participants to cooperate (PROMIT, 2009). For example, reluctance to share relevant data with colleagues prevents some transport intermediaries from operating in open ELM (Electronic Logistics Marketplace). In addition, a lack of security of online transactions and accountability issues related to the data to be exchanged can be seen as a barrier to the adoption of online applications. Policy barriers are related to the coordination and harmonization of different policy levels, which can be an effective tool to facilitate the implementation of some new technologies or methods. Tsamboulas et al. (2007) analyzed the potential impact of related policies on intermodal transport from a European perspective. They pointed out that policies should be designed to improve the efficiency and productivity of intermodal transport by improving organizational and technological aspects. Due to the multimodal nature of transport, which is mainly related to international freight transport, it is likely that each country will have its own policies that may have an impact on the deployment of ICT. INTEGRITY (2011), KOMODA (2009) identify different obstacles:

- Related to politics
- Different rules for each mode of transport
- Different security and safety standards or(and) rules for modes of transport
- Different legal systems according to freight categories
- Different standards and administrative procedures between countries
- Difference between national and European transport policy (Harris et al., 2015)

Moreover, standardized interfaces and open mechanisms of communication for the deployment of ITS in multimodal transport also need to promote and assistance related politics both at the national and European Union level (PROMIT, 2009). There is therefore an urgent need to harmonize and coordinate this isolated and fragmented policy, which is having a negative impact on the deployment of ITS in the various countries, to ensure reliable and efficient operations of international freight. Although the European Commission published several action plans and policy packages related to ITS and intermodality (EC, 2011), there is still a lack of coordination and synergies between stakeholders and the Member States concerned (KOMODA, 2009).

18.4 Systems of Decision Support and Big Data for Intermodal Transport Management

Even though these technological changes have increased the volume and completeness of the information collected by organizations, more important is the ability to use and take advantage of the value of these "big data". Future Decision Support Systems (DSS) manage intermodal transport, taking into account the technological trends under discussion and depending on real-time dynamic and integrated decision-making and better capabilities. Complex analysis can significantly improve decision-making. The analysis of the existing scientific literature focused on areas such as terminal work planning, vehicle work planning, loading, route optimization, and network design. For example, Caris et al. (2008) provide an overview of decision planning and decision methods related to intermodal transportation of freight at decision levels, where the review includes integrated applications and decision support systems. Macharis et al. (2011) propose a solution support framework for the analysis of intermodal transport industry support policies and for the applicable location analysis of intermodal terminals. There are three models of system: (1) NODUS, (2) LAMBIT, (3) SIMBA. These models help to evaluate different solutions: the market area, the new terminal optimal location, the new terminal potential and ultimately the impact on the performance of the waterway network. Kengpol et al. (2012) provide decision for framework of support for logistics service providers and SMEs to choose a multimodal transport route.

For example, the system can optimize a multimodal route in the countries of the Greater Mekong Subregion by integrating programming of analytical hierarchy process and zero-target. Trends in IoT, cloud computing, and business analytics will further enhance the capabilities of DSS and improve user access to respective features. For example, (Harris et al., 2015):

- Real-time dynamic model with management approach is developed, which allows to reduce empty vehicle journeys, to extend freight consolidation, and manage the dynamic disruption of transport networks of freight forwarders (Bock, 2010). The model combines chains of multimodal transport and multiple reload.
- An integrated system metamodeling system has been developed for the management of intermodal transport networks (ITNs) at the both levels: operational (real-time) and tactical (offline) (Boschian et al., 2011). The system has two main modules: the ITN Reference Model (Knowledge Base) and the Modelling Module, which anticipates system behaviour and allows integrated systems to align the proposed strategies and options for management.
- An DSS for intermodal transport is proposed for real-time multi-route planning taking into account several conflicting criteria (e.g. time, cost, emissions) based on a distributed multi-agent system (Dotoli et al., 2013). In order to obtain combinations of the user and the vehicles, a genetic algorithm must be used according to the wishes of the user. This system offers solutions when transportation is not possible and is affected by external factors (e.g. strikes).

The use and capturing of the value that DSS derives from detailed and real-time information will ensure competitiveness for intermodal carriers and will likely have an impact on their business performance. It can be argued that greater visibility would encourage organizations to be more active if something goes wrong. Beneficiaries would be informed of the consequences of developing alternative options to overcome uncertainty and disruption. Using telematics to monitor the position in relay time would make it possible to understand the behaviour of truck drivers and route optimization. Automating operations would reduce the need for manual labour and delivery times. However, the challenge is how quickly to analyze the growing amount of data, which is often poorly structured. Traditional DSS would often not be able to manage and analyze that non-traditional data. One of the solutions adopted by the industry is to use Hadoop, an open source software system, designed to work with different big data sets. According, Harris et al. (2015) it breaks down a large data set into smaller groups, processes them separately and then combines the results into a smaller data set that is easier to analyze.

The quality of road, rail and water transport and of the services differs. This is partly due to fundamental differences and peculiarities of the mode of transport, and differences in infrastructure. The user considers road freight carriage as a benchmark for freight carriage and does not see infrastructure constraints or terminal problems as a reason to move from road to an intermodal alternative. Congestions and freight transit points remain the main infrastructure problems. The majority of freight corridors also introducing innovations or new concepts, but intermodality usually stops at the entrance to the terminal. Usually, activities of an intermodal terminal require different experts, such as gate inspectors, dispatchers, local operators, information managers and others. Moreover, the inspection of operations and equipment, maintenance and repair, supply and waste management must all work effectively together.

Therefore, the management of complex congestion at terminals requires very sophisticated computer technology, not in terminal itself, but also increasingly more often at transport units, to be able to manage the arrival and departure of vehicles in terminals. The scientific literature often does not delve into the problem of information management, which is a key factor in the success of intermodal transport together with the equipment management software. Most components require computer technology to manage operations.

Identification and tracking technologies have been used increasingly more often at terminal. Information at the gate has been collected electronically, while computers receive information from labels to be able to decide on the freight handling processes, place of storage and all the necessary administrative information.

18.5 Conclusions

The successful use and integrity of intermodal and multimodal transport require not only specialized terminal equipment but also a unified IT system (or at least the successful interoperability of different systems). These solutions are needed to ensure that information is processed in a timely manner and that both the technological and management processes are not disrupted. However, with intermodality, this issue becomes a challenge. Failure to resolve it will not ensure the efficient operation of the transport system, proportionate capacity in the network, and so on. Therefore, it must be stated that in this context there is a problem that does not allow to ensure the integrity of information systems, which slows down the whole process. In view of these challenges, a number of EU projects are being developed to match supply and demand at minimal cost. However, of these projects, they focus more on the specificities of each mode of transport than on ensuring the interoperability of IT between applications and uses in different modes of transport. This goes further and further, due to the big data sets, which becomes a challenge for all involved in the process.

References

Baublys, A. (2007). *Transporto rūšių sąveika*. Technika.

Bock, F. (2010). Real-time control of freight forwarder transportation networks by integrating multimodal transport chains. *European Journal of Operational Research, 200*(3), 733–746.

Boris, J., & Bureika, G. (2017). Lietuvos intermodalinių terminalų veiklos vertinimas. 20-osios jaunųjų mokslininkų konferencijos „Mokslas – Lietuvos ateitis" teminės konferencijos. http://jmk.transportas.vgtu.lt/index.php/tran2017/tran2017/paper/viewFile/119/143

Boschian, V., Dotoli, M., Fanti, M. P., Iacobellis, G., & Ukovich, W. (2011). A metamodelling approach to the management of intermodal transportation networks. *IEEE Transactions on Automation Science and Engineering, 8*(3), 457–469.

Caris, A., Macharis, C., & Janssens, G. K. (2008). Planning problems in intermodal freight transport: Accomplishments and prospects. *Transportation Planning and Technology, 31*(3), 277–302.

CHINOS. (2009). *Deliverable D0.4 Final report*. European Commission.

Choi, J. Y. (2011). *A survey of single window implementation*. World Customs Organization Research paper No. 17. Accessed June 12, 2020, from http://www.wcoomd.org/en/topics/research/activities-and-programmes/~/media/2DF5A36D3ECA46CCB7B17BDF77ACC021.ashx

Coronado, A. E., Lalwani, C. S., Coronado, E. S., & Coronado, C. E. (2009). Facilitating multimodal logistics and enabling information systems connectivity through wireless vehicular networks. *International Journal of Production Economics, 122*(1), 229–240.

D2D. (2005). *D2D: Demonstration of an integrated management and communication system for door-to-door intermodal freight transport operations*. Final Technical Report. Accessed September 2, 2021, from http://www.transport-research.info/Upload/Documents/200607/20060727_143636_83260_D2D_Final_Report.pdf

Dotoli, M., Hammadi, S., Jeribi, K., Russo, C., & Zgaya, H. (2013). *A multi-agent decision support system for optimization of co-modal transportation route planning services*. Proceedings of the 52nd IEEE CDC (Conference on Decision and Control), Florence, Italy, December 10–13.

Dotoli, M., Fanti, M., Mangini, A., Stecco, G., & Ukovich, W. (2010). The impact of ICT on intermodal transportation systems: A modelling approach by Petri nets. *Control Engineering Practice, 18*(8), 893–903.

Dullaert, W., Neutens, T., Vanden, G., Berghe, T., Vermeulen, B., Vernimmen, F., & Witlox. (2009). MamMoeT: An intelligent agent-based communication support platform for multimodal transport. *Expert Systems with Applications, 36*(7), 10280–10287.

EC. (2011). *European Commission. Roadmap to a Single European Transport Area – Towards a competitive and resource efficient transport system*. Accessed July 25, 2021, from http://eur-lex.europa.eu/LexUriServ/LexUriServ.do?uri=COM:2011:0144:FIN:EN:PDF

e-FREIGHT. (2011). *e-Freight: European e-Freight capabilities for co-modal transport*. Accessed March 1, 2021, from http://www.efreightproject.eu

European Commission: Communication from the Commission COM. (2007). *The EU's freight transport agenda: Boosting the efficiency, integration and sustainability of freight transport in Europe*. 606 final, p. 4.

European Commission: White Paper, Roadmap to a Single European Transport Area, COM. (2011). 144 final, p. 2ff.

Evangelista, P., & Sweeney, E. (2006). Technology usage in the supply chain: The case of small 3PLs. *International Journal of Logistics Management, 17*(1), 55–74.

F-MAN. (2005). *F-MAN: Rail car asset management*. Deliverable D16: Final report. Accessed January 15, 2020, from http://www.transport-research.info/Upload/Documents/200604/20060411_172123_25402_F-MAN%20Final%20Report.pdf

Giannopoulos, G. A. (2004). The application of information and communication technologies in transport. *European Journal of Operational Research, 152*(2), 302–320.

GIFTS. (2004). *GIFTS: Global Intermodal Freight Transport System*. Deliverable D03. User needs identification and analysis. Accessed August 21, 2021, from http://gifts.newapplication.it/gifts/project_deliverables/public/GIFTS-REP-D03-1_0.zip

Grieger, M. (2003). Electronic marketplaces: A literature review and a call for supply chain management research. *European Journal of Operational Research, 144*(2), 280–294.

Gurnak, V., Khomenko, I., & Volynets, L. (2018). Mperatives of multimodal transport development in accordance with the tendencies of the transformational changes of the national economy in the conditions of europeanintegration. *Economics. Ecology. Socium*, *2*(1). http://www.ees-journal.com/index.php/journal/article/view/52/23

Harris, I., Wang, Y., & Wang, H. (2015). ICT in multimodal transport and technological trends: Unleashing potential for the future. *International Journal of Production Economics, 159*, 88–103, . ISSN 0925-5273. https://doi.org/10.1016/j.ijpe.2014.09.005

Hollenstein, H. (2004). Determinants of the adoption of information and communication technologies (ICT): An empirical analysis based on firm-level data for the Swiss business sector. *Structural Change and Economic Dynamics, 15*(3), 315–342.

INFOLOG. (1999). Final report. Accessed March 3, 2021, from https://trimis.ec.europa.eu/sites/default/files/project/documents/infolog.pdf

INTEGRITY. (2011). *INTEGRITY: Intermodal global door-to-door container supply chain visibility: FP7 project*. Deliverable D0.3. Final report. Accessed January 20, 2021, from http://www.integrity-supplychain.eu/

Jakobs, K., Pils, C., & Wallbaum, M. (2001). Using the internet in transport logistics – the example of a track and trace system. *Lecture Notes in Computer Science, 2093*, 194–203.

Jankauskaitė, A. (2010). Jūrų transporto ir uostų valdymo informacinės technologijos. Magistro baigiamasis darbas. VGTU.

Jarašūnienė, A., & Čižiūnienė, K. (2019). Krovinių vežimo geležinkelių ir jūrų transportu sąveikos probleminių aspektų vertinimas Lietuvoje. Vilniaus Gedimino technikos universitetas.

Kengpol, A., Meethom, W., & Tuominen, M. (2012). The development of a decision support system in multimodal transportation routing within Greater Mekong sub-region countries. *International Journal of Production Economics, 140*(2), 691–701.

KOMODA. (2009). *KOMODA: Co-modality – towards optimised integrated chains in freight transport logistics*. Deliverable D3.2: Action Plan. European Commission within the Seventh Framework Programme (2007–2013).

Leal Matamala, E., & Perez Salas, G. (2012). *Port–rail integration: Challenges and opportunities for Latin America* (p. 11). Bulletin FAL. No. 310.

Macharis, C., Caris, A., Jourquin, B., & Pekin, E. (2011). A decision support framework for intermodal transport policy. *European Transport Research Review, 3*(4), 167–178.

MarNIS. (2009). *Maritime navigation and information services*. Accessed January 15, 2020, from http://www.transport-research.info/web/projects/project_details.cfm?id=11127

MIT. (2011). *MIT – Metrocargo Intermodal Transport*. Accessed January 15, 2020, from http://www.mitproject.eu/

M-TRADE. (2007). *M-TRADE*. Accessed January 15, 2021, from http://www.newapplication.it/mtrade/mtrade_home.html

Palšaitis, R. (2011). *Tarptautinio verslo transportinis logistinis aptarnavimas*. Technika.

Perego, A., Perotti, S., & Mangiaracina, R. (2011). ICT for logistics and freight transportation: A literature review and research agenda. *International Journal of Physical Distribution and Logistics Management, 41*(5), 457–483.

Piplani, R., Pokharel, S., & Tan, A. (2004). Perspectives on the use of information technology at third party logistics service providers in Singapore. *Asia Pacific Journal of Marketing and Logistics, 16*(1), 27–41.

Pokharel, S. (2005). Perception on information and communication technology: Perspective in logistics. *Journal of Enterprise Information Management, 18*(2), 136–149.

PROMIT. (2009). *D 3.1-3.3 Best Practice Year 1, 2 and 3*. Accessed August 17, 2021, from http://www.promit-project.net/UploadedFiles/Deliverables/PROMIT_BPH3_April09_cp_MSR.pdf

Schilk, G., & Seemann, L. (2012). Use of ITS Technologies for Multimodal Transport Operations – River Information Services (RIS) Transport Logistics Services. *Procedia – Social and Behavioral Sciences, 48*, 622–631. https://doi.org/10.1016/j.sbspro.2012.06.1040

SMART-CM. (2011). *SMART-CM: Smart Container Chain Management*. Accessed January 10, 2021, from http://www.transport-research.info/web/projects/project_details.cfm?id=37567

TAP. (2000). *TAP (Telematics Applications Programme): Freight Intermodal: Final area report*. 4th Framework Programme for RDT&D. Accessed August 28, 2020, from ftp://ftp.cordis.europa.eu/pub/telematics/docs/tap_transport/freight_report.pdf

Tsamboulas, D., Vrenken, H., Lekka, A., & M. (2007). Assessment of a transport policy potential for intermodal mode shift on a European scale. *Transportation Research Part A: Policy and Practice, 4*(8), 715–733.

Vasilis Vasiliauskas, A. (2013). *Krovinių vežimo technologijos*. Klaipėda.

Wiegmans, B., & Behdani, B. (2018). A review and analysis of the investment, and coststructure of, intermodal rail terminals. *Journal Transport Reviews*.

Zeimpekis, V., & Giaglis, G. M. (2006). Urban dynamic real-time distribution services – insights from SMEs. *Journal of Enterprise Information Management, 19*(4), 367–388.

Chapter 19
Development of Intelligent Service System for Multimodal Transport Management

Dalė Dzemydienė

Abstract The main objective of this part of the research is to describe the development of a Smart-SDS for multimodal freight transport management. Components of decision-making processes are integrated into the Smart-SDS structures. Computerized decision support processes involve many different components of the infrastructure. The assessment of situations is based on the criteria and priorities applied in the practice of specialist experts. The conceptual models applied in a computer-based system must allow the representation of dynamically changing and complex processes. Logistics and transport management processes must be in line with the requirements and goals of the 2030 Strategy for Sustainable Development. The Sustainable Development Agenda's goals (SDG) and targets are interrelated with all levels of policy implementation. This section describes the results of research related to the object—multimodal freight transportation. However, the data show that the conditions of the war (the brutal aggressive actions of Russia against Ukraine and the crimes against of humanity) are very serious, and that the military invasion severely violates the lives and rights of many people and is completely incompatible with the GSD. This unpredictable war, which began in February of 2022 and early in 2014, is affecting major changes in Europe and around the world. Developing a computer system to enable the operational management solutions becomes a daunting task. Decision support models can be developed under normal, well-established routine conditions. We understand that the concept of "logistics" originated from the regulation of hostilities, given the changes in this concept in the historical development of this type of science. However, the process of building a knowledge base (KB) becomes complex by creating rules for representation of activities describing formally—how to act quickly in unforeseen, extreme, asynchronous situations. The entire Smart-SDS system is integrated with the ICT infrastructure and with embedded systems. The development methodology

D. Dzemydienė (✉)
Vilnius Gediminas Technical University Vilnius Tech, Vilnius, Lithuania

Vilnius University, Vilnius, Lithuania
e-mail: dale.dzemydiene@mif.vu.lt

D. Dzemydienė et al., *Development of Smart Context-Aware Services for Cargo Transportation*, International Series in Operations Research & Management Science 330, https://doi.org/10.1007/978-3-031-07199-7_19

of this type of Smart-SDS system (for multimodal transport management) combines modeling components, simulation models for subsystem behavior analysis and experimental studies with prototypical subsystems. Some of the results we get show outdated situation analysis. In war conditions, events and conditions can occur unpredictably, and forecasts become unpredictable. Once the environmental monitoring infrastructure is destroyed, the performance management system will not work. Rules that reflect fair operations and affect the environment in which transport should behave in unpredictable conditions will not work during the destruction of infrastructure in geographical areas. Risk assessment is possible, but there are no enough mathematical models to help in unpredictable situations. Under such contextual situations, the decision-making subsystems will not reflect real management and help in operative control of processes.

Keywords Multimodal transportation · Decision support system · Smart service delivery system · Conceptual model · Knowledge-based system

19.1 Introduction to Multimodal Transport Specifics

The research works which are analyzed the multimodal transport at primary stages are concerned the detailing analysis of inventory of transported goods and hazardous materials. A statistical analysis provides the spectral picture of traffics and accidents at those directions and locations (Batarlienė & Baublys, 2007; Barysienė, 2012; Batarlienė, 2007; Baublys & Jarašūnienė, 2010; Kaklauskas et al., 2009; Jakimavičius & Burinskienė, 2009). EU efforts continue for implementation of projects for ensuring safer transport. The projects are concerned development of safety means and systems which started early, like eCall (eSafety initiatives, 2007), services and components for development of intelligent transport systems (Wibisono et al., 2009; Du et al., 2021, ITS Platform, 2016; Bielskis et al., 2009; Dzemydiene & Burinskiene, 2021).

Emerging areas of the multimodal transportation of goods are already being analyzed using a combination of radio frequency technology (RFID) and a mobile sensor-based control systems. The functions of such systems are devoted for ensuring the seamless monitoring from source to destination (nodes) and visibility from global to local intermodal transport management, assessing the potential risks associated with the transport of goods and especially of dangerous goods (Dzemydiene et al., 2019; Knospe and Poh, 2004; Kang et al., 2008; Andziulis et al., 2010). Scientific results show that the integration of risk management practices into transport processes through a mobile management system would increase the mobility and overall efficiency of intermodal container transport with the optimal number of transport security checkpoints required to reduce the pre-determined loss of value during a given route (Dzemydienė & Dzindzalieta, 2009; Dzemydiene et al., 2016). Interesting example of system with sensor parameters is presented in Shariat-Mohaymany and Babaei (2010). The main goal of intermodal container transport managers is ensuring a high level of information and transportation security. The aims can be increased by achieving of ICT efficiency and by improving the transport strategy and operations. For improvements of capabilities of IS, ICT, and common infrastructure systems are needful the integration of all efforts and the implementation of an innovative mobile management systems (Bojković et al, 2010; Deveci et al., 2018; Faccin & Nunes, 2015; Osmundsen et al., 2018; Dzemydiene et al., 2022).

Examples in worldwide market show that new kinds of mobile channels and ICT help in modern modes implementation for operative management of multimodal transportation of goods following the recommendations of Grean Deal (European Commission, 2019, 2020). Attempts to develop the analytical systems for planning of the intermodal transportation of goods, which include harmful (hazardous) materials by rail truck are presented as well by following the requirements of (Evaluation and Classification of Chemical Wastes, 2010; Hazardous Substances Ordinance, 2004). Aspects of implementation of multi-objective optimization models and infrastructure for planning and managing of intermodal shipments are analyzed with inclusion of intermodal services in the US East (Verma & Verter, 2010). Risk analysis of intermodal transportation in EU regions was provided (Fabiano et al., 2005; Dzemydiene & Dzindzalieta, 2012). Intermodal transport route optimization problems are analyzed (Macharis et al., 2010; Ishfaq & Sox, 2010; Macharis & Pekin, 2009; Limbourg & Jourquin, 2009; Andziulis et al., 2016). Logistic chain analysis is considered in works (Shariat-Mohaymany & Babaei, 2010; Bazar et al., 2010).

Many authors pay attention for management of intermodal transport itself (Thill & Lim, 2010; Kreutzberger, 2008; Ruiz-Garcia et al., 2007). Analytical view insight of implementation of sensor-based RFID technology in intermodal container transport we can found in works (Vergara et al., 2007; (Amador & Emond, 2010; Chavali et al., 2008). Analysis of some security aspects and implementation of RFID data transmissions are analyzed in (Kaya et al., 2009; Piramuthu, 2007). The research results were obtained by describing the usage of IS for intermodal container transport (Oztekin et al., 2010; Dias et al., 2009; Ngai et al., 2007; Ferrer et al., 2010).

The RFID-based reverse logistics system and implemented a genetic algorithm to optimize the locations of product return collection points to maximize customer coverage, allowing for cost-effective and environmentally sound recycling is presented in Lee and Chan (2009). The new sensors are analyzed (Jedermann et al., 2006) by integrating the communication and software technologies to expand the premises of food tracking and tracing systems when sensor data are collected from a wireless network detects an embedded evaluation unit. The empirical research results were obtained in Kurmis et al. (2015), Dzemydiene and Dzindzalieta (2012), Dzemydiene et al. (2020).

The aims of this research part are going forward for analyzing possibilities of development of Smart-SDS and formulating methods for the assesment of context data and for reflecting results of evaluation surroundings in the system. The empirical research methodology includes the review of transport specialists experts opinions and results of imitational models and their evaluations. Our efforts are forwarded for achieving the goals of sustainable transport.

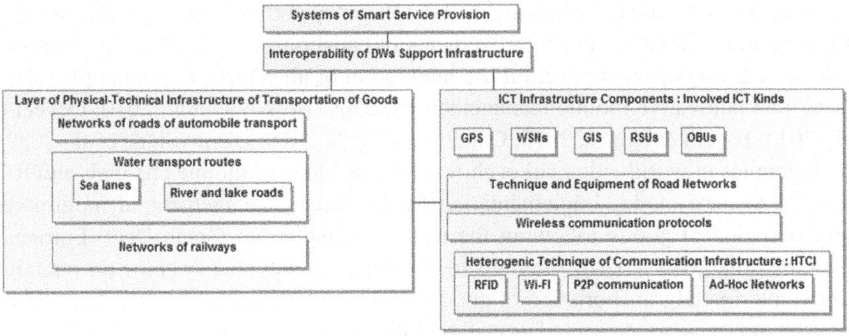

Fig. 19.1 Schema of representation of interoperability of two layers of infrastructure for the Smart-SDS

19.2 Multilayered Architecture of Smart-SDS with Integration of MC-DSS Components for Multimodal Transportation

The research challenges of developing the smart service delivery system (Smart-SDS) with integration of multi-criteria decision-making subsystems (MC-DSS) are related with creation of automated control structures that support and integrate all necessary information systems (IS), data warehouses (DWs), including possibilities of communication channels. Smart-SDS have operated in direct access mode (i.e., online). Multimodal freight transport as the main managing object is analyzed by multi-dimensional aspects and by multi-criteria estimations.

We are developing few models for description of dynamic characteristics of multimodal transportation processes. The specific areas of the elaboration are needful to integrate into the analyzing process at certain stages of the implementation of the methodology. An ontology can be created by using some computer-based ontology construction tools. The means, like UML, Petri nets and programming of imitational models are selected at the stages of development of ontology of domain (Pavliska, 2006; Šutienė et al., 2010). The multimodal transport infrastructure is assessed as dynamically changing and complex phenomenon (Fig. 19.1).

Decisions are based on many different factors:

- Description of transport technical infrastructure that is involved in the chains of logistics and the system for decision support.
- ICT needful to support transportation processes and context awareness, and the work of Smart-SDS.
- Management infrastructure, i.e., assessment of organizational priorities and dynamic aspects of control and governance.

- Comparison of actual transportation processes and emerging situations and reports with the goal of developing the action plans across the multiple levels of management in a more efficient and adaptable manner.
- Risk assessment and identification of risky situations and operational control of these actions.

The operative control activities are supported directly by the concrete actions provided by means of the Smart-SDS, which influence behavior of drivers. Algorithms that enable the generation of such actions are integrated with decision-making components and influence the operational management activities. Notifications may announce the corrections of actions on concerns.

Management actions are interlinked and affect transportation activities. The areas are included into the management area like management of the physical and technical infrastructure of transport (PTIT), the monitoring of the environment and weather conditions, the maintenance of "green" transport corridors, etc. The PTIT is connected with terminals of reloads of goods, cross-boarding posts and other interrelated structures.

The multi-layer structures of ISs are designed for description of all components. For example, the terminal description structure have integrated capacities, traffics, and scheduling, the dynamical changes of analyzed objects with involving of the set of processes around.

We can classify the set of activities for representation in ISs, like the activities of preparation of traffics, scheduling of reloading terminals. Into this analysis are included:

- Reloading from the auto tracks, trailers, containers, etc. to wagons of rails and opposite
- Reloading from wagons of rails into the water transport means and opposite
- Reloading from automobiles (auto tracks, etc.) to means of water body transport and opposite

As well as, the important means and modules are developed for more detailed representation of such activities:

- Modules for development of traffics and timetables
- Modules for optimization of workflows
- Modules for assessment of procedures of monitoring of reloading activities and enabling the possibilities of synchronization of activities
- Methods for evaluation of risks and recognition of real transportation situations

The assessment of activities of all types of junction-nodes is related to methods of evaluation of time consuming, macro environment status, impact of political situations, and similar.

The cycles of multimodal transport include such important features like:

- The goods are transported in the same transport unit using several different modes of transport.

- The sufficiently long freight transfer cycles are formed, which are planned and re-planned according to different real situations.
- The conditions for the consolidation of forces, which are required for the international and multimodal transportation of goods are considered, and the algorithms for the obtaining such procedures for activation are developed.
- Complex links between individual transport interoperability constituents and are related to individual business chains.
- Transport processes have a strong relationship with time and geographical location.

In order to evaluate and identify heterogeneous and complex subsystems, we need to ensure the functioning of all the complex mechanisms involved. The entire management structure transmitted through an intelligent system are related to management process optimization solutions.

To ensure these processes, the system have had ability to display:

- The general information about multimedia services
- The measures of capacities of WSNs and all components of ICT infrastructure

Based on the results of the evaluation, which should reflect real-life situations, templates for past work should be used for analytical analysis and reports.

The studies on the attractiveness of alternative corridors for multimodal transport are related with evaluation criteria. The review of opinions of participating in these processes actors (drivers, transportation experts and specialists) are obtained.

Multi-criteria assessment process starts from indicating the list of assessment parameters. The list of important assessment parameters (criteria) for evaluation of multimodal transportation are created by opinions of specialist experts of multimodal transportation:

$\Delta t_{t,i,j}$—the parameter for expression of time duration of concrete transportation trip [in hours]

$\text{Cost}_{t,i,j}$—the parameter for expression of cost of the concrete cycle of transportation [Euros by trip]

$\text{Rel}_{t,i,j}$—the parameter for expression of transportation corridor reliability, expressed by qualitative value evaluated in range from 1 to 5, by applying the Likert scale

$\text{Security}_{t,i,j}$—the parameter for expression of transportation corridor security, expressed by qualitative value from 1 to 5 by applying the Likert scale

$\text{Qtr}_{t,i,j}$—the parameter for expression of quality of transport system and others

The route stretch reliability ($\text{Rel}_{t,i,j}$) can be express as the Product of the Cartesian, including:

$$\text{Rel}_{t,i,j} = \{\text{Parameter of probability of possible attacks}\}$$

$$\times\{\text{Parameter of ensuring of cargo security}\}$$

$$\times\{\text{Parameter of ensuring of ecological safety}\}.$$

The ecological characteristics of goods are important for such evaluations too. Some parts of these parameters are provided by the CARNET-TIR cards information (e-documents of Transport International Routers). The classes of dangerous and hazardous materials and goods are implemented by special Directive.

The transferring processes are expressed by models with dynamic characteristics (Table 19.1). Transport technical infrastructure is involved in the transport logistics chain:

- Information communication infrastructure provided to describe typical transportation processes.
- Management infrastructure, i.e., assessment of organizational priorities and dynamic aspects of control and governance.
- Comparison of actual transportation processes and emerging situations and reports with the goal of developing action plans across multiple levels of management in a more efficient and adaptable manner.
- Risk assessment and identification of risk situations, which are a direct driver of decision-making and operational management to take corrective action.

Multimodal transport cycles are characterized by such type activities:

- Cargo is transported in the same transport unit using several and different modes of transport.
- A sufficiently long freight transfer cycle is formed, which should be evaluated.
- The conditions for the consolidation of forces required for the international transport of goods are activated.
- Complex links are forming between individual transport, interoperability constituents and individual business chains.
- Transport processes have a strong relationship with time control and geographical location.

The functioning of all complex mechanisms, which are involved in multimodal transportation are related to each other and are interacting as internal and external parts. The entire management structure transmitted through an intelligent system should enable the optimization solutions.

The Smart-SDS has general information about multimedia services and others components, which are described in Table 19.1. Additionally, the provision of management services and analyzing functions are included. The results of the evaluation of reflected real-life situations are provided for template-based analysis and are implemented for next work steps.

To define the interaction of objects in transport processes, when analyzing ground conditions, roads must be covered as structures that do not change so quickly over time, including land networks: road networks for motorized and/or autonomous vehicles, road networks for railways, and routes to water bodies (seas, rivers, lakes, etc.). Route selection requires specific information, which is especially

Table 19.1 Multi-layers of infrastructure of multimodal transportation which are involved in the Smart-SDS

Short description of layers	Detailing of components	Supporting infrastructure	Subsystems with AI methods
Road networks I Layer, including physical—technical infrastructure (PTI) of multimodal freight transport	An overland road network for road transport; Waterways (marine and other water bodies), means of transport, routes; Railway networks and facilities.	• Road network equipment and types of equipment; •Wireless networks; • Wireless Sensor Networks (WSN); • Global Positioning Systems (GPS); •Operating online geographic information systems (GIS); • Roadside units and management infrastructure (RSUs); • e-Service delivery systems; • Interoperability of IS with cloud computing services, IoT, etc. technologies	• Communication network infrastructure management and situation assessment, interfaces with solutions of PTST change planning; • Performance monitoring systems; • Supporting means for the quality of road networks; • Decisions on the development of "green" transport corridors; • Dynamic weather assessment and forecasting and warning systems; • Assessment of contextual information, including weather forecasts, assessment of freezing conditions for water bodies and other types of roads; • Information assistance in informing about real situations and conditions of road sections.
II Layer infrastructure for transshipment of cargo in multimodal transportation processes with enabling infrastructure and transshipment terminal works to support multimodal transport	Management of material, financial, and information flows through terminal infrastructure, including transshipment, warehousing (i.e., consolidation) processes; Operation of congestion terminals combining road, rail, and water vehicles.	Equipment for monitoring terminals of transshipment processes of transport units (trailers, containers) and other objects; existing online geographic information systems (GIS); • Roadside Infrastructure (RSU); • Smart service delivery systems; • Integration of	• Transportation planning systems; • Decision-making systems; • Traffic management and optimization tools; • Wireless sensor networks (WSNs) and monitoring systems. • Synchronous and asynchronous process control systems; • Risk assessment,

(continued)

Table 19.1 (continued)

Short description of layers	Detailing of components	Supporting infrastructure	Subsystems with AI methods
		information systems with cloud computing services, IoT, etc.	forecasting, and rapid response systems. • Workflow schedules; • Unification of e-document flows and recognition of such information to support all activities; • Automated management systems for customs procedures.
Layer III for transport management at transport corridor road sections from junction (e.g., border crossing points, road crossings, cities) to junction, where multimodal transport routes are formed and modified	Ongoing management processes at international access points by analyzing the road network between the parties involved	Prolonged processes related to customs procedures, accompanying documents and e-document processing systems, border crossing procedures, etc.	
IV Infrastructure layer of vehicle companies, location of vehicle companies, stories, etc.	Transport logistics systems, Organizational systems for various vehicles; Systems for assessing and monitoring the technical condition of vehicle types	Vehicle types and equipment in the vehicle; Types of computer communication networks, communication systems, standardized wireless communication protocols.	• Maintaining a relatively high quality of work in vehicles and transport fleets; • Traffic regulation subsystems; • Driver support subsystems; • Work and activity planning involving all parties involved; • Risk assessment and management subsystems.

important when developing intelligent service systems. What data are considered contextual information has been discussed in the first chapters and they influence decision-making.

The technical conditions of the road must allow to determine the technical needs (or other characteristics) of the road surface of the selected route for a specific seasonal period. They may also affect the package of requirements for freight transport during a certain period of the year and relate to meteorological forecasts. In the next stage of system development, it is important to define how the required parameters will be entered and how reliable expert sources will help to determine the specific required values of these parameters.

19.3 Representation of Subsystems with Knowledge Base Structures

This subsection is devoted for the aim to develop the decision support subsystem with functionality of multi-criteria evaluation and AI methods. The methods are used with possibilities to modify the planning and re-planning activities of freight transportation. In the dynamically changing situations, the usage of fixed offline plans is not suitable. Therefore issue in the trip points there, it is possible to make different alternative decisions are analyzed. Special artificial methods (AI) are included into the designing of the subsystems, i.e., multi-agent working engines, artificial planning components, estimation of possible alternatives, and road choosing algorithms. The conceptual models are used for expression of static and dynamic characteristics of transportation. An examples of developing the conceptual schemas of DBs and knowledge base (KB) are presented as well.

19.3.1 Structure of Decision Support Processes for Multimodal Transportation

An abstract subsystem architecture for planning solutions is given in Fig. 19.2. This subsystem deals with the planning and management of multimodal transport corridors and consists of:

- The activity planning level
- The planning of processes involved in the cooperation in order to carry out the transportation processes expeditiously
- The systems for monitoring of situations

The planning processes are performed by using hierarchical planning manner. The compliance or non-compliance with the initial plan is identified.

The architecture consists of two interacting components: the Operational management subsystem and the Field subsystem (Fig. 19.2).

The decisions are supported by the Operational subsystem and are focused on the highest level of planning, i.e., the means for finding of optimal combinations of transportation modes. The results of operational planning are obtained at the middle level of planning processes. The field description modules are focused on transport mode. The type of communication is V2X (i.e., Vehicle-to-Everything). Higher-level operational plans contain the means for path constructions, i.e., transporting corridor planning, re-planning, and scheduling.

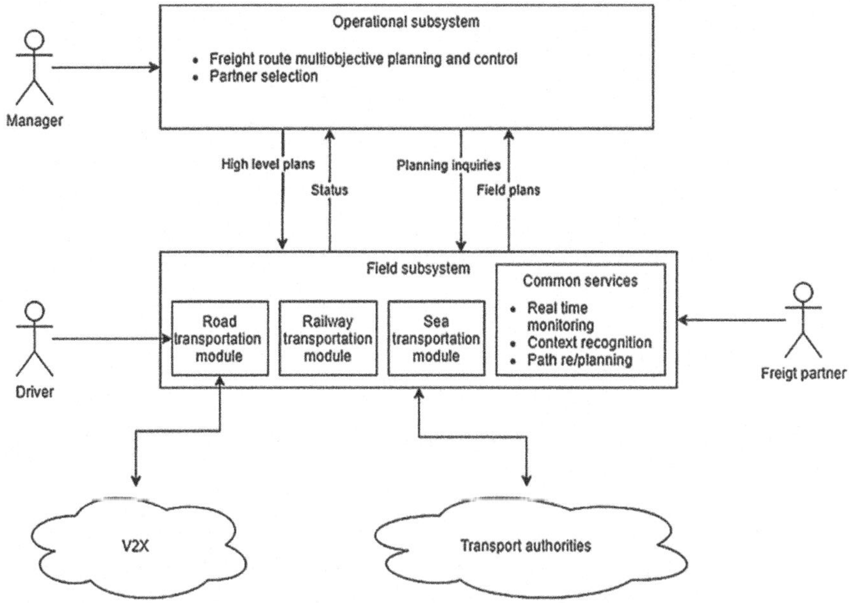

Fig. 19.2 Abstract structure of subsystems for planning solutions (Source: Dzemydiene et al., 2020)

19.3.2 Agent-Based Architecture of Subsystem for Planning of Transport Corridors

The agent-based approach is one of most popular in development of AI systems for complex and dynamic domains (Dzemydiene et al., 2020). The type of artificial agent's design method that is defined as the Belief-Desire-Intention (BDI) agent's structure practically used method. The BDI is usually implemented for autonomous working software.

An example of the architecture of such a subsystem is presented in Fig. 19.3. The operative work of such subsystem is activated by several artificial agents that operate according to the BDI model and help in the multimodal transport field for choosing the right plan at right time moments of considering situations.

The architecture implements some of the concepts that are expressed in the model by defining the connections between concepts: beliefs, desire, intentions. "Beliefs" are facts related to the knowledge base (KB) that allow a smart agent to act. KB was created using multiple layers of multi-component structures to describe the multimodal transport domain. "Desires" are the specific states of the world's agent that artificial agents seek to achieve. For example, the library of available travel plan templates contains a set of plans with goals and target states. "Intentions" indicates the current status of the agent. The agent's goals oblige him to reach a certain target

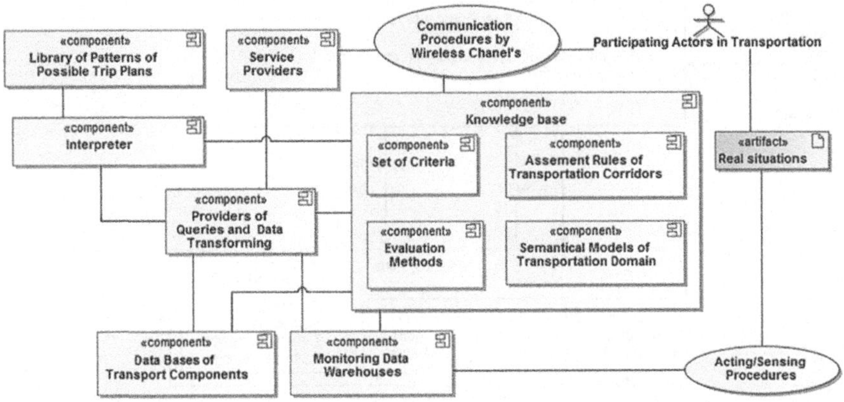

Fig. 19.3 Componential architecture of the subsystem for planning tasks of multimodal transportation based on BDI agents (Source: Dzemydiene et al., 2020)

state. The core of this subsystem is the subsystem moderator, which implements the cycles of considering alternative solutions and connects other parts of the subsystem.

The reflection and decision cycle allows the agent to choose the most appropriate follow-up plan for the situation under analysis. Of course, more precise explanations are needed to describe the situation. The agent undertakes to pursue the chosen goal with certain intentions. However, the reasoning cycle provides algorithms and a formal description of the actions that allow the creation of situational agents. The agent constantly monitors and changes commitments as they achieve the chosen goal, these commitments become insignificant.

For interaction with the real environment is created the artifact—"Real situations." The assessment of such artifact is carried out using the Communication Procedures and Acting/Sensing modules (Dzemydiene et al., 2020). Acting/Sensing procedures are merged since the software agent interacts with the world only by using other web services based on REST architectural style. The sensing can occur only by sending the request and receiving responses (Fig. 19.3).

When formally describing the performance of an artificial agent, its working algorithms need to be specified according to a formal description of the BDI method. Its working mechanism incorporates elements from KB. One part of the KB as Assessment Methods is constructed as at least a graph. A formal description is presented using concepts that are described in a unified Resource Description Framework (RDF) and enables abstract syntax for the data model. The work of agent actions is described in several n-interactions. One part of the KB has a formally described evaluation method.

Each graph consists of n-tuples, where each tuple has a subject, an object, and a predicate:

$$m = (n_m,\ t_m,\ p_m,\ b_m\ ),\ \text{where :}$$

n_m is the method name
t_m is the task name
p_m is the pre-conditions and they construct the set of queries for KB
b_m is the type of algorithm (i.e. the sequence of steps (tasks or actions) of agent)

The agent transition function is described as a four-member interaction $\Sigma = (S,\ A,\ \gamma,\ q)$, where:

- S—represent the set of states (a finite set)
- A—represent the set of actions
  The action can have one of states $a_i \in \{\text{draft, executing, done, failed}\}$.
  As well as the set of actions is divided into two subsets:
- A_k—is the set of information gathering form actions. These actions do not change the world state.
- A_s—is the set of state changing actions.
  $\gamma : S \times A_k \rightarrow S$ is a formula that describes the state-transition function.
  $q : S \times A_k \rightarrow Q$ is a formula that describes a quality assignment function. Where each $q_i \in Q$ is a vector.

The planning problem can be formally defined as the triple:

$$P = (\Sigma, s_0, g),\quad \text{where :}$$

- s_0—is the initial state;
- g—is the goal.

The plan is a sequence of actions that lead to a goal state, i.e., $\pi = a_0, \ldots, a_n$ such realizing the state from $\gamma(s_0, \pi) \in S_g$.

19.3.3 Representation of Semantical Models in KB

The multimodal evaluation of transportation corridors is based on semantical models of KB. The DWs contain semantical models for description of parameters and attribute structures of object classes. The classes include the semantical structures about the freightage junctions, the border crossing points, cities with different population densities, ports, and similar objects. The conceptual models are developed for all needful components of the KB. The meta-models are developed for working repositories of the DWs.

When assessing the technical conditions of the road, the technical characteristics of the pavement of the selected road section, which change in different periods of the year, are assessed. These requirements of the Road specifications may also affect the package of requirements for the carriage of goods during a certain period of the year.

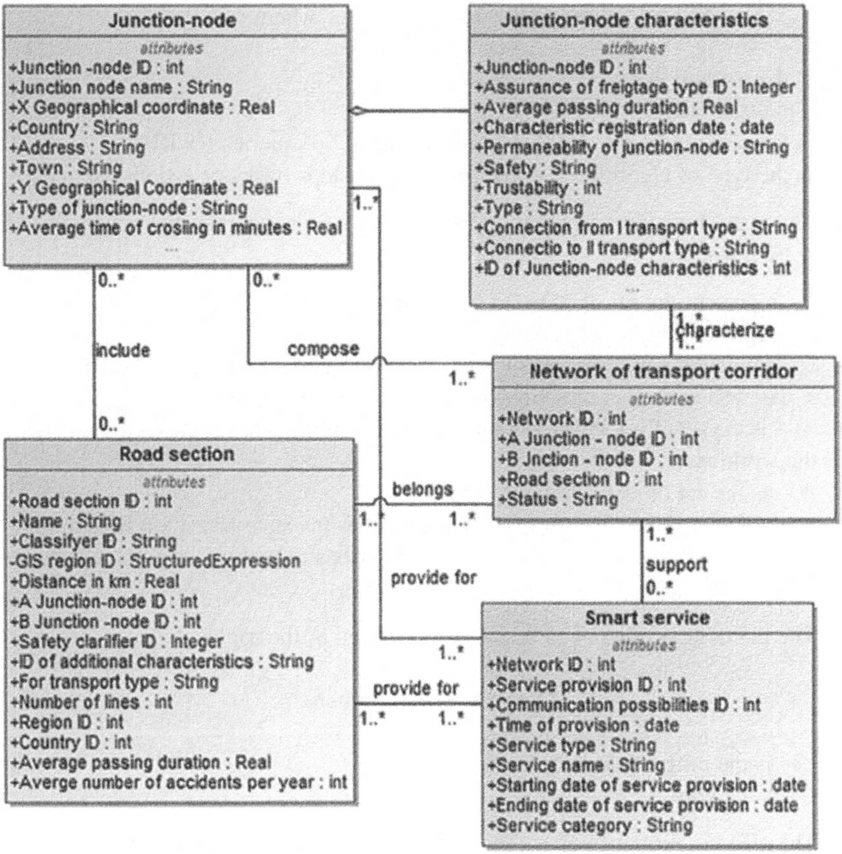

Fig. 19.4 Representation of road's network characteristics

The assessment of meteorological conditions is included in the decision selection subsystem as well. The example of description of the junction-nodes and relations with transportation networks is presented in Fig. 19.4.

The dynamic aspects are representing in other models for achieving of coordination of activities during transportation of freights (example is presented in Fig. 19.5). These models are implemented for synchronous work of all participating actors: like senders and receivers, agents responsible for cargo forwarding, operators of reloading terminals, drivers of sea ferry bridges, etc.

For representation of river transport, the ships, ferries, bridges of rivers and lakes are included.

In order to make the right routing decision, all transport corridors are assessed by including alternative routes from the same destination to the same end point. To achieve such an assessment, the modeling (simulation, emulation) methods are applied. The case study of the multimodal transport corridor workflow model evaluation model is presented (Fig. 19.5).

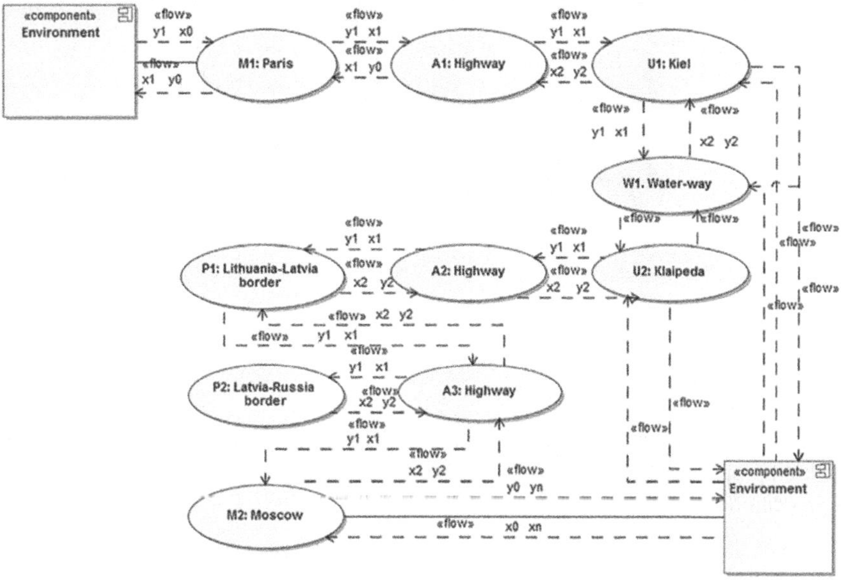

Fig. 19.5 The representation of workflow model of analyzed multimodal transport corridor

Cooperative and non-cooperative behavioral aspects are included. In the selection process of routes involved the agent-user of vehicle (or other transport mean) and some criteria like cost and duration of assessing corridors are expressed by: tr_cost_i, Δt_i, tr_rel_i.

The decision is made in such a way as to minimize the parameters of the cargo transportation (price, travel time, security and reliability). However, it is not possible to reduce the total cost of the network, and decisions are made to partially increase the efficiency of traffic.

The assessment of road sections in multimodal corridors includes the overall model under assessment and possible delays related to traffic movements in urban areas, which are more difficult to estimate accurately over time. Traffic movement includes traffic on streets and intersections. Congestion occurs when traffic is extremely high and its speed is extremely low. In some sections of the road, traffic flow can reach maximum peak time. An estimate of the flow rate that may be present in situations with the lowest possible velocity can be predicted.

Flow analysis, commonly referred to as standard traffic flow analysis, is most commonly performed. It is based on static single-period valuation models. Static single-period models are used to estimate congestion costs. It turned out that the static models did not capture congestion for some time. Because congestion varies, it is important to simulate traffic patterns over time. Simulation can help you create traffic schedules for a specific time period and location. Freight delivery uncertainty can also be modeled based on estimates of time and road segment characteristics in such models. Some authors are looking for possible options to reduce traffic

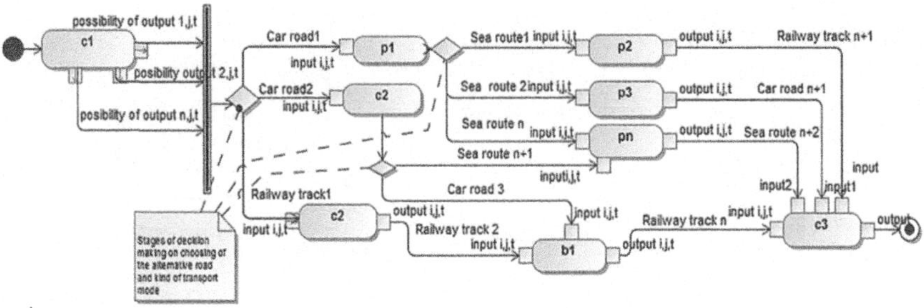

Fig. 19.6 The representation of the MC-DS processes for choosing alternatives of different multimodal transport corridors

congestion or pollution. The historical traffic data collected is used for analysis, including optimization models. Dynamic optimization models used to control traffic lights can help reduce traffic congestion.

19.3.4 Formal Specification of Decision Support Processes

Discrete time model's components and certain time-aggregated positions are used to describe the processes. For ensuring the simulation results of continuous processes are used aggregated positions for representation and simulation of continuous processes. Some of the possibilities of simulation models are based on making decisions that are more applicable in practice. When the calculations cannot express the processes simulated with clear set constructions, aggregation functions, analysis of additional indicators, vague means of fuzzy logic are used.

In order to illustrate the possible decision-making processes in which the appropriate alternatives are selected, we provide their description at the possible stages of travel planning, taking into account the specificities of multimodal transport corridors. An indicative model for the evaluation of such transport components is presented in Fig. 19.6, by using the UML diagram notation for representation of activities. Rhombic positions represent decision points where all possible versions of alternatives (solutions) are evaluated at specific locations in the corridors of the multimodal chain. Possible versions are analyzed under conditions where this is possible.

In the simulation model, two sets of data flow monitoring for each process (operation) are applied, namely the set of inputs—In $= \{\text{input}_{i,j,t}\}$ and the set of outputs—Out $= \{\text{output}_{i,j,t}\}$. This is important information to accompany the cargo. Formally describing $i = 1,\ldots,n$ are transport identifiers; $j = 1,\ldots, m$—vehicle identifiers; t_k—time moments during their recording in the monitoring data recording system (and in the retrospective analysis of such data, which is extracted from the monitoring data repositories).

Monitoring data is recorded when the transport vehicle reaches the corresponding node n_l. Recording moments of time is important in monitoring processes, such data is stored in data warehouses (DWs).

Accumulated tracking DW can help both operationally manage and analyze transportation and logistics processes and provide intelligent services based on MC DSS (Alazzawi & Žak, 2020; Anton et al., 2004; Mitropoulos & Prevedouros, 2016; Zavadskas et al., 2007a, b; Brauers & Zavadskas, 2009).

Some possibilities for formal expression are based on the principle of more adaptable solutions by analogy. When estimates are based on fuzzy set designs, an inappropriate aggregation function may be used or generalizations may be estimated with deviations, summaries can be based on outliers.

The simulation model uses several types of node-defining activities that are important and must be identified in the decision support process simulation model to recognize:

$C = \{c_i\}$ is the set for description of activities in cities;
$P = \{p_j\}$ is the set for description of activities in ports;
$B = \{b_k\}$ is the set for description of behavior board crossing points.

The applied aggregate values of total quality assessment can justify the possibilities of simulating decisions, i.e., they are explained by comparisons as "lower or equal" quality values taken from comparisons described above. Applying the expressions of summary quality indicators in our decision-making model, we will follow the formal description described in (Castillo & Pitfield, 2010). The following formalized description of the decision-making process helps to construct algorithms of the decision-making processes in the system:

$$ls_1 \leq_Q ls_2 \Leftrightarrow M_1(ls_1) \leq M_1(ls_2) \wedge M_2(ls_1) \leq M_2(ls_2) \wedge \ldots \wedge M_n(ls_1) \leq M_n(ls_2). \quad (19.1)$$

Performing the algorithm steps to select the best solution from the presented set of multimodal transport corridor alternatives $\{ls_k\}$. At each of the possible points, a comparative analysis of all possible alternatives ls_k and the parameters $\{M_i\}$ describing them is performed. Only after determining that all parameters meet certain conditions of a better solution (assessing the context), it is concluded that the aggregate quality of the solution ls_{k+1} is better than ls_k. This step of the algorithm works when the parameters evaluated according to the whole ls_{k+1} alternative are better than the ls_k (according to the established estimates of the priority evaluations of the criteria). Sometimes measures M_i can be contradictory and partly redundant. There is room here for bipolar and asymmetric connection.

Following the recommendations of Pamučar et al. (2020) and Pamučar and Savin (2020) in the mathematical description of validity, coverage, and deviations, which help to evaluate the best solution for choosing one alternative in case of deviations:

- If the validity degree v is very small or very high.
- And if the sufficient coverage C is very small.

Following (Pamučar et al., 2020) the outliers can be calculated by:

$$O = t(s(v, 1 - v), (1 - C)). \tag{19.2}$$

T is a t-norm, s is a t-conorm, v is validity, and C is covered. The formula can express the coverage:

$$i_c = \frac{\sum_{i=1}^{n} t(\mu_S(x_i), \mu_R(x_i))}{n}, \tag{19.3}$$

where i_c is the average evaluation of the founding criteria i by its coverage c that is influencing the forming process of concrete possible decision; $\mu_S(x_i)$—the relative importance of the criterion i that was assessed in a decision-making situation dulled by the average calculation by the relevant specialists experts; and $\mu_R(x_i)$—the relative importance of criterion i according to the structure of R part importance.

The adaptable solution (AS) with restriction usually covers a relatively small subset of the considered data (Pamučar et al., 2020), and the formula can define its coverage C:

$$C = f(i_c) = \begin{cases} 0 & i_c \le r_1 \\ 2\left(\frac{i_c - r_1}{r_2 - r_1}\right)^2 & r_1 > i_c \le \frac{r_1 + r_2}{2} \\ 1 - 2\left(\frac{r_2 - i_c}{r_2 - r_1}\right)^2 & \frac{r_1 + r_2}{2} < i_c < r_2 \\ 1 & i_c \ge r_2 \end{cases}. \tag{19.4}$$

There, adjusting r_1 and r_2 according to the structure of R part.

By evaluation of quality, we can simplify that AS is of high quality when validity and non-outliers of indirect coverage are high:

$$Q_C = t(v, 1 - O) = t(v, s(t(1 - v, v), C)). \tag{19.5}$$

Alternatively, coverage is defined as significant when the coverage value is 0.5 or greater:

$$Q_C = \begin{cases} t(v, C) & C \ge 0.5 \\ 0 & \text{otherwise} \end{cases}.$$

The product of t-norm makes this distinction, but the aggregated value is low. It is not a problem when searching for the best AS, but it may indicate that, even though ls_2 is the best, its quality is not high.

$$Qc < (C \text{ and } v), \text{when } (C \text{ and } v) \text{ are lower than } 1. \tag{19.6}$$

For the mining AS, according to the recommendation provided in Pamučar and Savin (2020), it is possible to find such an alternative by considering the linguistic description of possible solutions of interest. We can reveal all relevant summaries evaluating high validity. The finding of Q, S, R by Pamučar and Savin (2020) recommendations on the subject could be formulated as such.

$$\begin{aligned} &\text{find} \quad Q,\ S,\ R \\ &\text{subject to} \\ &Q \in \_Q; \quad S \in \_S; \quad R \in \_R; \quad v(Q,S,R) \geq v_k. \end{aligned} \tag{19.7}$$

Q is a set of quantifiers of interest; R and S are sets of relevant linguistic expressions for restriction and present summary, respectively; v_k is the threshold value from the (0, 1] interval. It is important to state that all feasible solutions produce such linguistic summary (Q^* R^* are S^*).

By mining of AS of interest regarding quality measures, we can reveal all relevant summaries considering quality measures, the algorithm structure is expressed by Eq. (19.8):

$$\begin{aligned} &\text{find} \quad Q,\ S,\ R \\ &\text{subject to} \\ &Q \in \_Q; \quad S \in \_S; \quad R \in \_R; \quad v(Q,S,R) \geq v_k; \\ &c(Q,S,R) \geq c_k; o(Q,S,R) \geq o_k, \end{aligned} \tag{19.8}$$

where v_k, c_k, and o_k are threshold value from the (0, 1] interval.

The pseudo algorithm for comparison may be converted into

$$\begin{aligned} &\text{find} \quad Q,\ S,\ R \\ &\text{subject to} \\ &Q \in \_Q; \quad S \in \_S; \quad R \in \_R; \quad QV_c(Q,S,R) \geq qv_k, \end{aligned} \tag{19.9}$$

where qv_k is threshold value of the aggregated quality measure.

Values that are assigned for measures is from the set of S = {small, medium, high}.

Values of terms of the set for expressing quantifiers are assigned from Q = {few, about half, most of}.

The assessment procedure can be expressed as a relaxed conjunction, if attributes should be satisfied:

$$\nu(z) = \wedge_{j=1}^{p} \mu_{R_i}(z) \otimes \mu_Q \left(\frac{1}{q} \sum_{i=1}^{q} \mu_{P_i}(z) \right) \in [0, 1], \tag{19.10}$$

where p is the number of hard conditions, q is the number of soft conditions and (x) is an aggregation by expression of conjunction or suitable of unformal conditions.

19.4 Representation of Functional Requirements of Multimodal Transportation in Simulation Models

When designing a multi-component evaluation model for the assessment of multimodal freight corridors, we are applied the piece-wise linear aggregation-based simulation modeling method (PWLASM) (Pranevičius, 2003; Šutienė et al., 2010). This method allows to express both discrete and continuous characteristics of simulated processes and imitate their properties. Because transportation processes can have linear and discrete character, PWLASM method enables us to express discrete and continuous processes and their interactions. PWLASM became the main methods for simulation modeling of multimodal transport.

The formalism of Petri nets, which was very suitable for the formal specification of processes, is also applied. This type of Petri nets belongs to a class of time automata with continuous process and evaluation components. Transport processes are depicted as sets of interactions between transitions and positions that interact with each other. Transformation (or evaluation) procedures and associated components for activating transitions. The discrete state component includes a set of properties to be evaluated in that transition process. The subset of transition-type positions can perform continuous processes. This is related to the expression of continuous dynamic processes and can be related to the specific characteristics of the transport process (e.g., for expressing of location co-ordinates, obtained during transportation references, etc.).

The components of multimodal transport are defined mathematically by several sets: state sets Z with input signals X and output signals Y. The state of an aggregate can change only in two cases: when an input signal enters the aggregate or when a continuous component acquires a defined value. The state $z \in Z$ of piecewise-linear aggregate has analogy with the formalism of description of the state of piece-linear Markoff process. Functions $z(t) = (\nu(t), z_\nu(t))$, represent transitions where $\nu(t)$ is discrete state component taking values on countable set of values and $z_\nu(t)$ is continuous components comprising of $z_{\nu_1}(t)$, $z_{\nu_2}(t)$, …, $z_{\nu_k}(t)$ co-ordinates (Fig. 19.7).

In the imitational model, all transition performances are considered in dynamics of the processes expressing them in the set of time moments t_i from the used scale of time measurements T. We are following all requirements of dynamic processes representation implementing and following of axioms of temporal logics.

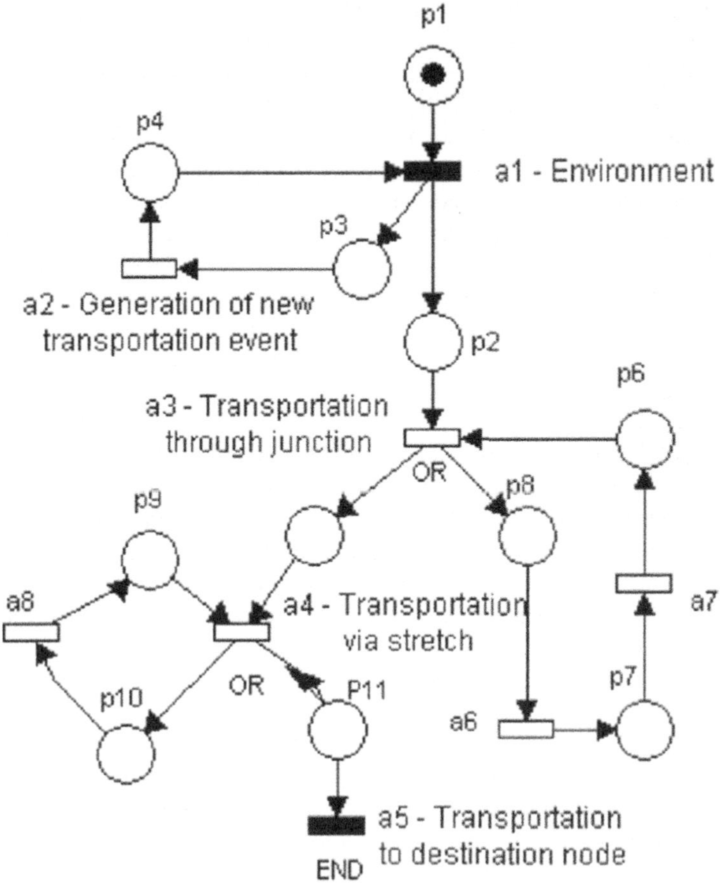

Fig. 19.7 Representation of processes of case study of multimodal transportation by using notation of Petri nets

Also, each "Intersection Node" as the subsystem of the whole imitational system has input channels with a "Chain" subsystem on the right and left (except for the first and last "Intersection Nodes" and with "Environment" channels) through which the Junction-Node receives an incoming signal for the transport of goods through the completion functionality of the "Chain." The length of the delay time is (waiting for the transport time schedule, stops for the driver to relax). The "Chain" subsystem is intended for a model of a specific transport route chain, i.e., concrete stretch of ground road, sea or rail. When goods are arrived at the "Chain," they have to wait or not wait for another transport's medium-term schedule. Data for the Environment subsystem has two incoming channels with the first and the last node of the subsystem.

The modeling characteristics of the model are a certain route transportation time, cost, and confidence (Pavliska, 2006). For example, in imitational models are expressed such things like answering the question for decision-making: how long the goods can be delayed calculating such value from minimum to the maximum, based on the average transportation time of before obtained example.

The aggregate components of intermodal transportation are defined by sets of states Z with, input signals X and output signals Y. The state of an aggregate can change in two cases only: when an input signal arrives at the aggregate or when the continuous component acquires a definite value. The aggregate functioning is considered in set of time moments $t \in T$. The state $z \in Z$ of piecewise-linear aggregate is the same as the state of *piece-linear Markoff process*, i.e. $z(t) = (\nu(t), z_\nu(t))$, where $\nu(t)$ is discrete state component taking values on countable set of values and $z_\nu(t)$ is continuous components comprising of $z_{\nu_1}(t), z_{\nu_2}(t), \ldots, z_{\nu_k}(\mathrm{t})$ co-ordinates (Fig. 19.7).

Also each "Junction-Node" subsystem has incoming channels with "Chain" subsystem to the right and to the left side (except the first and the last "Junction-Nodes" which have channels with the "Environment"), through which the "Junction-Node" gets the incoming signal about the completeness of transportation of goods through the "Chain." The delaying time duration are (waiting for a transport time schedule, stops for the driver relaxation). The subsystem "Chain" is dedicated to model of the certain chain of the transportation route, i.e., a land-road, a sea road or a rail. When the goods arrive to the "Chain" they have to wait or not for the next transport mean time schedule. Data about subsystem "Environment" has two incoming channels with the first and the last node of subsystem.

Modeling characteristics of the model are certain route transportation time, price and confidant (i.e., how long the goods can maximally be late with respect to the medium transportation time). A set of input locations represent signals for transportation through junction-node. For example, the set of positions (locations) is $PX_k = \{p_2, p_6\}$, where p_2 is a signal stating that the load has arrived to node from route in one or another direction (if the transportation from destination or source node is described), $i = 1, 2$. The set of token characteristics is expressed like $\xi j_i = (\xi j_{i1}, \xi_{i2}, \xi_{i3}, \xi_{i4})$, and they can express, for example row data parameters, where ξ_{i1} can be the code of the load (cargo); ξ_{i2} express concrete time moment when the load (cargo) has arrived to the route; ξ_{i3} is a time interval during which the load (cargo) is transported through the route; ξ_{i4} is transportation expenses, evaluating a passing part of the route.

The subsystem "Chain" is dedicated to the model of certain chain of transportation routes. When the goods arrive at the "Chain," they have either to wait or not for the next transport means according to the time schedules of transport means. The needed data on chains of the route are certain list of route chains and the good code.

The set of output signals from junction-node is initialization signal for starting transportation thought route stretch. The transition a_4 has input positions $PY_k = \{p_3, p_6\}$, where p_3 is a signal stating that the transportation of load has finished through this route and the load drive into the node to one or another direction (destination/

source); $i = 1, 2$; $\xi y_i = (\xi y_{i1}, \xi y_{i2}, \xi y_{i3}, \xi y_{i4})$, where ξy_{i1} is the code of the load; ξy_{i2} is a time moment when the load has arrived to the system; ξy_{i3} is a time interval during which load is transported through the route; ξy_{i4} is a transportation cost, evaluating passing part of the route.

The set of external events: $E' = \{e'_1, e'_2\}$, where e'_i denotes the finishing of attendance in the junction—node and the load arrived in the chain; $i = 1, 2$.

The set of internal events: $E'' = \{e''_1, e''_2\}$, where e''_i—transportation of load has finished in the chain of route and the load drive into the junction—node, where $i = 1, 2$.

Controlling sequences of internal events:

$$e''_i \Rightarrow \{\varepsilon_{ij}\}_{j=1}^{\text{Const}},$$

where $i = 1, 2$ ε_{ij} is duration of transportation in chain of the route, which distributed continuously in interval $[a,b]$; j corresponds to direction (forward/backward).

The macro transition of Petri net has the same properties as the aggregate state: $Z_\nu(t_m) = \{W(t_m), \surd(t_m)\}$, where $W(t_m) = \{w(e''_1, t_m), w(e''_2, t_m)\}$ is a continuous aggregate state component; $w(e''_1, t_m)$ is a time instant when the event e''_i has to occur; $i = 1, 2$; $\surd(t_m) = \{\nu_1, \nu_2, \nu_3, \nu_4(t_m), \nu_5(t_m)\}$ is discrete aggregate state component, where ν_1 is a parameter corresponding to the expenses of load transportation in analyzed chain; ν_2 is a parameter corresponding to the initial time moment of appearance of load in the route; ν_3 is a parameter corresponding to duration of transportation in certain chain of route; $\nu_4(t_m)$ is equal to 0, if there are not load in the route, and equal to 1, if load is transported by the route; $\nu_5(t_m)$ is a parameter corresponding to delaying of transportation (time duration τ).

The dynamic properties of transportation processes can be described by applying Petri nets (Fig. 19.7).

Initial state: $w(e''_i, t_0) := \infty$, $i = 1, 2$; $t_0 = 0$;

$$\nu_4(t_0) := 0, \quad \nu_5(t_0) := \infty.$$

The algorithm for description of behavior of the "Chain" system is expressed by the set of transitions and output operators:

```
H(e'i) : i = 1, 2;
ν4 (tm+1) := 1; τp:= tm mod24; Start X:= ν2;
Wwhile Start X < τp Do Start X := Start X + ν3;
   j = 1,..., Const;
If Start X=τp
Then begin w(e"i, tm + 1) := tm + εj ; ν5 (tm + 1) := 0;
   end
Else begin
   ν5 (tm+1) := Start X − τp;
   w(e"i, tm + 1) := tm + εj + ν5 (tm + 1);
   end
   ttr := xi3 + εj + ν5 (tm+1);
```

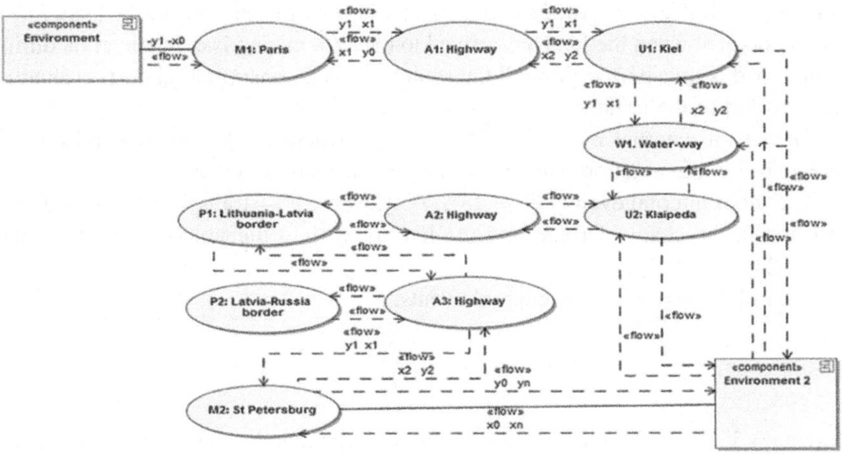

Fig. 19.8 Example of description of alternative corridor in West-East directions of multimodal transportation in Europe

$k_{tr} := x_{i4} + \nu_1;$
$STD(\nu_5(t_{m+1}));$
$G(e'_i): i = 1, 2.$
$Y := \varnothing;$
$H(e''_i): i = 1, 2;$
$w(e''_i, t_{m+1}) := \infty;$
$\nu_4(t_{m+1}) := 0;$
$\nu_5(t_{m+1}) := \infty;$
$G(e''_i): i = 0, 1, 2;$
$y_i := (y_{i1}, y_{i2}, t_{tr}, k_{tr}).$

The proposed formal description is used for assessment of imitational processes used for the Smart-SDS development. In real world, various multimodal transport corridors could be named. For the application of the methodology, the set of transport corridors is initially defined. Later, the number of transportation corridors is narrowed to the ones that fulfill initial selection objectives. Later on, the multi-objective function is used for the comparison and the ranking of possible alternatives.

A deep representation of knowledge by the semantic model can help choose priorities of the principles and rules provided for specialist experts of different target areas. The adequate imitational models of behavioral analysis allow us to predict further evolution of alternatives and increase the quality of decision-making. The representation of one of the analyzed corridors is presented in Fig. 19.8.

The concrete corridor's real environment applied for multimodal transportation includes all information flows from all participating multimodal components. Data transfer channels have relations in two directions through them are passing the signals about starting and ending processes in each of participating ways (i.e., A1–A3, W1).

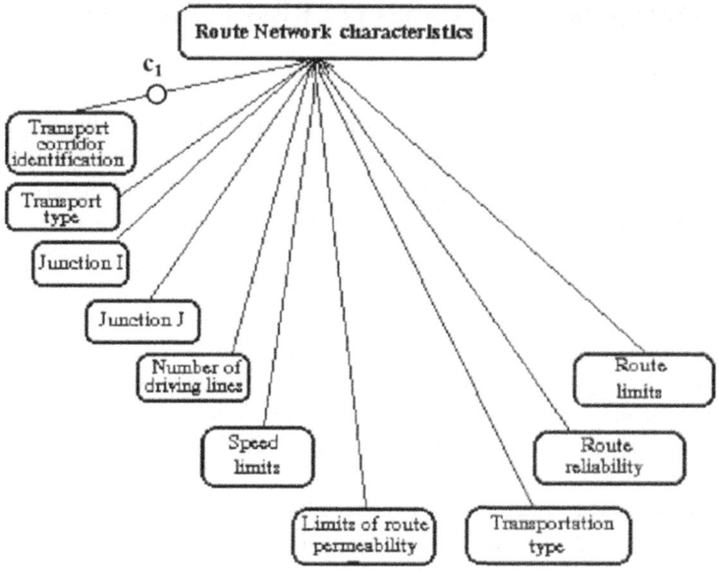

Fig. 19.9 Main parameters of evaluation of road network characteristics

The information of starting and completeness of processes and including the unsuspected and accident events, delays are transmitted fraught these channels accordingly in the West-East or East-West directions, by analyzing the Europe region (Fig. 19.9).

Rules for making decisions are of the form: *IF<conditions>-THEN <actions or conclusions>*. Such rules can be expressed by more detailed form:

IF X_1 is A_1 AND…AND X_n is A_n THEN Y is B, where A_1,.., A_n and B are certain predicates characterizing the variables X_1, ..., X_n and Y. The set of *IF-THEN* rules forms linguistic description:

$$\begin{gathered} R_1 := \text{IF } X_1 \text{ is } \mathrm{A}_{11} \text{ AND} \ldots \text{AND } X_n \text{ is } \mathrm{A}_{1n} \text{ THEN } Y \text{ is } B_1 \\ \cdots\cdots\cdots\cdots\cdots\cdots\cdots\cdots \\ R_m := \text{IF } X_1 \text{ is } \mathrm{A}_{m1} \text{ AND} \ldots \text{AND } X_n \text{ is } \mathrm{A}_{mn} \text{ THEN } Y \text{ is } B_m. \end{gathered} \tag{19.11}$$

There each transition of node of Petri net corresponds to one rule or more rules and can be described by linguistic description.

19.5 Possibilities of Assessment of Risks in Multimodal Transportation

The risk assessment should consist of many factors. The route is divided into road sections, and each has different characteristics. Risks are associated with accident scenarios that are affected by the types of dangerous goods and the environment. Methods for multiple description of complex scenarios influence their classification by type and may be based on an ontology of this phenomenon. Federal and provincial legislation acts are provided an extensive list of products, substances, or organisms classified as hazardous. The products are classified into nine classes: explosive, flammable, radioactive, and others. The transport risk model proposed by the authors focuses on the probabilistic estimation of the frequency of accidents, describing the parameters of cargo tracking by describing certain scenarios of risky events following works (Fabiano et al., 2001, 2005; Dzemydiene & Dzindzalieta, 2012).

One of approaches that helps in representing the formal presentation of risky situations and their assessment is based on evaluation of the set $S = \{s_k\}$ of types of scenarios of accident events of transportation, which we can to recognize (where $k = \overline{1, n}$. is number of obtained types).

The estimated number of fatalities due to accidents that may occur on a given road section r and may develop under the s_k, scenario can be expressed as follows:

$$B_r = \sum_{k=1}^{n} f_r N_{r,s_k} P(s_k), \tag{19.12}$$

where f_r is the function of frequency of accident in the r-th road section [number of accidents per previous year (year-1)]

N_{r,s_k}—for expression of the number of fatalities according to a scenario s_k in the r-th road stretch [accident fatalities-1]

$P(s_k)$—for expression of the probability of evolving scenarios of type s_k, following the accident (i.e. collision; roll-over; failure, etc.)

The transportation network can be considered as a number of road stretch and junctions (nodes) linked one to another by a number of arcs (Fig. 19.10).

The junctions represent the cross roads, towns, tool-gates, storage areas, etc. in the transportation network. An arc between two junctions can be characterized by a different number of road sections. The expected number of fatalities for the road section (arc) can be expressed as:

$$B = \sum_{r} \sum_{s_k} f_r N_{r,s_k} P(s_k). \tag{19.13}$$

The frequency of an accident involving in the scenario s_k, on the r-th road section, can be expressed as:

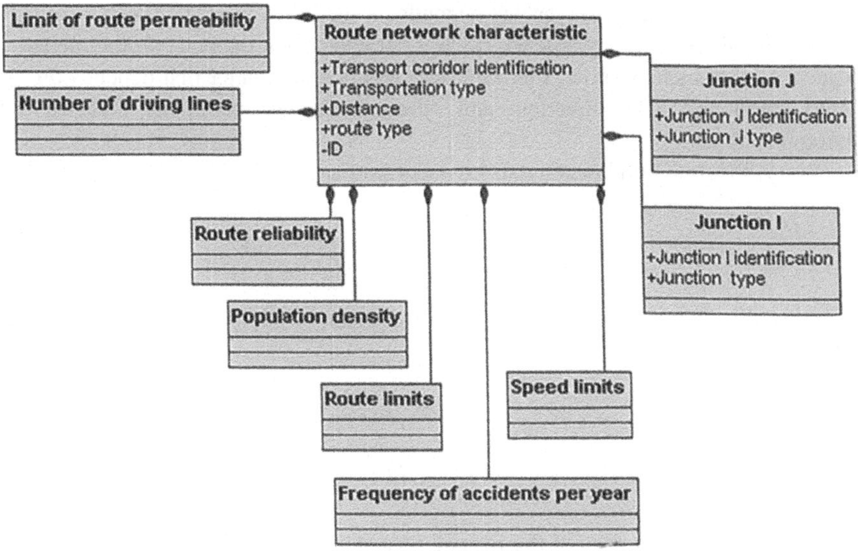

Fig. 19.10 Conceptual representation of route stretch characteristics

$$f_{r,s_k} = f_r P(s_k); \tag{19.14}$$

$$f_r = \gamma_r L_r n_r; \tag{19.15}$$

$$\gamma_r = \gamma_{0,r} G,, \tag{19.16}$$

Expressed in formulas γ_r is expected frequency on the r-th road section [number of accidents per km–1·according to the number of vehicle-1·per previous year-1],

L_r—express the road length [km],
n_r—is the number of vehicles through the road r-th section [vehicle],
$\gamma_{0,\,r}$—is the regional accident frequency [number of accidents per distance of km-1 of number of·vehicle-1·per previous year-1].

The probabilistic parameter G can be characterized as a common evaluation parameter of risks of such road section environment. Various factors are influenced and can be implemented for assessment of the accident events: environmental, behavioral, physical, mechanical, road intrinsic descriptors. The common probabilistic parameter is evaluated by these parameters:

$$G = \prod_{j=1}^{m} G_j, \tag{19.17}$$

The probability G is the local enhancing/mitigating parameter. For expressing more detailed structure of G components, the model can implement, for example the $G1$ as a parameter expressing the risk of accident events dependence from temperature, $G2$ is a parameter that depends on the inherent factor (such as tunnel, bend radii, slope, height gradient), $G3$ is a parameter that depends on the meteorological factor (such as snow, sun, rain, ice), $G4$ is a parameter that depends on the wind speed and wind direction, and others until such parameter that we can recognize Gm:

N_{r,s_k} is the total number of fatalities according to Eq. (19.18):

$$N_{r,s_k} = \left(\Phi_{s_k}^{\text{in}} o^{\Delta t} v_r + \Phi_{s_k}^{\text{off}} d_r\right) P(F, s_k). \tag{19.18}$$

Being this, the in-road and the off-road number of fatalities can be calculated, respectively, as:

$$N_{r,s_k}^{\text{in}} = \Phi_{s_k}^{\text{in}} o^{\Delta t} v_r P(F, s_k); \tag{19.19}$$

$$N_{r,s_k}^{\text{off}} = \Phi_{s_k}^{\text{off}} d_r P(F, s_k), \tag{19.20}$$

where $\Phi_{s_k}^{\text{in}}$ is a consequence of the in-road area associated with scenario s_k [m^2].

$\Phi_{s_k}^{\text{off}}$—express a consequence of the off-road area associated with scenario s_k [km^2].

$P(F, s_k)$—express a probability of fatality F for accident scenarios$_k$.

$o^{\Delta t}$—is the average vehicle occupation factor during specific time period Δt, which can depend on the seasons or day time.

v_r—is the vehicle density on the road area [number of vehicles per m^2].

d_r—is the population density of the r-th road area environment [number of inhabitants per km^2].

The formal expression of risky situations and accident evens can help in inclusion of such type services in the Smart-SDS, for prediction of drivers during the cycles of multimodal transportation.

19.6 Conclusions

The approach for representing of static and dynamic aspects of a decision support processes and reflecting them in the Smart-SDS by using knowledge representation methods of dynamic domain of multimodal transportation is presented. We are focused on management aspects of logistics of multimodal transportation of cargo. Some aspects of risk evaluation of transportation are analyzed, as well as the possibilities of representation them by imitational models are analyzed. Possibilities of revealing of the formal expressions for assessment of overall value of loss-free cargo route planning method are analyzed by adding new elements to the whole framework of multimodal transportation. The conceptual presentation explains the key variables through proposed mobile control system. The benefits are revealed and explained of transportation operations by evaluating the proposed general information. Models for description of services evaluate the some types of risks, context parameters and for description of risks in whole transportation route are described.

Due to the high risks involved in new RFID and mobile sensor-based mobile control system, implementation is advised. An approach for careful management with the forecasting possibilities of fleet replacement decreasing situations, when installing the new system in future replaced containers and with the check location number and placements over the transportation routes.

The descriptions of some important parts of the decision support system are presented. The main components of supporting infrastructure for providing smart services for the management of multimodal transportation processes were described. We hope that our approach will be helpful in the operative management processes of transportation. The main structures of the subsystems were presented and enable implement the components of KB. Integration of methods of decision-making and multi-criteria evaluation is analyzed. Methods for more real construction of KB and methods for the provision of the intelligent planning tools for evaluating multimodal transportation corridors are helpful in choosing of alternatives at the real conditions of transportation.

Several options are focusing on problems of solving sustainable transportation. We provide the more adaptable multi-criteria decision support methods for choosing the best transportation corridors between the East and West directions of Europe. The multi-componential system is integrated into the overall work of the ICT infrastructure and enables intellectualize the work of e-service provision. Such type systems are associated with communication channels between all participating parts. The types of multimodal transportation were analyzed on the base of the underground Earth means. The proposed approach of Smart-SDS development enables context recognition procedures, provides methodology for service delivery infrastructure development, and shows a way how to build KB for operative decision support. Wide spectrum of context awareness is reviewed. Inclusion of context awareness in smart system enables provide broader and more adaptable solutions during logistical cycles of multimodal freight transportation in real situations.

References

Alazzawi, A., & Żak, J. (2020). MCDM/A based design of sustainable logistics corridors combined with suppliers selection. The case study of freight movement to Iraq. *Transportation Research Procedia, 47*, 577–584. https://doi.org/10.1016/j.trpro.2020.03.134

Amador, C., & Emond, J. P. (2010). Evaluation of sensor readability and thermal relevance for RFID temperature tracking. *Computers and Electronics in Agriculture, 73*, 84–90.

Andziulis, A., Jakovlev, S., Adomaitis, D., Steponavičius, R., Kurmis, M., & Pareigis, V. (2010). Integration of information system models in intermodal container transportation systems. In *Proceedings of 14th International Conference „Transport Means"* (pp. 127–130).

Andziulis, A., Eglynas, T., Bogdevicius, M., Jusis, M., & Senulis, A. (2016). Multibody dynamic simulation and transient analysis of quay crane spreader and lifting mechanism. *Advances in Mechanical Engineering, 8*(9). https://doi.org/10.1177/1687814016670803

Anton, J. M., Grau, J. B., & Andina, D. (2004). ELECTRE and AHP MCDM methods versus CP method and the official choice applied to high-speed railway layout alternative election. *WSEAS Transactions on Business and Economics, 1*, 64–69.

Barysienė, J. (2012). A multi-criteria evaluation of container terminal technologies applying the COPRAS-G method. *Transport, 27*, 364–372. https://doi.org/10.3846/16484142.2012.750624

Batarlienė, N. (2007). Implementation of advanced technologies and other means in dangerous freight transportation. *Transport, 22*(4), 290–295.

Batarlienė, N., & Baublys, A. (2007). Mobile solutions in road transport. *Transport, 22*(1), 55–60.

Baublys, A., & Jarašūnienė, A. (2010). Statistical probability evaluation of operating ITS. *Transport, 25*(2), 163–170.

Bazaras, D., Ledauskaitė, K., & Palšaitis, R. (2010). Comparative analysis of information and communication technologies in Lithuania and Finland. *Transport, 25*(1), 98–104.

Bielskis, A. A., Dzemydienė, D., Denisovas, V., Andziulis, A., & Drungilas, D. (2009). An approach of multi-agent control of bio-robots using intelligent recognition diagnosis of persons with moving disabilities. *Technological and Economic Development of Economy, 15*(3), 377–394.

Bojković, N., Anić, I., & Pejčić-Tarle, S. (2010). One solution for cross-country transport-sustainability evaluation using a modified ELECTRE method. *Ecological Economics, 69*, 1176–1186. https://doi.org/10.1016/j.ecolecon.2010.01.006

Brauers, W. K., & Zavadskas, E. K. (2009). Multi-objective optimization with discrete alternatives on the basis of ratio analysis. *Intellectual Economics, 2*(6), 30–41.

C-ITS Platform. C-ITS Directive (2010/40/EU) Supporting Framework and Enabling Conditions. C-ITS platform final report & Annexes available at: http://ec.europa.eu/transport/themes/its/c-its_en.htm

Castillo, H., & Pitfield, D. E. (2010). ELASTIC – A methodological framework for identifying and selecting sustainable transport indicators. *Transportation Research Part D: Transport and Environment, 15*, 179–188. https://doi.org/10.1016/j.trd.2009.09.002

Chavali, M., Lin, T. H., Wu, R. J., Luk, H. N., & Hung, S. L. (2008). Active 433 MHz-W UHF RF-powered chip integrated with a nanocomposite m-MWCNT/polypyrrole sensor for wireless monitoring of volatile anesthetic agent sevoflurane. *Sensors and Actuators A, 141*, 109–119.

Deveci, M., Canıtez, F., & Gökaşar, I. (2018). WASPAS and TOPSIS based interval type-2 fuzzy MCDM method for a selection of a car sharing station. *Sustainable Cities and Society, 41*, 777–791. https://doi.org/10.1016/j.scs.2018.05.034

Dias, J. C. Q., Calado, J. M. F., Luıs, O. A., & Morgado, L. F. (2009). RFID together with multi-agent systems to control global value chains. *Annual Reviews in Control, 33*, 185–195.

Dzemydienė, D., & Burinskienė, A. (2021). Integration of context awareness in smart service provision system based on wireless sensor networks for sustainable cargo transportation. *Sensors, 21*(15), 1–22. https://doi.org/10.3390/s21155140. Special Issue Artificial Intelligence and Internet of Things in Autonomous Vehicles. Basel: MDPI. Art. no. 5140.

Dzemydiene, D., Burinskiene, A., & Miliauskas, A. (2020). An assessment of provision of heterogeneous services for sustainable cargo transportation process management by roads. *Sustainability, 2020*(12), 8405. https://doi.org/10.3390/su12208405

Dzemydienė, D., Dzemydaitė, G., & Gopisetti, D. (2022). Application of multicriteria decision aid for evaluation of ICT usage in business. *Central European Journal of Operations Research, 30*, 323–343. https://doi.org/10.1007/s10100-020-00691-9

Dzemydienė, D., & Dzindzalieta, R. (2009). Development of decision support system for risk evaluation of transportation of dangerous goods using mobile technologies. In M. Grasserbauer, L. Sakalauskas, & E. K. Zavadskas (Eds.), *Knowledge-based technologies and OR methodologies for strategic decisions of sustainable development* (pp. 108–113).

Dzemydienė, D., & Dzindzalieta, R. (2012). Multi-layered architecture of decision support system for monitoring of dangerous good transportation. In A. Čaplinskas, G. Dzemyda, A. Lupeikienė, & O. Vasilecas (Eds.), *Databases and information systems: 10th international Baltic conference; local proceedings*, July 8–11, 2012, Lithuania. Žara.

Dzemydienė, D., Kurmis, M., Baikštienė, V., & Dzindzalieta, R. (2019). Daugialypių mobiliųjų paslaugų teikimo galimybės autotransporto komunikacijos tinkluose = Possibilities to support heterogeneous mobile services in vehicle communication networks. *Informacijos mokslai, 85*, 176–192. https://doi.org/10.15388/Im.2019.85.22

Dzemydiene, D., Maskeliunas, S., Dzemydaite, G., & Miliauskas, A. (2016). Semi-automatic service provision based on interaction of data warehouses for evaluation of water resources. *Informatica, 27*, 709–722. https://doi.org/10.15388/Informatica.2016.107

Du, Y., Wang, J., Rizos, C., & El-Mowafy, A. (2021). Vulnerabilities and integrity of precise point positioning for intelligent transport systems: Overview and analysis. *Satellite Navigation, 2*(3). https://doi.org/10.1186/s43020-020-00034-8

eSafety Initiative. (2007). *Safer roads for European citizens*. http://www.esafetysupport.org/en/esafety_activities/

European Commission. (2019). *A European green deal: Striving to be the first climate-neutral continent 2019*.

European Commission. (2020). *Communication from the Commission to the European Parliament, the Council, the European economic and social committee and the committee of the regions: A European strategy for data 2020*.

Evaluation and Classification of Chemical Wastes. (2010). http://www.oc-praktikum.de/en/articles/pdf/WasteClassification_en.pdf

Fabiano, B., Currò, F., Reverberi, A. P., & Pastorino, R. (2005). Dangerous good transportation by road: From risk analysis to emergency planning. *Journal of Loss Prevention in the Process Industries, 18*(4–6), 403–413.

Fabiano, B., Palazzi, E., Currò, F., & Pastorino, R. (2001). *Risk assessment and decision-making strategies in dangerous good transport. From an Italian case-study to a general framework. Loss Prevention and Safety Promotion in the Process Industries. Proceedings 2* (pp. 955–966). Elsevier Science B.V..

Faccin, J., & Nunes, I. (2015). BDI-agent plan selection based on prediction of plan outcomes. In *Proceedings of the 2015 IEEE/WIC/ACM International Conference on Web Intelligence and Intelligent Agent Technology (WI-IAT)* (Vol. 2, pp. 166–173).

Ferrer, G., Dew, N., & Apte, U. (2010). When is RFID right for your service. *International Journal of Production Economics, 124*, 414–425.

Hazardous Substances Ordinance. (2004). *GefStoffV (BGBl. I p. 3758) as amended by Article 2 of the Ordinance of 23 December 2004 (BGBl. I p. 3855)*. www.zuv.uni-eidelberg.de/sw/gefahrstoffe/gesetze/HazardousSubstances-Ordinance.pdf

Ishfaq, R., & Sox, C. R. (2010). Intermodal logistics: The interplay of financial, operational and service issues. *Transportation Research Part E, 46*, 926–949.

Jakimavičius, M., & Burinskienė, M. (2009). A GIS and multi-criteria-based analysis and ranking of transportation zones of Vilnius city. *Technological and Economic Development of Economy, 15*(1), 39–48.

Jedermann, R., Behrens, C., Westphal, D., & Lang, W. (2006). Applying autonomous sensor systems in logistics—Combining sensor networks, RFIDs and software agents. *Sensors and Actuators A, 132*, 370–375.

Kaklauskas, A., Zavadskas, E. K., & Šaparauskas, J. (2009). Conceptual modelling of sustainable Vilnius development. *Technological and Economic Development of Economy, 15*(1), 154–177.

Kang, S. Y., Lee, D. G., & Lee, I. Y. (2008). A study on secure RFID mutual authentication scheme in pervasive computing environment. *Computer Communications, 31*, 4248–4254.

Kaya, S. V., Savas, E., Levi, A., & Ercetin, O. (2009). Public key cryptography based privacy preserving multi-context RFID infrastructure. *Ad Hoc Networks, 7*, 136–152.

Knospe, H., & Poh, H. (2004). RFID security. *Information Security Technical Report, 9*(4), 39–50.

Kreutzberger, E. D. (2008). Distance and time in intermodal goods transport networks in Europe: A generic approach. *Transportation Research Part A, 42*, 973–993.

Kurmis, M., Andziulis, A., Dzemydiene, D., Jakovlev, S., Voznak, M., & Gricius, G. (2015). Cooperative context data acquisition and dissemination for situation identification in vehicular communication networks. *Wireless Personal Communications, 85*(1), 49–62. https://doi.org/10.1007/s11277-015-2727-1

Lee, C. K. M., & Chan, T. M. (2009). Development of RFID-based reverse logistics system. *Expert Systems with Applications, 36*, 9299–9307.

Limbourg, S., & Jourquin, B. (2009). Optimal rail-road container terminal locations on the European network. *Transportation Research Part E, 45*, 551–563.

Macharis, C., Hoeck, E., Pekin, E., & Lier, T. (2010). A decision analysis framework for intermodal transport: Comparing fuel price increases and the internalization of external costs. *Transportation Research Part A, 44*, 550–561.

Macharis, C., & Pekin, E. (2009). Assessing policy measures for the stimulation of intermodal transport: A GIS-based policy analysis. *Journal of Transport Geography, 17*, 500–508.

Mitropoulos, L. K., & Prevedouros, P. D. (2016). Incorporating sustainability assessment in transportation planning: An urban transportation vehicle-based approach. *Transportation Planning and Technology, 39*, 439–463. https://doi.org/10.1080/03081060.2016.1174363

Ngai, E. W. T., Cheng, T. C. E., Auc, S., & Lai, K. H. (2007). Mobile commerce integrated with RFID technology in a container depot. *Decision Support Systems, 43*, 62–76.

Osmundsen, K., Iden, J., & Bygstad, B. (2018). Digital transformation: Drivers, success factors, and implications. In *Proceedings of the MCIS* (p. 37).

Oztekin, A., Pajouh, F. M., Delen, D., & Swim, L. K. (2010). An RFID network design methodology for asset tracking in healthcare. *Decision Support Systems, 49*, 100–109.

Pamučar, D., Ecer, F., Cirovic, G., & Arlasheedi, M. A. (2020). Application of improved best worst method (BWM) in real-world problems. *Mathematics, 8*. https://doi.org/10.3390/math8081342

Pamučar, D. S., & Savin, L. M. (2020). Multiple-criteria model for optimal off-road vehicle selection for passenger transportation: BWM-COPRAS model. *Vojnotehnicki Glasnik, 68*, 28–64. https://doi.org/10.5937/vojtehg68-22916

Pavliska, V. (2006). *Petri nets as fuzzy modelling tool*. http://irafm.osu.cz/research_report/84_rep84.pdf

Piramuthu, S. (2007). Protocols for RFID tag/reader authentication. *Decision Support Systems, 43*, 897–914.

Pranevicius, H. (2003). The use of aggregate approach for simulation of logistics systems. 1st International Industrial Simulation Conference, Valencia, Spain, pp. 271–275.

Ruiz-Garcia, L., Barreiro, P., Rodriguez-Bermejo, J., & Robla, J. I. (2007). Review. Monitoring the intermodal, refrigerated transport of fruit using sensor networks. *Spanish Journal of Agricultural Research, 5*(2), 142–156.

Shariat-Mohaymany, A., & Babaei, M. (2010). An approximate reliability evaluation method for improving transportation network performance. *Transport, 25*(2), 193–202.

Šutienė, K., Makackas, D., & Pranevičius, H. (2010). Multistage K-means clustering for scenario tree construction. *Informatica, 21*(1), 123–138.

Thill, J. C., & Lim, H. (2010). Intermodal containerized shipping in foreign trade and regional accessibility advantages. *Journal of Transport Geography, 18*, 530–547.

Vergara, A., Llobet, E., Ramírez, J. L., Ivanov, P., Fonseca, L., Zampolli, S., Scorzoni, A., Becker, T., Marcoe, S., & Wöllenstein, J. (2007). An RFID reader with onboard sensing capability for monitoring fruit quality. *Sensors and Actuators B, 127*, 143–149.

Verma, M., & Verter, V. (2010). A lead-time based approach for planning rail–truck intermodal transportation of dangerous goods. *European Journal of Operational Research, 202*, 696–706.

Wibisono, W., Zaslavsky, A., & Ling, S. (2009). Improving situation awareness for intelligent on-board vehicle management system using context middleware. In *2009 IEEE Intelligent Vehicles Symposium* (pp. 1109–1114). IEEE. https://doi.org/10.1109/IVS.2009.5164437

World Shipping Council. (2009). *Partners in trade* [online]. Accessed September 10, 2010, from http://www.worldshipping.org/about-the-industry/global-trade/trade-statistics

Zavadskas, E. K., Kaklauskas, A., & Kaklauskienė, J. (2007a). Modelling and forecasting of a rational and sustainable development of Vilnius, emphasis on pollution. *International Journal of Environment and Pollution, 30*(3–4), 485–500.

Zavadskas, E. K., Kaklauskas, A., Peldschus, F., & Turskis, Z. (2007b). Multi-attribute assessment of road design solutions by using the copras method. *Baltic Journal of Road and Bridge Engineering, 2*, 195–203.

Glossary

2S-LRP (Two-stage location routing problem) Optimal number, the capacity, and the location of facilities are determined, and the optimal set of vehicle routes into two stages from each facility through the intermediate warehouse is planned and implemented.

5G (The fifth generation) Denoting developments in computer design to produce machines with artificial intelligence

ABC (Artificial Bee Colony) Optimization algorithm based on the intelligent foraging behaviour of honey bee swarm

ACL (Agent Communication Language) Standard language for agent communications

ADAS (Intelligent Driver Assistance Systems) Groups of electronic technologies that assist drivers in driving and parking functions. Through a safe human–machine interface, ADAS increase car and road safety.

Ad-Hoc A wireless communication network connecting 1, ..., n destination points and created according to the need to achieve a specific goal.

AHP (Analytical hierarchy process) Structured technique for organizing and analyzing complex decisions, based on mathematics

AI (Artificial Intelligence) The science area for development, creation, and implementation of systems with abilities to imitate human activities by artificial intelligence methods and equipment.

AIFS (Arbitration Inter-Frame Space) An arbitrary outer frame space. This is an access class prioritization method.

AIS (Automatic Identification System) An automated tracking system that displays other vessels in the vicinity.

ALSN (The Automatic Locomotive Signalling) Type of cab signalling system that provides track status information to the train cab and uses the rails as a continuous communication channel between track and train.

ANP (Analytical network process) General form of the analytic hierarchy process

D. Dzemydienė et al., *Development of Smart Context-Aware Services for Cargo Transportation*, International Series in Operations Research & Management Science 330, https://doi.org/10.1007/978-3-031-07199-7

AODV (On-Demand Distance Vector) An on-demand distance vector that can act as a reactive protocol in which routes are created only when data needs to be transmitted.

API (Application programming interface) Software intermediary that allows two applications to talk to each other.

ARAS (Additive ratio estimation method) Utility function value determining the complex efficiency of a feasible alternative is directly proportional to the relative effect of values and weights of the main criteria considered in an analysis.

AVL (Automatic vehicle locating) Automatic vehicle location is a means for automatically determining and transmitting the geographic location of a vehicle. This vehicle location data, from one or more vehicles, may then be collected by a vehicle tracking system to manage an overview of vehicle travel.

B Internal company boundaries

B2(B2B) (Business-to-Business-to-business) Multi-tier electronic documentation transfer services within business enterprises

B2B (Business-to-business) Exchange of products, services, or information between businesses

BFO (Basic formal ontology) Small, upper-level ontology that is designed for use in supporting information retrieval, analysis and integration in scientific and other domains

BI (Business intelligence) Combines business analytics, data mining, data visualization, data tools and infrastructure, and best practices to help organizations to make more data-driven decisions

BigData Large set of data that can be revised computationally to identify patterns and dynamics

BPM (Business process management) Organizational discipline where a company takes a step back and looks at all of these processes in total and individually.

BPMMFisher Business process management modelling proposed by L. Fisher (2004)

BPMMM (Business process management maturity model) Prospering approach to improving a company's processes

BPR (Process transformation) Changes made in a business or other organization which improve the company's employee and management participation in the development of the business

BSC (Balanced scorecard) Strategic management performance metric that helps companies identify and improve their internal operations to help their external outcomes.

BWM (Best worst method) Approach for weighing a collection of options against a set of decision criteria

C2C-CC (CAR 2 CAR Communication Consortium) Car-to-car communication consortium.

C3 (Car-to-Car Cooperation) Infrastructure of information communication technology enabling car-to-car cooperation.

CA (Car Agent) A car agent is a component like a car that is equipped with systems using artificial intelligence methods.

CAMD (Context management) Dynamic computer process that uses "subjects" of data in one application, to point to data resident in a separate application also containing the same subject

CAN (Network of controller area) Another type of serial communications protocol that was developed within the automotive industry to allow several electronic units on a single vehicle to share essential control data

CAR Controlled Automobile Reporting System

CCL Container Control Logistics System

CDMA (Code-Division Multiple Access Technology) Code division multiple access technologies

CEPEC (Coordinated External Peer Communication) Coordinated channel access communication for external network nodes and a protocol that ensures its operation.

CEPT (Conference of European Postal and Telecommunications Administrations) European Conference of Postal and Telecommunications Administrations.

CIM (Rail Consignment Note or *Rail* Transport Document) Document proving the conclusion of a transport contract with a *railway* undertaking.

CIS (Center for Internet Security) Nonprofit organisation that harnesses the power of a global IT community to safeguard public and private organizations against cyber threats

Cluster Group of customers assigned to one route

CMR Convention on the Contract for the International Carriage of Goods by Road

CO_2 (Carbon dioxide emission) Missions stemming from the burning of fossil fuels and the manufacture of cement; include carbon dioxide produced during consumption of solid, liquid, and gas fuels as well as gas flaring

COMPOSS Computer-aided Methods for Planning and Designing the System

COPRAS (Complex proportional evaluation) Method uses step-wise ranking and evaluating procedure of alternatives in terms of significance and utility degree

COPRAS-G Complex proportional evaluation of grey ratio alternatives

CRN (Cognitive Radio Networks) Cognitive Radio Networks with radio diapasons can enable cognitive functions.

CSMA/CD (Carrier Sense Multiple Access with Collision Avoidance) A protocol that provides a way to avoid collisions, the essence of which is the use of additional data packets.

CTS (Clear-To-Send) Confirmation to send.

CVRP (Limited capacity vehicle routing problem) Vehicles with limited carrying capacity need to pick up or deliver items at various locations

DAKOSY Data Communication systems

DAVIS Transport data processing system

DE (Differential evolution algorithm) Method that optimizes a problem by iteratively trying to improve a candidate solution concerning a given measure of quality.

DEA (Data envelopment analysis) Nonparametric method in operations research and economics for the estimation of production frontiers

Delivery Process of transporting goods from a source location to a predefined destination

DES (Discrete event simulation) Models the operation of a system as a (discrete) sequence of events in time

DG (Dangerous goods) Substances that when transported are a risk to health, safety, property, or the environment

DG MOVE The department for mobility and transport is responsible for the EU Commission's policy on transport for private and professional purposes.

DMS (Document Management System) System used to receive, track, manage and store documents and reduce paper.

DSM (Distributed Sorting Mechanism) Distributed sorting mechanism

DSRC (Dedicated Short Range Communications) Type of communication with dedicated functions and working in short-range communication areas.

DSS (Decision Support Systems) Decision support system is an information system that supports business or organizational decision-making activities.

DTLF Forum of Digital Transportation and logistics

DTRA (Dynamic Transmission Range Assignment) Transmission Range Assignments that can work in dynamic conditions of moving objects.

DW (Data Warehouse) Big data bases (qubes), i.e. large store of data accumulated from a wide range of sources and used to guide management decisions

EA (Evolutionary algorithm) Algorithm that uses mechanisms inspired by nature and solves problems through processes that emulate the behaviours of living organisms

EATWOS A method consisting of an analysis of entropy and efficiency with the satisfaction of the output.

EC (European Commission) Executive branch of the European Union.

eCMR Electronic version of the document issued based on Convention on the Contract for the International Carriage of Goods by Road

ECMT (European Conference of Ministers of Transport) The European Conference of Ministers of Transport is an intergovernmental organisation set up in 1953 at the initiative of the Committee of Transport of the OECE (Organisation for Economic Co-operation in Europe) which, in the early 1960s, was to become the OECD. Owing to its existence to the drive for mutual assistance and integration in post-war Europe, it has firmly established itself as a forum for consultation among ministers responsible for inland transport. It is also a think-tank for forward-looking policies to address the challenges and problems of transport systems on a pan-European scale. One of the major tasks the organisation is now facing is the shaping and implementation of an all-European transport policy.

EDCA (Enhanced Distributed Channel Access) Extended Distributed Channel Access.

EDI (Electronic Data Interchange) Concept of businesses electronically communicating information that was traditionally communicated on paper, such as purchase orders and invoices.

EDI (Electronic data interchanges) Concept of businesses electronically communicating information that was traditionally communicated on paper

EGNOS (European Satellite Signalling System) Europe's regional satellite-based augmentation system (SBAS) that is used to improve the performance of global navigation satellite systems (GNSSs), such as GPS and Galileo.

ELM Electronic Logistics Marketplace

EOQ (Economic order quantity) Transport order quantity that minimizes the total transportation costs and other costs associated with freight delivery management

ERP (Enterprise resource planning system) Type of software that organizations use to manage day-to-day business activities such as accounting, procurement, project management, risk management and compliance, and supply chain operations.

ERTMS (The European Rail Traffic Management System) Single European signalling and speed control system that ensures interoperability of the national railway systems, reducing the purchasing and maintenance costs of the signalling systems as well as increasing the speed of trains, the capacity of infrastructure and the level of safety in rail transport.

ERTRAC Horizon Europe partnerships and the European Commission in identifying future research needs for upcoming R&I programmes, to further facilitate a sustainable and efficient road transport system in Europe, while also fostering international cooperation.

ETA (Estimated time of arrival) Time when a ship, vehicle, aircraft, cargo, emergency service, or person is expected to arrive at a certain place.

Ethernet Computer networking technology for local area networks

eTIR Electronic version of customs transit document used to prove the existence of the international guarantee for duties and taxes for the goods transported under the TIR system

EU (European Union) Political and economic union of 27 member states that are located primarily in Europe.

EU (European Union) The European Union is a political and economic union of 27 member states that are located primarily in Europe.

EvDo (Evolution-Data only) Evolutionary data only.

EvDv (Evolution-Data/Voice) Evolutionary data, voice.

FADR Modified fuzzy method for preference order similarity to the ideal solution

F-ANP Applied for the improvement of ANP by combining the fuzzy set theory

FIDEUS Project on freight innovative delivery in European urban space

FIPA (Foundation for Intelligent Physical Agents) Body for developing and setting computer software standards for heterogeneous and interacting agents and agent-based systems

FLP (Facility location problem) Classic optimization problem that determines the best location for a factory or warehouse to be placed based on geographical demands, facility costs, and transportation distances.

FMP (Fuzzy multi-criteria group decision support system) Model aggregates both subjective and objective information under multi-level hierarchies of criteria and evaluators.

Freight ID (Unique number of transport unit) Unique identifier of freight unit

FTE (Full-time equivalent) Refers to the unit of measurement equivalent to an individual

FTL (Full Truckload) Type of shipping mode whereby a truck carries one dedicated shipment.

FTTP (Fiber to the Premises) Equipment used in fibre access deployments where fibres extend to the end-user premises and the equipment is designed and optimized for use in residential applications

F-VIKOR Aggregating fuzzy merit that represents the distance of an alternative to the ideal solution.

GA (Genetic algorithm) Metaheuristic inspired by the process of natural selection that are commonly used to generate high-quality solutions to optimization and search problems by relying on biologically inspired operators such as mutation, crossover, and selection

GDP (Gross domestic product) Measures the monetary value of final goods and services—that is, those that are bought by the final user—produced in a country in a given period

GenCLOn (Ontology for city logistics) Ontology for the domain of City

GIAI (Global Individual Asset Identifier) Part of the GS1 system of standards. It is a simple tool to identify an asset by type and can identify assets uniquely where required.

GIFTS (Global Intermodal Freight System) *Global Intermodal Freight* Transport *System* rarely.

GIS (Geographic Information System) Geographic information system is a type of database containing geographic data, combined with software tools for managing, analyzing, and visualizing those data.

GPS (Global Positioning System) Global positioning system that allows to determine the coordinates of an object anywhere in the world and working on the infrastructure of satellite equipment and communication channels.

GPS (Global Positioning System) Space-based radio-navigation system consisting of a constellation of satellites broadcasting navigation signals and a network of ground stations and satellite control stations used for monitoring and control.

GPSR (Greedy Perimeter Stateless Routing) Route-finding algorithm that does not secure a perimeter.

GS1 (Global standards organization) International organization developing and maintaining standards including barcodes

GSIN (Global shipment identification number) Number assigned by a seller and shipper of goods to identify a shipment comprised of one or more logistic units that are intended to be delivered together.

GSK (Group Session Key) The key that can be assigned for the session of a defined group of included in this group objects.

HSPA (High-Speed Packet Access) High-speed packet access that provides 42 Mb/s reception and 22 Mb/s transmission speeds over mobile networks.

HTTP (Hypertext Transfer Protocol) Application protocol for distributed, collaborative, hypermedia information systems that allow users to communicate data on the World Wide Web.

I2V (Infrastructure to Vehicle) Type of infrastructure that enables the interaction (communication) of vehicles equipment or systems of the whole infrastructure.

IASDS (Intelligent Adaptive Service Delivery System) Modifying their runtime behaviour to achieve system objectives

ICT (Information and communication technologies) Extensional term for information technology (IT) that stresses the role of unified communications[1] and the integration of telecommunications (telephone lines and wireless signals) and computers, as well as necessary enterprise software, middleware, storage and audiovisual, that enable users to access, store, transmit, understand, and manipulate information.

ICT (Information Communication Technology) Wide area of means (software and hardware), which enable storage and work with data, information and enable a wide spectrum of communication means with methods and involving people in these processes.

IEEE (Institute of Electrical and Electronics Engineers) International professional organization dedicated to technological advancement.

IEEE 1609 Family of standards for wireless access in automotive environments being developed by the IEEE.

IEEE 802.11 Set of standards for the deployment of wireless local area networks in the 2.4, 3.6, and 5 GHz frequency bands.

IEEE 802.11p Improvement on the IEEE 802.11 standards for wireless access in automotive environments in the 5.9 GHz frequency band. The standard is based on the higher level IEEE 1609.

IEEE 802.15.4 A standard that describes the physical and MAC layers in low-speed wireless personal area networks.

IEEE 802.16 Set of wireless broadband access standards, otherwise known as WiMAX.

Intermodality Transportation involving more than one form of carrier, as truck and rail, or truck, ship, and rail

IoT (Internet of Things) The Internet of things describes physical objects that are embedded with sensors, processing ability, software, and other technologies that connect and exchange data with other devices and systems over the Internet or other communications networks.

IPTV Television service for the transmission and reception of television signals by packet data over the Internet (IP).

IRP (Inventory routing problem) Integration and coordination of two components of the logistics value chain: inventory management and vehicle routing

IRU (International Road Transport Union) Global road transport organization, which upholds the interests of bus, coach, taxi and truck operators to ensure economic growth and prosperity via the sustainable mobility of people and goods by road transport

IS (Information system) Formal, sociotechnical, organizational system designed to collect, process, store, and distribute information.

ISO (International Organization for Standardization) Organisation sets standards in many businesses and technologies, including computing and communications

ISO 31000, Risk management Guidelines, provides principles, a framework and a process for managing risk

ITS (Intelligent Transportation System) Intelligent transportation system. According to EU Directive 2010/40/EU (7 July 2010), ITS is defined as systems that apply information and communication technologies to road transport, including infrastructure, cars, consumers, traffic management and interfaces with other modes of transport.

IT2FS Interval type-2 of fuzzy sets

ITMS (Intelligent transport management systems) Subset of ITS, which comprises of a set of broad and diverse technologies including information processing, network communications, electronics/instrumentation control & software engineering

ITN Intermodal transport networks

ITS (Intelligent transport systems) Systems that apply information and communication technologies to road transport, including infrastructure, cars, consumers, traffic management and interfaces with other modes of transport.

JIT ("Just in Time") Inventory system is a management strategy that aligns raw-material orders from suppliers directly with production schedules.

JSON (Useful data serialization and messaging format) Open standard file format and data interchange format that uses human-readable text to store and transmit data objects

JSON-LD Lightweight syntax to serialize Linked Data in JSON

KB (Knowledge base) Set of facts, assumptions, and rules which a computer system has available to solve a problem

KIF (Knowledge interchange format) Computer language designed to enable systems to share and re-use information from knowledge-based systems

KOMODA EU-funded project "Co-modality—towards optimised integrated chains in freight transport logistics"

KPI (Key performance indicators) Measurable value that demonstrates how effectively a company is achieving key business objectives

KQML (Knowledge Query and Manipulation Language) Language and protocol for communication among software agents and knowledge-based systems

LDP (Location distribution problem) Concerned with how to select a location from the potential set so that the total relevant cost is minimized

LISP (Standard research tool of artificial intelligence scientists) Family of programming languages, which involves a computer; consequently, programming languages are usually defined and studied this way

LLC (Logical Link Control) Logical communication control that describes the operation of the second layer of the OSI model for data communication. This is a sublayer that enables simple Bridge-type routing between wireless and wired Ethernet networks.

LOTSE (Logistic-Tele-Service) Common system that acts as an adapter for hardware.

LRIT (Long-Range Identification and Tracking) System is a designated International Maritime Organization (IMO) system designed to collect and disseminate vessel position information received from IMO member States ships that are subject to the International Convention for the Safety of Life at Sea (SOLAS).

LRP (Location routing problem) Optimal number, the capacity, and the location of facilities are determined, and the optimal set of vehicle routes from each facility is also sought

MAC (Medium Access Control) Media Access Control (Layer 2 of the OSI model).

MADM (Multiple Attribute Decision Making) Dedicated for making preference decisions (such as evaluation, prioritization, selection) over the available alternatives that are characterized by multiple, usually conflicting, attributes.

MANET (Mobile Ad-Hoc Network) Mobile Ad-Hoc network.

MAS (Multi-agent system) Computerized system composed of multiple interacting intelligent agents.

MCF (Minimum cost flow model) Optimization and decision problem to find the cheapest possible way of sending a certain amount of flow through a flow network.

MCP (Maximum Coverage Problem) Problem of maximum coverage.

MEC (Multi-access edge computing) MEC brings technology resources closer to the end-user.

MERCURIO Project on a European analysis of the Road Injuries Management System concerning the social and economic impact of emergency and post-injuries services on national finances and households

MIMO (Multiple Input Multiple Output) The use of multiple antennas for transmission and reception to improve device performance.

MIS (Management Information Systems) Computer system consisting of hardware and software that serves as the backbone of an organization's operations.

Mobile computing Mobile computing is human–computer interaction in which a computer is expected to be transported during normal usage, which allows for the transmission of data

MOLP (Multiple objective linear programming) Linear program with more than one objective function

MOORA (Multi-purpose optimization based on ratio analysis) Process of simultaneously optimizing two or more conflicting attributes (goals) subject to certain restrictions

MRP (Materials and production resource planning) System for calculating the materials and components needed to manufacture a product. It consists of three primary steps: taking inventory of the materials and components on hand, identifying which additional ones are needed and then scheduling their production or purchase.

Multimedia Type of possibilities to express a multimedia medium information that includes several types of information (like audio, video, text visualization, etc.).

Multi-modality Freight delivery under a single contract, but performed with at least two different modes of transport

MXF (Maximum flow model) Network of pipelines that transports crude oil from oil wells to refineries

N3 (Notation3) Format being developed by Tim Berners-Lee and others from the Semantic Web community

NLOS (Non-Line-of-Sight) Type of Indirect Visibility.

NLP (Nonlinear programming) Process of solving an optimization problem where some of the constraints or the objective function are nonlinear

Node-movement-scenario configuration Configuration of the movement script

N-Quads Widely accepted format to represent named graphs for data exchange on the Web

N-Triples Format for storing and transmitting data

OBU (On-Board Unit) Means and is equipped on-board of vehicle.

OD Delivery from the origin point to destination point

OFDM (Orthogonal Frequency-Division Multiplexing) Rectangular Frequency Density

OIS Operational Information Systems

OPKIS (Operational Computer Information System for Carriage) System that allows exchanging information between the main stations of the Baltic States and the railway information centres of the CIS countries, and collecting information on freight train composition, traffic, assembling, disassembling, arrival and departure of rolling stock, the crossing of border stations and all data on the wagon fleet.

OSI (Open Systems Interconnection Reference Model) Abstract description of communication protocols used in the communication of computer networks and systems which are referred by International Standards Organization (ISO) for describing of the multi-layered infrastructure of communication of internet and other communication channels.

OTIF (On-time and in-full) Measurement of delivery performance in a supply chain and measures how often the customer gets what they want at the time they want it.

OTRP (Overlay Token Ring Protocol) Covered network protocol.

OWL (Ontology web language) Family of knowledge representation languages for authoring ontologies

OWL-S (Web Ontology Language for Services) High-level language (XML-based) used for describing Web service properties

PER (Packet Error Rate) Number of packet errors.

PML (Process Modelling Language) Process class diagrams that describe methods and resources for process modelling

PROMETHEE (Preference Ranking for Enrichment Assessment) Significant method for evaluating alternatives concerning criteria in multi-criteria decision-making problems

PSO (Particle swarm optimisation) Computational method that optimizes a problem by iteratively trying to improve a candidate solution concerning a given measure of quality

QoS (Quality of Service) Term for service quality indicator.

RDF (Resource description framework) World Wide Web Consortium (W3C) standard originally designed as a data model for metadata

RDF/XML Corresponding syntax when a node element about a resource has multiple property elements

REHODES Hierarchical Optimized Distributed Efficient Real-Time System

REST (Representational state transfer services) Architectural style for developing web services

RFID (Radio Frequency Identification) Identification of devices operating and communicating based on radiofrequency technology.

RFID (Radio frequency identification) Technology that uses radio waves to passively identify a tagged object

RIS (River Information Services) Concept whereby information services in inland navigation support traffic and transport management in inland navigation, including interfaces with other modes of transport.

Route ID (Assigned route number) Within such a class, routes are distinguished from each other by a route number

Router Device that transmits or distributes data packets over computer networks.

RREP (Route Reply) Route response packet sent by the destination node to the original node on the reverse route.

RREQ (Route Request) Route request packet sent by the originating node to initiate the routing protocol overflow of the network.

RS (Rank sum) Procedure the weights, are the individual ranks normalized by

RSU (Road-Side Unit) Devices (equipment) and systems integrated into the roadside.

RTS (Request-to-Send) Message for the request to send.

SA (Signal Agent) Component, working like system-agent with special signalling functionality.

SAW (Simple additive weighting) Method is finding a weighted sum of the performance on each alternative on each attribute

SC (Supply chain) Network of all the individuals, organizations, resources, activities and technology involved in the creation and sale of a product

SCATS Sydney Coordinated Adaptive Motion System

SCOOT Split-cycle bias optimization method

SD (System dynamics) Computer-aided approach to policy analysis and design. Applies to dynamic problems in social, managerial, economic, and ecological

SDG (Single Digital Gateway) European gateway for Europeans to information and procedures

SDN (Software-defined network) Technology is an approach to network management that enables dynamic, programmatically efficient network configuration to improve network performance and monitoring, making it more like cloud computing than traditional network management

SDS Solution Delivery Systems

SE (Simulation Engine) System that enables functions of the simulation engine.

Smart-SDS (Smart Service Delivery System) System for smart services delivery with integrated AI components.

SOA (Service-oriented architecture) Designed to connect distributed, separately maintained and deployed software components

SOAP (Simple object access protocol) Message protocol that allows distributed elements of an application to communicate

SON (Self-Organizing Networks) Type of neuro-nets (the type of systems with a specific type of artificial intelligent methods for reasoning, learning) with the functionality of self-organizing networks.

SP Service provider for sending and receiving electronic freight documentation

SP Set splitting

SP2SP Service provider-to-service provider is transferring electronic freight documentation

SRMA/PA (Soft Reservation Multiple Access with Priority Assignment) Systems with an easy reservation with priority assignment.

SWARA (Step-wise Weight Assessment Ratio Analysis) Criterion weights and alternative selection

Synchro-modality Synchronized intermodality which employs multiple transport modes in a flexible, dynamic way to induce a modal shift towards more environmentally friendly transport modes like rail or inland waterways, without compromising on responsiveness and quality of service

T Period during which something occurs or is expected to occur

TC (Transaction costs) Expenses incurred when buying or selling a good or service

TCA (Transaction cost analysis) Method, focusing on transaction costs calculation

TCP (Transmission Control Protocol) Main protocol of Internet that enables transmission control functions and provides reliable data transmission (Layer 4 of the OSI model).

TEU (Twenty-foot Equivalent Unit) Exact unit of measurement used to determine cargo capacity for container ships and terminals

TICUR Transport infrastructure capacity and use ratio

TIR (International Road Transport) Customs transit document used to prove the existence of the international guarantee for duties and taxes for the goods transported under the TIR system, within the limit of the amounts specified by the contracting parties and under conditions stipulated in the TIR Convention.

TOPSIS (Technique for Order of Preference by Similarity to Ideal Solution) Concept that the chosen alternative should have the shortest geometric distance from the positive ideal solution and the longest geometric distance from the negative ideal solution

Truck License Plates Vehicle registration numbers visible to all can be easily removed or moved

TSLS (Two stages least square) Regression uses instrumental variables that are uncorrelated with the error terms to compute estimated values of the problematic predictor(s) (the first stage), and then uses those computed values to estimate a linear regression model of the dependent variable (the second stage).

TTRP (Routing truck and trailer problem) Generic name was given to a whole class of problems where it is needed to design a set of routes.

UDP (User Datagram Protocol) User data transmission protocol that is faster than the TCP transmission protocol because it does not provide reliable data transmission and does not provide an error correction mechanism (operates in Layer 4 of the OSI model).

UFO (Unified Foundational Ontology) Developed based on several theories from Formal Ontology, Philosophical logic, Philosophy of Language, etc.

UMTS (Universal Mobile Telecommunications System) Universal mobile telecommunications system.

UN (United Nations) Universal global organization

USA The United States of America

UTCS Urban traffic control system

UVRP (Unlimited capacity vehicle routing problem) Vehicles pick up or deliver items at various locations

V2I (Vehicle to Infrastructure) Special type of communication infrastructure that enables the interaction of the vehicle with infrastructure.

V2V (Vehicle to Vehicle) Special type of communication that enables vehicle-to-vehicle interaction.

VANET (Vehicular Ad-Hoc Network) Special type of network for the creation of interaction possibilities between cars (vehicles) like Ad-Hoc networks.

VAR (Vector Autoregression) Statistical model used to capture the relationship between multiple quantities as they change over time

VEC (Vector error correction) Restricted vector autoregression designed for use with nonstationary series that are known to be cointegrated

Vehicle Platooning Situation when a huge amount of vehicles are displayed in a short area.

VIKOR Serbian term "VIekriterijumsko KOmpromisno Rangiranje" for "multi-criteria optimization and compromise solution"

VIN (Vehicle identification number) 17-digit number provided by the producer

VMI (Vendor-managed inventory) Supply chain agreement where the manufacturer or supplier takes control of the inventory management decisions for the seller or retailer.

VOT (Value of time) The opportunity cost of the time that a traveller spends on their journey

VTMIS (Vessel Traffic Management and Information System) Extension of the Vessel Traffic Service (VTS), in the form of an Integrated Maritime Surveillance, which incorporates other telematics resources to allow allied services and other interested agencies in the direct sharing of VTS data or access to certain subsystems to increase the effectiveness of port or maritime activity operations as a whole, but that does not relate to the purpose of the VTS itself.

WASPAS (Weighted aggregate product estimation sum) Method exploits the advantages of the weighted sum model and weighted product model

WAVE (Wireless Access for Vehicular Environments) Wireless access in an automotive environment.

WCO (World Customs Organisation) Intergovernmental organization established as the Customs Co-operation Council (CCC) in 1952 to improve the efficiency and effectiveness of customs administrations worldwide

Wi-Fi Wireless technology brand owned by the Wi-Fi Alliance.

WiMAX (Worldwide Interoperability for Microwave Access) Broadband technology used for long-distance data transmission.

WLAN (Wireless Local Area Network) Wireless local area network.

WMN (Wireless Mesh Network) Wireless distribution network (mesh) network.

WS-BPEL (Web services in the Business process execution language) Standard or language for automating business processes

WSDL (Web service description language) Interface description language that is used for describing the functionality offered by a web service

WSM (Weighted sum method) Multi-criterion decision-making method in which there will be multiple alternatives and we have to determine the best alternative based on multiple criteria

WSMO Meta-model for Semantic Web services related aspects (Web service modelling ontology)

WSN (Wireless Sensor Network) Network of wireless sensors.

WSN (Wireless sensor networks) Group of devices that communicates monitoring and recording conditions at diverse locations through wireless links

XML (Extensible Markup Language) Language used to describe data

ZOR (Zone of Danger) Description of the zone with special parameters, which indicate the dangerous conditions of that zone.

The manufacturer's authorised representative in the EU is Springer Nature Customer Service Centre GmbH, Europaplatz 3, 69115 Heidelberg, Germany. If you have any concerns regarding our products, please contact ProductSafety@springernature.com

Printed and bound by CPI Group (UK) Ltd, Croydon, CR0 4YY

28/09/2026

02224061-0002